# BORON
## Volume 2
### Preparation, Properties, and Applications

# BORON

## Volume 2
### Preparation, Properties, and Applications

Based on papers presented at the 1964
Paris International Symposium on Boron

Edited by

## Gerhart K. Gaulé

Institute for Exploratory Research
U. S. Army Electronics Command
Fort Monmouth, New Jersey

Springer Science+Business Media, LLC

1965

ISBN 978-1-4899-6266-9      ISBN 978-1-4899-6574-5 (eBook)
DOI 10.1007/978-1-4899-6574-5

Library of Congress Catalog Card Number 60-13945

# Foreword

In one of the most awe-inspiring parts of the Mojave desert in southeastern California, near a small town appropriately named Boron, are the richest boron ores thus far discovered. The passing tourist is stunned by the view of the large mining establishments bustling with human activity under the cruel desert sun in the midst of the barren, seemingly uninhabitable wastes. The desert mining operations based in Boron symbolize to this editor the research effort based on boron; boron research seems to take place in productive oases of knowledge within a vast desert of uncertainty, still dominated by an apparently uncooperative nature. That some of these oases of knowledge are now rapidly expanding and even beginning to merge became evident at the Second International Symposium on Boron held in Paris on July 17—18, 1964.

Most of the chapters of this book originated from the scientific presentations given at the Paris symposium; additional papers (from authors who could not attend the symposium) were included in order to give a complete presentation of those recent developments in boron research which fall within the scope of this volume, namely, preparation, purification, and analysis (Part I), structure (Part II), electronic and mechanical properties (Part III), and applications (Part IV).

In order to judge advances made in these fields during the past few years, it is helpful to compare this volume with its predecessor, Volume 1, which originated from the first symposium on boron held in Asbury Park in 1959.. In the chapters of the first part of this volume, the reader will find new methods of preparation, such as that of alpha-rhombohedral boron from the iodide and that of beta-rhombohedral boron from diborane. New methods for the growth of larger single crystals of the alpha, beta, and tetragonal polymorphs are described, and new approaches to the production of boron films and of boron filaments are outlined. The erstwhile application to boron of such powerful analysis tools as gas chromatography, neutron activation, and mass spectrometry is reported. Those concerned with semiconductor effects from impurities will welcome the new information given on diffusion rates, activation energies, and doping effectiveness of several foreign elements in boron. While papers predominantly on metal borides or other boron compounds are not included in this volume, a number of compounds are treated in conjunction with purification and structure problems.

The reader will note with satisfaction that in Part II the structure of beta-rhombohedral boron, which had resisted clarification for so many years, is now fully determined. The description of the arrangement of the 105 boron atoms within their unit cell is fascinating, and it demonstrates most clearly the unique bonding behavior of the boron atom. New information is also presented on the structure of tetragonal boron, and a new method to obtain single crystals of this polymorph via a substitution reaction is described. One chapter is devoted to a new polymorph with semiconductor properties which forms under high pressure at high temperatures. Relationships among the polymorphs of boron have been clarified; it appears that beta-boron, though not the densest form, has by far the widest range (in temperature and pressure) of stability. The formation of the other polymorphs, on the other hand, is frequently favored by nonequilibrium conditions during the preparation of boron. Such conditions are discussed by several authors.

Eventually, it should become possible to link theoretically the unique bonding and, thus, the structural properties of boron with its unique semiconductor properties. The first step toward such a theory is presented in a chapter which outlines the structure of the energy bands of the electrons for a solid with rhombohedral crystal symmetry.

Most of the new subjects, and also a clarification of those already treated in Volume 1, will be found in Part III of this volume, which is devoted to electronic and mechanical properties. Here, of course, important details of sample preparation, analysis, and structure are included wherever necessary. Among the new phenomena discovered in beta-boron are the anisotropy of the optical absorption, conductivity and magnetoresistance, excess infrared absorption after illumination with visible light, other low-temperature trapping effects, and electron paramagnetic resonance. Birefringence and dichroism were discovered in alpha-boron. Recent studies on amorphous boron films reveal optical properties different from those of bulk boron and a very small conductivity which, however, increases rapidly at high electric fields, probably due to impact ionization effects. Nonlinear avalanche-current increases were also reported for doped beta-boron. The conventional rectification effect, which other semiconductors display at contacts to metals or at $p-n$ junctions, still has not been found with boron, probably due to the very small carrier mobilities in boron. It is generally agreed that boron is usually $p$-type and that its hole mobility in beta-boron does not exceed the value of $10^{-1}$ cm$^2$/V-sec and is probably much below this value. A more precise determination, which was the aim of the Hall-effect measurements carried out by several authors, has thus far been impossible because of the very small mobilities. Based on the mobility results and on the results of novel measurements of the conductivity at high frequencies, two authors propose a mechanism of conductivity via hopping of carriers between impurity sites for beta-boron. This mechanism would co-exist with the conventional conductivity via the energy bands, but would prevail under certain conditions, especially at lower temperatures. These hypotheses will undoubtedly stimulate future experimental and theoretical studies.

Papers on the mechanical properties of bulk boron and of boron filaments reveal unexpectedly high strength and elasticity and interesting surface phenomena.

Differences among data obtained from various research groups and differences in the interpretation of these data persist in some areas. Whenever possible, pertinent discussions have been included at the end of each chapter. To facilitate the search for detailed subjects and their treatment by all the authors concerned, a subject index has also been incorporated into this volume.

Particularly large differences are still found between reported values for the band gap of beta-boron. It is noted, however, that in some cases the differences are not due to measurement, but merely to a definition which bases the energy-gap value on a given measured curve. In most cases, these definitions are explicitly stated by the authors. The recent discovery of recombination radiation from beta-boron (Part III) should throw new light on the structures of the band edges and on the direct or indirect transitions betweeen them.

Many of the discrepancies still in existence are, of course, due to differences between the samples used—in particular, differences in the impurity concentration. This could be resolved to some degree by an exchange of samples between laboratories. Even so, the fact remains that the concentration of carbon, and possibly also that of concomitant structural imperfections (vacancies) in the boron used for semiconductor research today, is uncomfortably high in comparison with the impurity concentration of, for example, silicon used in semiconductor research. The high concentrations of impurities and other imperfections make it impossible to estimate how closely the properties measured or observed approximate those for ideally pure and perfect boron. In other words, many questions now pending must await answer until the continuing efforts of several groups to produce boron with impurities considerably below the parts-per-million level have been successful. However, this does not preclude the continuing development and application of boron devices; the well-controlled preparation and sophisticated analysis methods presented in the first part can certainly yield a boron which, though not extremely pure, has very reproducible properties.

A certain optimism with respect to device applications of boron is indeed justified by the fourth and last part of this book, which describes experiments with device prototypes. Generally, all these devices are made of beta-boron or of boron films having predominantly beta-structure. The devices behave essentially as ohmic resistors, except for the very strong and reproducible temperature dependence of the resistance (thermistors). This feature permits, for example, the building of switching circuits without moving contacts. A novel neutron detector was built using two thermistors made of the two stable isotopes of boron. The detector effect is based on the unusually large difference between the neutron-capture cross sections of the two isotopes. The two boron isotopes also have a relatively large mass difference, whose influence is discussed in the chapter on thermal conductivity. With the exception of the two topics just mentioned, nuclear properties or applications of boron are not treated in this volume. Applications of avalanche and of electro-optical effects in boron and structural application of bulk and filamentary boron are discussed in the summaries of some of the earlier chapters.

The editor wishes to express his thanks to all those who contributed to this volume, either by submitting manuscripts or by providing stimulating discussions. The editor also shares the gratitude of all the participants to the prime movers of the Second International Symposium on Boron—Dr. Jürgen Smidt and Dr. Wolfgang Dietz (both of Consortium für elektrochemische Industrie, Munich, Germany). The completion of the book would have been impossible without the able and patient help of the staff of Plenum Press—in particular, Miss Evelyn Grossberg.

The following pages, so the editor hopes, will convey to the reader not only useful (if incomplete) information, but also the challenge of the vast, ever-new field that is boron research.

Gerhart K. Gaulé

October, 1965

# Contents

## Part I

### Preparation, Purification, and Analysis

## Part II

### Structure

## Part IV

## Applications

# List of Contributors

# List of Contributors

# Synthesis of Red, $\alpha$-Rhombohedral Boron

E. Amberger and W. Dietze

*Institut für Anorganische Chemie*
*Universität München, Munich, Germany*

The results of investigations on the synthesis of large, single crystals of red, $\alpha$-rhombohedral boron are presented in this work. The synthesis was made by the pyrolytic decomposition in the gas phase of pure, recrystallized, zone-melted $BI_3$, on tantalum wire of diameter 0.1–2.0 mm in the temperature range of 850–1100°C. A discussion of six important experimental variables is presented.

## INTRODUCTION

Of the established modifications of boron — the so-called high-temperature or $\beta$-rhombohedral boron, the tetragonal boron, and the low-temperature or $\alpha$-rhombohedral boron, the last-named modification is of particular interest. Alpha-rhombohedral boron has a comparatively simple structure, containing in the unit cell twelve atoms which form an icosahedron. Alpha-rhombohedral boron was first prepared in 1958 [1,2], but it seems that the synthesis was not reproducible and only very tiny crystals were obtained. The object of our work has been the reproducible synthesis of large, single crystals of red, $\alpha$-rhombohedral boron. These single crystals are desired in order to carry out physical measurements.

Our first experiments were similar to those described in the literature. The pyrolysis of boron halides, e.g., boron bromide and boron chloride, on heated tantalum wires of 0.1–2.0-mm diameter was tried, but no traces of $\alpha$-rhombohedral boron were found in the thick boron deposits. Systematic experimentation on the subject then led to the discovery of the optimum conditions for obtaining red, $\alpha$-rhombohedral boron. The most important variables are: (1) type of starting compound, (2) purity of starting compound, (3) substrate (filament), (4) temperature of the filament, (5) diameter of the filament, and (6) region of formation of $\alpha$-boron.

1

## THE STARTING COMPOUND

Both boron halides and boron hydrides are potential starting compounds. At first, it would seem that the hydride, $B_{10}H_{14}$, with its intrinsic icosahedral structure of twelve boron atoms, would be a quite satisfactory starting compound, especially since it has a low decomposition temperature. However, since pure, carbon-free $B_{10}H_{14}$ is difficult to obtain, a boron halide was used instead. The iodide, rather than the bromide or chloride, was used because its practical decomposition equilibrium occurs at the lowest temperature of the three halides mentioned. The decomposition reaction is

$$BI_3 \overset{\Delta}{\rightleftharpoons} B + \tfrac{3}{2} I_2$$

## PURITY OF THE STARTING COMPOUND

We obtained $a$-rhombohedral boron in a reproducible form only when the $BI_3$ that was used was extremely pure, i.e., impurities of the order of parts per million. If $BI_3$ is prepared from the elements at a high temperature, contamination will inevitably result from impurities in the very materials of which the apparatus is made. Purification of the $BI_3$ would then be tedious and unsatisfactory. We, therefore, prepared $BI_3$ under more favorable conditions and at a lower temperature by the reaction of lithium boranate with iodine.

$$LiBH_4 + 4 \ I_2 \xrightarrow{\text{n-hexane}} BI_3 + LiI + 4 \ HI$$

Boron iodide was then recrystallized in n-hexane and repurified by the zone-melting method in tubes made of Pyrex glass. The efficiency of the zone-melting method was tested by adding 0.01–0.05% $SiI_4$, $AlI_3$, $MgI_2$, and $CI_4$ to the $BI_3$ to be zone-melted. Samples were taken at various locations in the zone-melting tube, and the impurity concentrations were then determined by colorimetric methods.

Figure 1 is a plot of the logarithm of final concentration $c$ divided by the initial concentration $c_0$ of the impurities as a function of position in the zone-melting tube. The steepest concentration gradient, that is, the best purification, is obtained with the smallest value of the parameter $k$. The purification was most effective for $SiI_4$ ($k = 0.3$), less effective for $AlI_3$ ($k = 0.5$), and least effective for $MgI_2$ ($k = 0.7$). Fortunately, it was not necessary to perform the quite difficult analysis for carbon in $BI_3$, since even small amounts

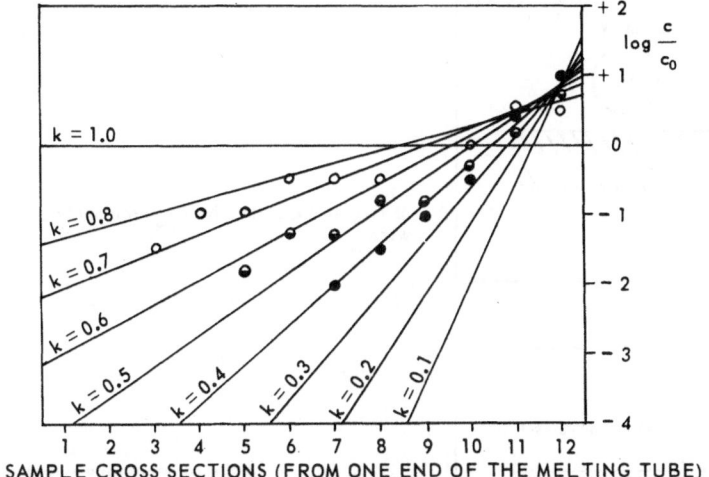

Fig. 1. Purification effect on highly pure $BI_3$ with added impurities. Ordinate: Logarithm of the contentration $c$ of impurities after zone-melting as a function of location in the $BI_3$ melting tube divided by the concentration $c_0$ of the added impurities. Abscissa: Distance from the input end of the melting tube at which samples were taken. Cross sections 1–8 indicate the "pure" end of the melting tube, and 9–12 the "impure" end. Open circles: $BI_3$ with $MgI_2$ as impurity. Half-open circles: $BI_3$ with $AII_3$ as impurity. Black circles: $BI_3$ with $SiI_4$ as impurity.

of $CI_4$ can be recognized through a color effect. $CI_4$ has a strong red color; $BI_3$ is usually colorless, but it becomes colored upon contamination with $CI_4$. The recrystallized and zone-melted $BI_3$ was colorless and remained colorless upon either heating or exposure to light.

The effect of purification is also reflected in the change of the melting point of the $BI_3$, which was 42.9°C before the zone-melting, and, afterwards, 48.0–48.6°C at the "pure" end (two-thirds the whole length) and 38°C at the "impure" end (one-third the whole length) of the zone-melting tube (Fig. 1).

Removal of occluded oxygen in the $BI_3$ crystals was accomplished by at least one sublimation of the crystals at high vacuum in the decomposition apparatus.

Pyrolysis experiments show unmistakably that the impurities in $BI_3$ have a great influence on the synthesis of $\alpha$-rhombohedral boron. In particular, it was found that (1) use of unrecrystallized, non-zone-melted $BI_3$ very seldom produces $\alpha$-boron, (2) use of recrystallized, non-zone-melted $BI_3$ produces more $\alpha$-boron, and (3) use of recrystallized, zone-melted $BI_3$ always produces $\alpha$-boron

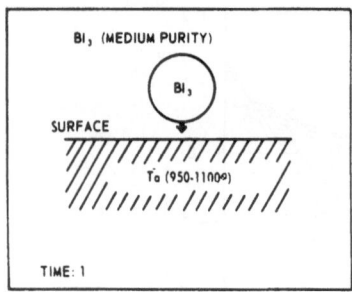

Fig. 2. Sketch depicting the onset of the pyrolysis of $BI_3$ at Time 1: The $BI_3$ approaches the tantalum wire.

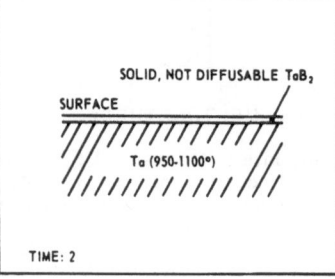

Fig. 3. Sketch of the progressing pyrolysis of $BI_3$. Time 2: The $BI_3$ has reached the outer surface of the tantalum wire on which there is a thin deposit of $TaB_2$. This $TaB_2$ does not react with the boron deposits, and there is no diffusion of $TaB_2$ into the boron to be deposited later.

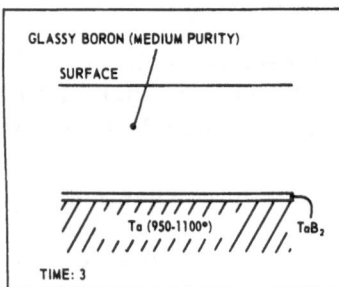

Fig. 4. Sketch of the pyrolytic surface after the events of Figs. 2 and 3. Time 3: Large quantities of glassy boron are deposited on the $Ta/TaB_2$ surface when the pyrolytic conditions are constant. The degree of purity of the glassy boron is the same as the degree of purity of $BI_3$.

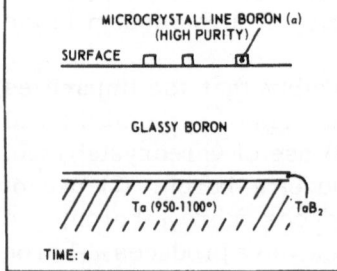

Fig. 5. Sketch of the progressing pyrolysis of $BI_3$ at Time 4: Probably because of the effect of transient pyrolytic conditions, there is production of a few microcrystals of $\alpha$-rhombohedral boron. These microcrystals have a degree of purity greater than that of the glassy boron.

when other experimental conditions are fulfilled. Some of the most important experimental variables — the substrate and the temperature and diameter of the filament — are discussed below.

### The Substrate (The Filament)

Good results were obtained with a glowing tantalum wire. (Unfortunately, a boron wire was not at our disposal.) At 950–1100°C, a thin layer of $TaB_2$ is formed, which apparently does not permeate into the boron. Oxygen was removed from the wire, and from the decomposition apparatus as well, by heating.

### Filament Temperature

Our experiments show no influence of temperature on the formation of $\alpha$-rhombohedral boron between 850–1100°C.

### Filament Diameter

A peculiar relationship between the diameter of the filament used and the quantity of $\alpha$-boron synthesized has been demonstrated. American investigators [2] obtained far less $\alpha$-boron on a 0.25-mm-diameter wire than on thicker wires. We observed this effect on the zone-melted $BI_3$ on a much smaller scale. An explanation of this thickness effect follows.

The boron deposits are heated mainly by Joule heat generated by a current through the filament core. For the boron, heat losses outweigh heat generation. Since the temperature of the outermost surface is kept constant during the experiments, the inner filament must be heated to a higher temperature as the deposited layer gets thicker. This would not be so for a thicker filament with a thinner deposited layer. The hotter filament of the first case will send more impurities (by diffusion) into the growing deposit. In our opinion, the $\alpha$-modification of boron is the one which offers the

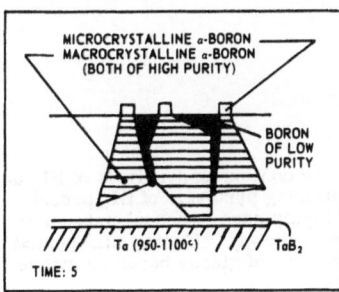

Fig. 6. Time 5: Highly pure $\alpha$-boron and boron of lower purity on the tantalum wire. The microcrystalline pure $\alpha$-boron provides crystallization centers for $\alpha$-boron from the not-so-pure glassy boron. The impurities are segregated between the large crystals of $\alpha$-boron (black areas) and cause the boron there to crystallize in a low-purity form.

least space for foreign atoms. The more foreign atoms present, the smaller the probability that $a$-rhombohedral boron will form. Because of the use, in our case, of very pure $BI_3$, the total number of impurities is small, and impurities from the filament are not of great import.

## SUGGESTED GROWTH MECHANISMS

We assume the following degrees of purity for the various substances under discussion:

recrystallized, zone-melted $BI_3$ . . . . . . medium purity
glassy boron. . . . . . . . . . . . . . . . . . . . medium purity
$a$-boron . . . . . . . . . . . . . . . . . . . . . . . .high purity
tetragonal boron . . . . . . . . . . . . . . . . . . low purity
rhombohedral boron. . . . . . . . . . . . . . . . low purity

The pyrolytic decomposition in the gas phase of pure, zone-melted $BI_3$ on a tantalum wire (diameter, 0.1–2.0 mm) produces microcrystalline $a$-boron and tetragonal boron in the temperature range 850–950°C. We presume that $a$-boron whiskers are formed, which are embedded in the tetragonal boron. However, in the temperature range of 950–approx. 1100°C, a thin skin of $TaB_2$ is produced, at first (Figs. 2 and 3). Glassy boron deposits on this skin from the gas phase (Fig. 4). This glassy boron probably contains icosahedrons of twelve boron atoms or fragments of these icosahedrons, as shown by the diffuse X-ray diagram. Completely constant experimental conditions, such as temperature, pressure of $BI_3$, and foreign-atom concentration, are needed to form this glassy, probably thermodynamically unstable modification of boron.

At some time during the deposition of boron at 950–1100°C, microcrystals of an $a$-rhombohedral boron are produced on the

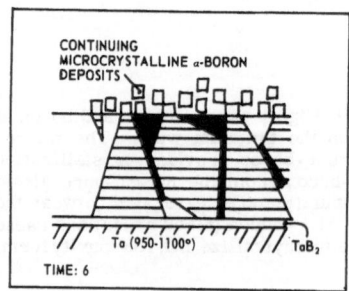

Fig. 7. Sketch of the continuing pyrolysis of $BI_3$ at Time 6: The continuing pyrolysis of $BI_3$ produces microcrystals of highly pure $a$-boron, embedded in impure boron (not shown). During this period, there is no production of glassy boron or macrocrystals of $a$-boron.

surface of the glassy boron deposit (Fig. 5). These microcrystals are formed when one of the experimental conditions varies.

The microcrystalline $\alpha$-boron at the surface of the glassy boron provides crystal growth centers. The microcrystals grown in the area of the glassy boron form large macrocrystals of $\alpha$-boron. Since these macrocrystals are of higher purity than those of the glassy boron, the impurities must settle between the macrocrystals of $\alpha$-boron (black areas on Figs. 6 and 7). The impurities in the glassy boron induce the boron to crystallize in the tetragonal or in the $\beta$-rhombohedral form, which forms are not as exacting, with respect to purity, as the $\alpha$-rhombohedral form.

## REFERENCES

1. McCarty, L. V., J. S. Kasper, F. H. Horn, B. F. Decker, and A. E. Newkirk, J. Am. Chem. Soc. 80:2592 (1958).
2. McCarty, L. V., and D. R. Carpenter, J. Electrochem. Soc. 107:38 (1960).

# Purity of Boron Produced by the Decomposition of Diborane and Subsequent Zone-Melting

Ingeborg Hinz and Heinz Wirth

*Consortium für elektrochemische Industrie GmbH*
*Munich, Germany*

Boron is prepared from diborane by the van Arkel process under special safety precautions and then purified by zone-refining. No impurities can be detected spectrochemically after the zone-refining. Nevertheless, the purity is controlled with a 2-m grating spectrograph using a DC arc in a nitrogen or argon atmosphere for excitation. Spectrochemical limits of detection are 1 ppm for silicon, and 0.1–0.5 ppm for copper, magnesium, calcium, and silver, for example. Neutron activation analysis is also discussed. By this method, the concentrations of tantalum, copper, and arsenic in the zone-refined boron were found to be $< 0.3$, $< 10^{-2}$–$10^{-3}$, and $< 10^{-4}$ ppm, respectively. No other elements could be identified after activation times up to 120 hr. The main impurity in our boron is carbon (30–50 ppm). Because of the unfavorable distribution coefficient for carbon in boron (approx. 4), removal by zone-refining is very time-consuming. A coulometric method to determine carbon in boron is discussed.

## INTRODUCTION

The van Arkel process of pyrolysis of gaseous compounds is used for the preparation of boron. The starting material, a boron–hydrogen compound $B_2H_6$, can be obtained in high purity in a relatively simple manner. However, the high inflammability and toxicity of this compound require special precautions. Quantities of $B_2H_6$ more than a few grams are processed outdoors only. Two special rooms serve for the decomposition of the $B_2H_6$: (1) The room housing the valves and the flood and pressure indicators must not be entered during operation. (2) The adjoining operating station contains a switchboard for the electrical supply. Here, the deposition apparatus is the only area that contains $B_2H_6$. The meters can be observed from the operating station through a safety-glass window in the partition separating the two rooms.

Figure 1 shows the control room equipped with two decomposi-

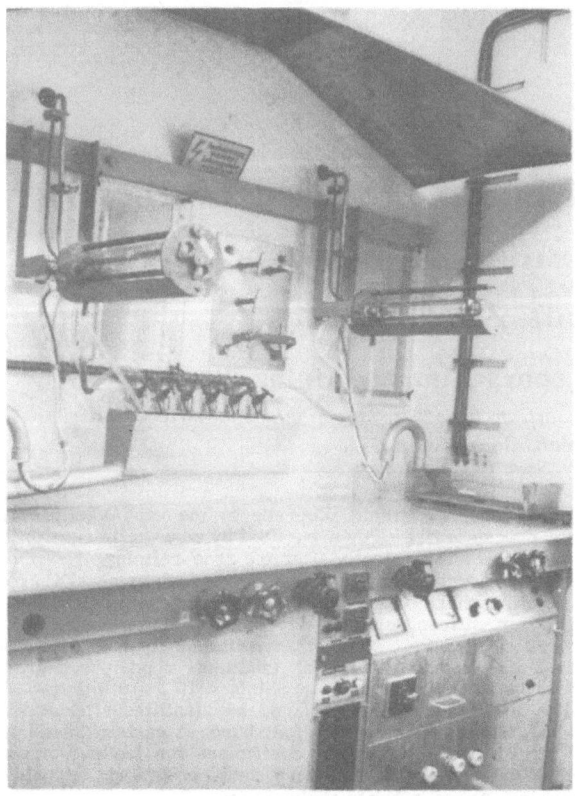

Fig. 1. Control room with decomposition apparatus for the preparation of boron from $B_2H_6$.

tion systems, a supply of cooling water, the valves, and the electrical controls. It is possible to observe through the two windows the piping installed at the back of this wall (Fig. 2). The diborane passes from a steel container installed outdoors through a pipe into a precision valve and a flood indicator, thus reaching the decomposition arrangement on the other side of the wall. The surplus gas is frozen in a steel trap. When the decomposition is finished, the gas, rarefied by nitrogen, is let out. The rest of the piping and the valves serve to clean the decomposition system with hydrogen, to feed nitrogen into the trap, and to rinse the piping with water. When the decomposition is finished, the entire pipe system is cleaned with water in order to hydrolyze possibly precipitated explosive, solid, decomposition products. By zone-melting after decomposition, the material (shown in Fig. 3) is degassed and purified.

Boron rods of 3–10 mm in diameter and up to 250 mm in length are obtained (Fig. 4). Apart from a thin surface layer ranging from 1–2 mm, the rods show a largely uniform orientation parallel to the hexagonal c-axis. Numerous crystal edges on rhombohedral faces are visible. The cores of the rods show — depending upon the diameter — more or less pronounced cracks running in various directions (Fig. 5).

## SPECTROCHEMICAL ANALYSIS

For the spectrochemical analysis of boron in the longwave UV range, we used a 2-m-grating spectrograph. Of all the methods of

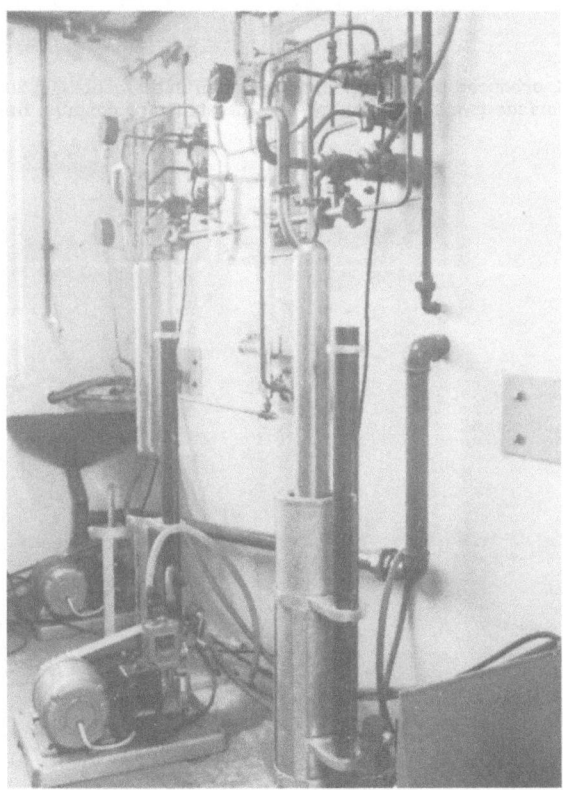

Fig. 2. Chemical section of decomposition arrangement.

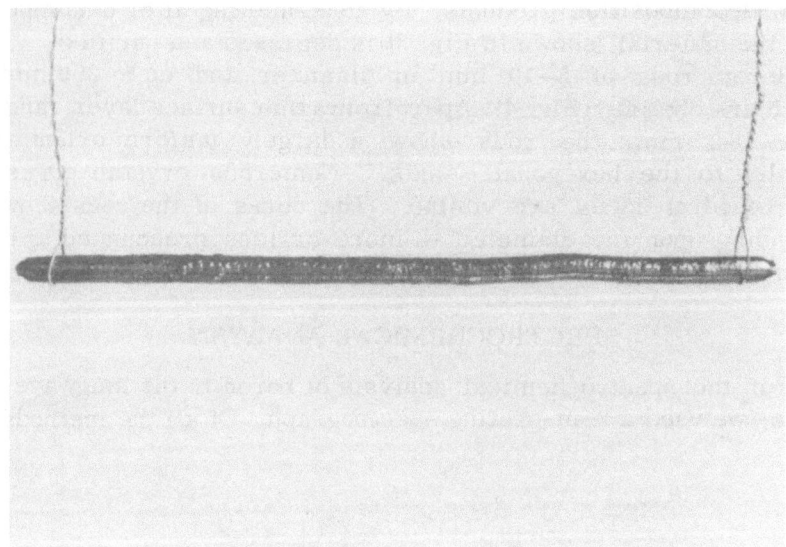

Fig. 3. Boron rod, produced by decomposition, 10 mm in diameter and 200 mm in length. The surface is black and shiny, and the structure contains flaws.

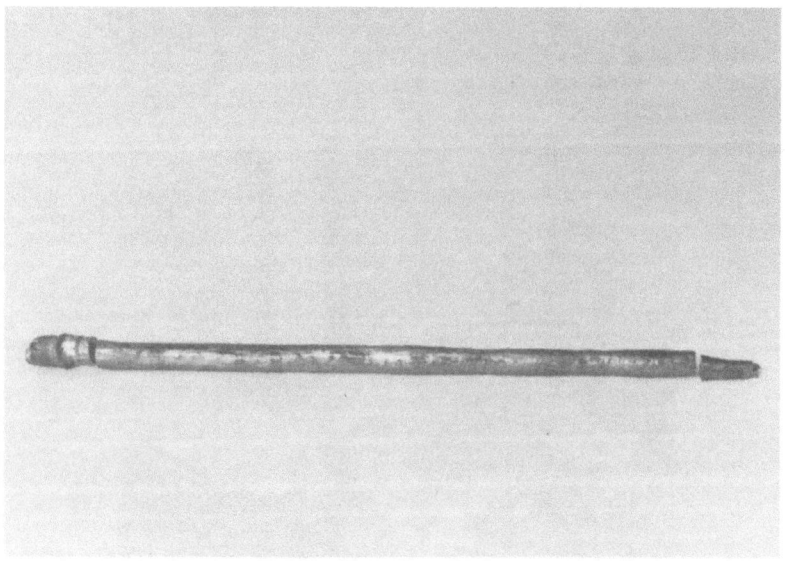

Fig. 4. Zone-refined boron rod, 8 mm in diameter and 200 mm in length.

excitation, a DC arc discharge has proved the most effective for trace impurities in boron. Discharging of the arc in air was out of the question because of the strong band background. For this reason, experiments were made to find a suitable, protective-gas atmosphere. Argon used as a protective gas offers the best sensitivity for the detection of some elements, because the background is continuous.

In a 30-A, DC arc in argon, some of the impurities evaporate almost completely within 15 sec, with the bulk part of the boron remaining. Thus, a certain effect of compilation is obtained, and the background intensity can be limited by a short exposure time.

Fig. 5. This face reveals the largely uniform orientation of the core. The reflecting face originated accidentally by cleavage.

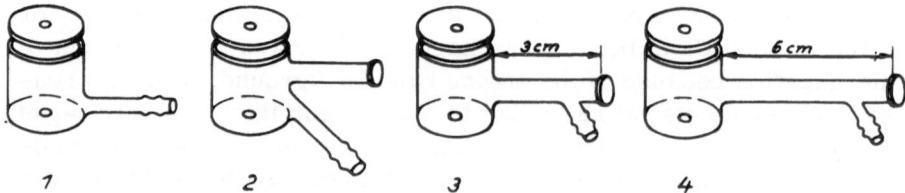

Fig. 6. Various forms of quartz-glass chambers for the spectrochemical analysis of boron.

For less volatile elements, such as silicon, the sensitivity of detection is best with nitrogen as the protective gas. In that case, only 10 A is needed, while the best exposure time varies from 60-90 sec.

A quartz-glass chamber that has a loosely applied cover, similar to the Schöntag chamber, is used for the discharge of the arc in a protective gas. The Schöntag chamber cannot be used in its original form, because its walls become coated, preventing any transmission of radiation. Figure 6 shows the various forms of quartz-glass chambers which were tested. Form 1 is the Schöntag chamber. Better results were obtained with the modifications of forms 2 and 3, but form 4 produced the most gratifying results. Our zone-melted boron samples, analyzed under these optimum conditions, usually showed no impurities. Nevertheless, they were continually examined spectrographically in order to control the purity. When spectral purity is ascertained, the contents of the impurity concentrations are always below spectrographical limits of detection, which have been approximately established as described below.

Since suitable standard samples were not available, they had to be made synthetically. Oxides of the elements in question were added in graduated quantities to the spectrographically almost-pure boron powder, produced from zone-melted boron. Mixing took place in small plastic tubes within a vibrator. The standard mixtures were then excited in argon. The same procedure was carried out in nitrogen after mixing the oxide–boron combinations with graphite powder in a 1:1 ratio. If not mixed with graphite powder, the boron spatters upon complete combustion, and results are not reproducible. Argon cannot be used for quantitative determinations of the elements because the testing conditions cannot be reproduced. We failed to evaporate the entire substance in a short enough time period to make the background tolerable. A short discharge in argon, rather than in nitrogen, results in a much greater sensitivity of detection for some elements.    A complete burning in nitrogen takes 4 min.   If the arcing time in argon is only 90 sec, however, the sensitivity of detection is still better than 4 min, due to the decreased background.  The sensitivity of the detection may be recognized from the first appearance of the line blackening of the elements and the increase of line intensity with concentration in the standard series.

As neither beryllium nor germanium could be used as an internal standard, a boron line or the background was used.  If exposed in argon, the established sensitivities of detection range from 2–3 ppm for arsenic, tin, mercury, and aluminum.  The sensitivity of detection of these elements in nitrogen is worse by the factor of at least ten.   For magnesium, copper, and silver, the sensitivity of detection is equally good in argon and nitrogen.  (The sensitivity of detection in nitrogen is 0.1–0.5 ppm for magnesium, copper, and silver and about 1 ppm for silicon.)  Of course, these results, determined with the help of the synthetic standard samples, offer only approximate values.   In the case of silicon, there were also at our disposal two natural standard samples with a colorimetrically determined silicon content of 8 and 20 ppm, which served to control the calibration curve.  The values of these natural standards corresponded quite well with the synthetic samples.  All patterns were taken using filters with transmittances of 10, 50, and 100%.   The main detection line of silicon was still visible in the sample with 8 ppm Si with the 10% filter.

For the routine spectrographical control of the purity of the boron rods, compact rods of 4-mm diameter and 4-mm length were used rather than boron powder, in order to avoid contamination. These small rods are etched with fluoric and nitric acids, and then excited in a DC arc under protective gas in a preformed crater electrode used as the anode.  Samples were arced in both argon (15 sec, 30 A) and nitrogen (90 sec, 10 A). The sensitivity of detection for the compact boron samples is better than that for the powder samples, since much more sample material is used.

## ACTIVATION ANALYSIS BY NEUTRONS

Activation analyses were made to investigate impurities in boron that cannot be detected by spectrochemical analysis.  Boron rods of about 5-mm diameter were sealed into quartz tubes and activated in the core of a reactor at a thermal neutron flux of approximately $1-2 \cdot 10^{13}$ n/cm²-sec for at least 40 hr. The advantage of this kind of activation is that there is no risk of contamination as the rods are powdered after the activation and thorough etching. Unfortunately, boron is an especially unsuitable element as far as quantitative activation analysis is concerned, since it absorbs thermal neutrons to a high degree. For a quantitative detection of impurities, the standards of the elements to be defined are usually activated together with the boron sample. In this case, however, the impurities receive a much smaller thermal neutron flux than the standard, since the boron absorbs many more thermal neutrons. Thus, they are less activated. Moreover, the neutron flux consists not only of thermal neutrons, but also shows a whole energy spectrum. Thus, the low-energy part of the neutron flux for all the elements in a boron rod is more attenuated than the high-energy part, and elements with resonance bands for epithermal neutrons are more activated than those responding to thermal neutrons only. We tried to determine a factor representing the attenuation of the neutron flux in boron for some elements. For a boron rod of 5-mm diameter, this factor is approximately 22.5 for copper and phosphorus and approximately 2.5 for tantalum and arsenic.

The factor for copper, for example, was determined as follows. A powdered boron sample with a colorimetrically determined copper content of 20 ppm was pressed into a small quartz-glass tube of 5-mm diameter, and then was activated together with a copper standard sample. After the activation, the two samples were chemically analyzed following the addition of a carrier. The copper was separated by dithizon extraction and was precipitated as Cu-I-rhodanide (CuSCN). The impulse rates were compared with the methane flowmeter. The copper content radiochemically found in the activated boron sample was too small by a factor of 22.5. The copper contained in boron as impurity received a neutron flux that was 22.5 times less than that through the standard copper sample. The factors for arsenic and tantalum were analogously determined. Arsenic and tantalum, which have strong resonance bands for epithermal neutrons, are (when in boron) more strongly activated than copper; then, the factor is less — as low as 2.5. These factors were determined only once and are only approximate values when applied to other activation analyses because: (1) they were determined in boron powder, whereas rods are usually used; (2) the diameter of the rods is not constant throughout the rod's length; and (3) when the activation is performed in a reactor, the conditions are never constant regarding the energy distribution of the neutrons. The conditions, therefore, are unfavorable and exact quantitative values cannot be expected when analyzing activated boron by neutron activation.

However, qualitative data and the orders of magnitude of the impurity concentrations can be obtained. By activation analysis, we could find only the elements tantalum, copper, and arsenic in our zone-refined boron samples. The activated samples were examined with a gamma-ray spectrometer. The radioactive isotopes, after chemical separation by inactive carriers, were identified by their gamma energies and their half-lives. The impulse rates measured by the methane flowmeter were compared with those of the simultaneously radiated standard samples of the elements in question, taking into account the above-mentioned factors for the attenuation of the neutron flux. In the zone-refined samples, the tantalum, copper, and arsenic contents (in ppm) were found to be $< 0.3$, $< 10^{-2}-10^{-3}$, and $< 10^{-4}$, respectively.

The maximum period of activation was 120 hr, but, even after such a long time, no further elements were detected by the usual, chemical-group separation after adding inactive carriers of a number of elements, such as molybdenum, tellurium, silver, nickel, bismuth, phosphorus, iron, and tin, to the dissolved boron sample. Traces of arsenic and phosphorus were discovered only in non-zone-refined boron samples. These elements are no longer detectable after zone-refining.

Table I shows the sensitivity of detection for various elements in boron for an activation period of 48 hr and a thermal neutron flux of $1 \cdot 10^{13}$ $n/cm^2$-sec. A factor of 22.5 for the diminution of the neutron flux was assumed for all elements, with the exception of tantalum and arsenic. This seems to be the most unfavorable value. For some elements with resonance bands for epithermal neutrons, the sensitivity of detection will be somewhat better. The sensitivity of detection is given in parts per million and calculated for a 1-g amount of boron. The sensitivity of detection of the elements copper, bromine, arsenic, gallium, gold, terbium, tungsten, palladium, iridium, sodium, antimony, tantalum, and mercury (group I) is below $5 \cdot 10^{-3}$ ppm or $5 \cdot 10^{-7}$ wt. %. The limitation of detection of the next group of elements is in the vicinity of 0.01–1 ppm, i.e., better than that for spectral analysis (magnesium excepted). For the third group of elements, the sensitivity of

## TABLE I

### Activation-Analysis Limits of Detection Sensitivity for Various Elements in Boron after 48-hr Activation Period and Thermal Neutron Flux of $1 \cdot 10^{13}$, Employing a Correction Factor of 22.5.

("d" signifies day; "a" signifies year; SA refers to the results of spectral analysis (in ppm); CA refers to the results of colorimetric analysis (in ppm); $T \frac{1}{2}$ refers to half-life time.)

**I. $< 5 \cdot 10^{-3}$ ppm**

| Element | Half-life | Detection limit, ppm | Remarks |
|---|---|---|---|
| Cu | 12,8 hr | $3.2 \cdot 10^{-5}$ | |
| Br | 3,3 hr / 36 hr | $4.4 \cdot 10^{-5}$ | |
| As | 26,7 hr | $6.8 \cdot 10^{-5}$ | (factor, 2.5) |
| Ga | 14,1 hr | $1.3 \cdot 10^{-4}$ | |
| Au | 2,7 d | $1.6 \cdot 10^{-4}$ | |
| Tb | 72 d | $< 4.9 \cdot 10^{-4}$ | |
| W | 24 hr | $8.0 \cdot 10^{-4}$ | |
| Pd | 13,6 hr | $1.1 \cdot 10^{-3}$ | |
| Ir | 74 d | $1.2 \cdot 10^{-3}$ | |
| Na | 15 hr | $1.4 \cdot 10^{-3}$ | |
| Sb | 2,8 d | $2.5 \cdot 10^{-3}$ | |
| Ta | 115 d | $2.8 \cdot 10^{-3}$ | (factor, 2.5) |
| Hg | 65 hr | $< 5.0 \cdot 10^{-3}$ | |

**II. 0.01–1 ppm***

| Element | Half-life | Detection limit, ppm | Remarks |
|---|---|---|---|
| K | 12,5 hr | 0.02 | |
| Ge | 11 d | 0.0225 | |
| Cd | 54 hr | 0.024 | |
| Cr | 27,8 d | 0.024 | |
| Pt | 19 hr | 0.037 | |
| Mn | 2,6 hr | 0.04 | after $5 T \frac{1}{2}$ |
| In | 50 d | 0.053 | |
| Ce | 33 hr | 0.063 | |
| Rb | 18,6 d | 0.07 | |
| Mo | 67 hr | 0.09 | |
| Zn | 13,9 hr | 0.11 | |
| Hf | 45 d | 0.11 | |
| Sn | 27 hr | 0.115 | |
| Co | 5,2 a | 0.126 | |
| Ag | 253 d | 0.4 | |
| Ni | 71 d(Co) | 0.61 | |
| Te | 58 d | 0.73 | |
| Se | 120 d | 0.87 | |
| Mg | 15 hr(Na) | 0.85 | SA, 0.1–0.5 |

**III. $> 1$ ppm**

| Element | Half-life | Detection limit, ppm | Remarks |
|---|---|---|---|
| Sr | 2.9 hr | 1 | after $5 T \frac{1}{2}$ |
| P | 14 d | 1.27 | |
| Bi | 5 d | 1.42 | |
| Al | 15 hr(Na) | 1.6 | |
| Ba | 12 d | 3 | |
| Si | 2.6 hr | 4 (after $4 T \frac{1}{2}$) | SA, ~1 |
| S | 87 d | 5.6 | |
| Zr | 65 d | 7.7 | |
| Pb | 3.3 hr | 8 (after $3 T \frac{1}{2}$) | SA, 1–5 |
| Fe | 2.6 a | 8.1 | |
| Ca | 153 d | 10.1 | SA, 0.1–0.5 |
| Ti | 1.8 d | 49 | CA, ~5 |

*Elements also classified in group II are Cl, Cs, Th, Tm, U, Os, Pr, Re, Ru, Sc, Sm, Gd, Y, and Yb.

detection by activation analysis is above 1 ppm. The spectrochemical analysis of the elements silicon, lead, and calcium proved to be more sensitive. The chemical analysis gave even better results with titanium. As mentioned above, the sensitivity of detection is calculated for an activation period of 2 days. For the elements with long half-lives, the sensitivity can be increased by long activation periods.

## DETERMINATION OF CARBON

Carbon is the main impurity present in our boron samples. To determine carbon content, the powdered boron samples are decomposed at 1020°C with lead borate in a porcelain vessel under oxygen flow. The $CO_2$ thus produced is coulometrically determined. The principle of the coulometric method is described below.

After combustion, the carbon (now in the form of $CO_2$) is absorbed in a $Ba(OH)_2$ and $BaCl_2$ solution of pH 9.5, thus forming insoluble $BaCO_3$, which process decreases the pH value. The amount of $Ba(OH)_2$ used up is continuously electrolytically renewed at the cathode by decomposition of $BaCl_2$ until the initial pH value is reached. From the amount of charge (measured in mA) and the amount of time (in sec) required for the regeneration of $Ba(OH)_2$, the consumed amount of $Ba(OH)_2$ is calculated. According to Faraday's law, 10 mA and 10 sec correspond to 6224 $\gamma$ of carbon.

Figure 7 shows a sketch of the arrangement. The electrolysis vessel consists of a cathode and an anode section separated by a diaphragm. Platinum sheets, connected with a stabilized DC source, serve as electrodes. The pH measurements are made with the help of a single rod glass electrode and a pH-meter. The $CO_2$ gas is conducted through a frit glass into the absorbing solution; a thorough mixing is effected by a magnetic stirrer. A Heraeus Co. tube furnace serves as the combustion furnace. The porcelain vessel is in a quartz tube and is pushed into the hot zone of the furnace by means of a quartz rod, moved by a magnet. This may not be done before the air is completely removed from the furnace. The oxygen for the combustion flows at a rate of 2–5 liter/hr. It is freed of hydrocarbons by passing it over glowing CuO and then over soda asbestos.

The coulometric carbon determination is exact and very sensitive (2 $\gamma$ of carbon are detectable); yet our sensitivity of detection is not better than 5–8 $\gamma$ due to the blank value of about 8 $\gamma$. This

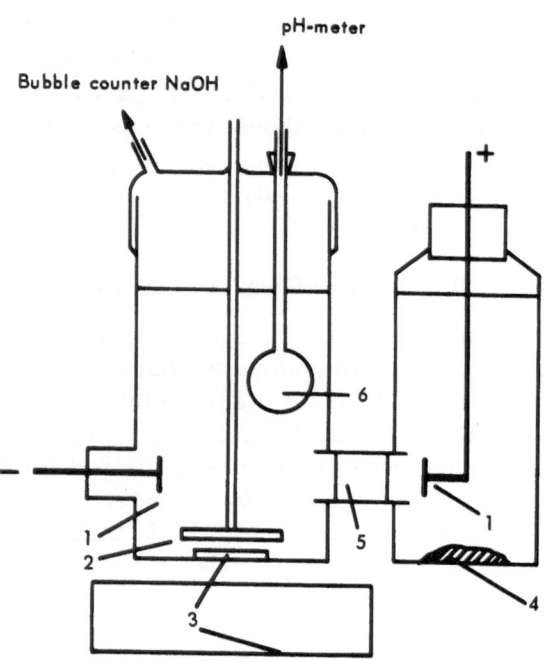

Fig. 7. Arrangement for carbon determination by decomposition of powdered boron with lead borate under oxygen flow. The $CO_2$ is coulometrically determined. (1) Platinum electrodes, (2) frit-glass $CO_2$ inlet, (3) magnetic stirrer and magnet, (4) $BaCO_3$, (5) diaphragm, (6) glass electrode.

blank value originates from the lead borate, the porcelain vessel, and the whole arrangement. The lead borate contributes almost nothing because it was previously kept at 1000°C for a long period in an oxygen flow. To decompose 250 mg boron, 2 g of lead borate is necessary. The melting takes place in a freshly outgassed porcelain vessel, which is put into a similar vessel to protect the quartz tube from the melt that creeps over the edge. We failed to reduce the total blank value with this arrangement.

Since the amount of boron required is 250–300 mg, the sensitivity of detection is, in this case, about 20–30 ppm. The carbon content of our samples is about 30 ppm. Decrease of the carbon content by zone-melting is possible only to a limited degree. By determining the carbon concentration after 1, 2, 5, and 8 zone passes at a travel rate of 2 mm/hr, an effective distribution coefficient of $k = 4$ was found. It follows that boron samples with an extremely low carbon content can be produced; but that zone-refining of large quantities requires much time.

# Ultrapure Boron from Halide Intermediates

## A. F. Armington
*Air Force Cambridge Research Laboratories*
*Office of Aerospace Research*
*L. G. Hanscom Field, Bedford, Massachusetts*

## J. T. Buford and R. J. Starks
*Eagle-Picher Company*
*Miami, Oklahoma*

The preparation, purification, and analysis of high-purity boron produced from iodide and bromide intermediates are discussed. The discussion begins with the preparation of these intermediates, predominantly from the elements. Of the several methods of purification considered, distillation appears to be the most satisfactory. Decomposition on boron and other substrates was performed using both hydrogen and vacuum procedures. In some cases, the final product was then zone-refined. Analytical results from emission and mass spectroscopy are given. Advantages and disadvantages of the bromide and iodide processes are compared.

## INTRODUCTION

Several methods of purification have been reported for boron in recent years, most of which report a final product with a purity better than 99%. The purest boron, however, has been produced employing chemical intermediates with relatively low melting and boiling points. The halide intermediates have been most frequently used, although the hydride might be a possible intermediate. The hydride has by far the most desirable boron percentage in the intermediate, but the explosive danger of the hydride presents a serious problem. The authors note with considerable interest a paper on this subject by Hinz and Wirth.* The high thermal stability of boron trifluoride creates a problem for the vapor-phase decomposition of this intermediate. The low boiling point (approx. - 100°C) also makes this material difficult to handle. Electrolysis of a fused salt solution of potassium fluoborate, potassium chloride, and boric oxide has been successful, however, in producing relatively high-

*See Hinz and Wirth, "Purity of Boron Produced by the Decomposition of Diborane and Subsequent Zone-Melting," this volume, p. 9.

purity boron [1]. Boron trichloride also has been investigated [2], although the low boiling point (12.5°C) makes this material difficult to handle, also. Boron trichloride is quite stable thermally and is usually hydrogen-reduced at 1000–1200°C. The resulting boron is generally of about 99.5% purity. Boron trichloride has been used by one of the authors for the preparation of ordered alpha-rhombohedral films on silicon [3]. In this case, the boron trichloride was distilled at low temperature prior to hydrogen reduction. The purity of the films is uncertain, but it is believed to be about 99.5%.

The bulk of our research work on boron purification involves bromide and iodide intermediates [4]. These intermediates have the most desirable physical properties when standard purification techniques are applied. Boron tribromide melts at -45°C and boils at 90°C, while boron triiodide melts at 48°C and boils at 211 or 212°C. Thus, while these materials are corrosive, particularly in the presence of moisture, they can be contained and handled easily.

## FORMATION OF THE INTERMEDIATES

Of the two intermediates discussed below, boron tribromide is the simplest to prepare. The established synthesis is made directly from the elements by passing bromine through powdered boron at 850°C using a fluidized bed reactor (Fig. 1). Using this apparatus, up to 12 liters can be produced in a single run. In experiments employing a horizontal fixed bed, the reaction still goes to completion at temperatures as low as 420°C [5]. Less contamination of the product appears to occur at the lower temperatures, but there has been no attempt as yet to produce large amounts of material in this manner.

Two principal methods have been employed to produce boron triiodide. The first method, which we have found useful in producing small amounts of material, is the reaction of iodine with sodium borohydride [6], usually employing a solvent [7]. A yield of about 40% is possible when small amounts of reactants are used, but attempts to scale-up the reaction result in a decreasing yield. This is possibly the result of sodium iodide, a by-product of the reaction, which coats the unreacted sodium borohydride and prevents its reaction. Direct combination of boron and iodine appears to be the best method of producing large amounts of boron triiodide [8–10]. The yield is not as good as in the case of boron tribromide, for which 90–100% is generally reported. The temperature for maximum

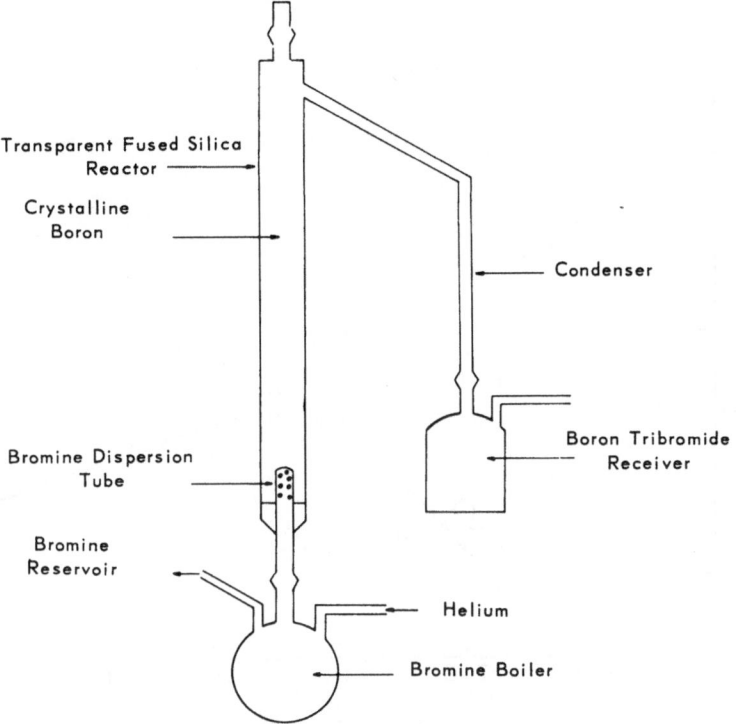

Fig. 1. Boron tribromide synthesis unit.

yield is 850–950°C, depending on the conditions. Figure 2 shows some results of yield versus formation temperature for a horizontal bed reactor using amorphous boron pellets. This reaction is quite sensitive to the conditions employed for the formation. Thus, using a horizontal fixed bed reactor, a yield of 50–60% can be produced with amorphous boron pellets, while a maximum yield of 10% can be produced using crystalline boron. However, using a fluidized bed reactor (vertical), a 60% yield can be obtained with crystalline boron. This is possibly a surface effect of some type. As in the case of boron tribromide, the formation temperature can affect the purity of the resulting intermediate, as shown in Table I. In the case of the iodide, however, reaction at a lower temperature has the distinct disadvantage of lowering the yield.

## PURIFICATION

The physical properties of the halide intermediates determine what purification techniques can be applied. Fractional crystal-

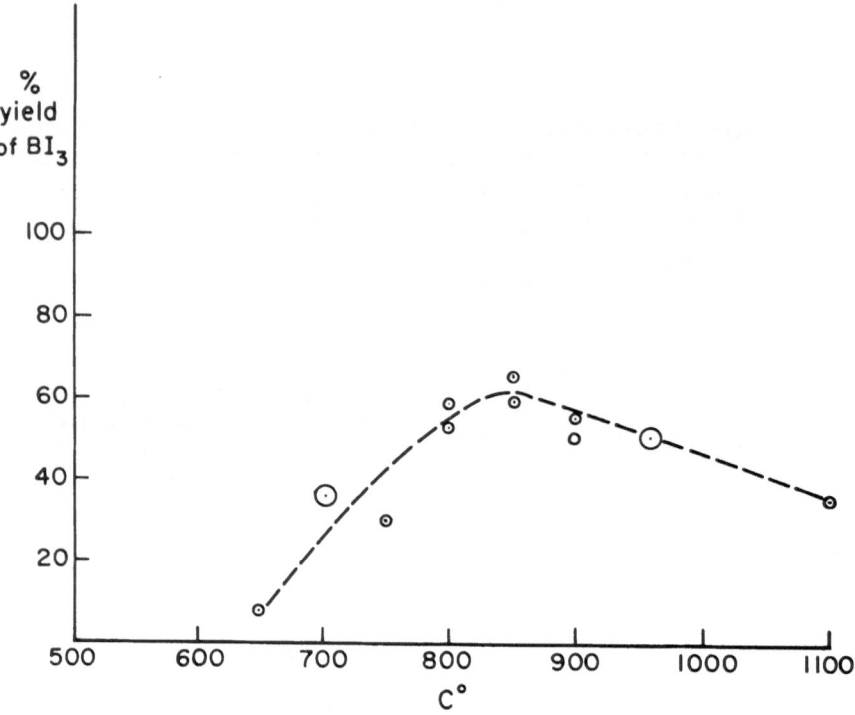

Fig. 2. Effect of temperature on $BI_3$ yield.  Carrier gas flow, 200 ml/min.

lization would not work for the bromide, since it is a liquid at room
temperature and down to -45°C.  Fractional recrystallization of the
iodide, using several organic solvents, has been attempted, but
found unsatisfactory, because of the high solubility of the iodide.  The
danger of carbon contamination from pyrolysis of occluded solvent
molecules during decomposition of the intermediate would also be
a serious drawback for this method, since often extensive purification procedures are required to eliminate this possibility.  Vapor-
phase chromatography of both intermediates has been unsuccessful
due to their high reactivity.  This would be a marginal purification
technique, at any rate, due to the low throughput of presently
available preparative columns.

    Three standard purification methods can be applied to the
purification of boron triiodide, while purification of the bromide is
limited to one or possibly two techniques. Thus, we can successfully
apply sublimation, zone-refining, and distillation to the iodide, while

## TABLE I

### Effect of Formation Temperature on Purity of $BI_3$

| Temperature, °C | Impurity, ppm | | | | | |
|---|---|---|---|---|---|---|
| | Si | Mg | Sn | Fe | Cu | Pt* |
| 650 | 100 | 1 | 5 | < 1 | < 0.5 | 10 |
| 750 | 300 | 1 | 10 | < 1 | < 0.5 | 30 |
| 800 | 100 | ND† | ND | < 1 | < 0.5 | 10 |
| 850 | 300 | 0.5 | < 2 | < 1 | < 0.5 | 10 |
| 900 | 600 | 11 | < 2 | < 1 | < 0.5 | 10 |
| 1000 | 1000 | 0.5 | < 2 | < 1 | < 0.5 | 20 |
| 1100 | 1000 | 0.5 | 5 | < 1 | < 0.5 | 10 |
| Detection limit | 1 | 0.5 | 2.0 | 1 | 0.5 | 10 |

*Platinum believed present due to drying of $B_2O_3$ in platinum boats at red heat.
†ND indicates not detected.

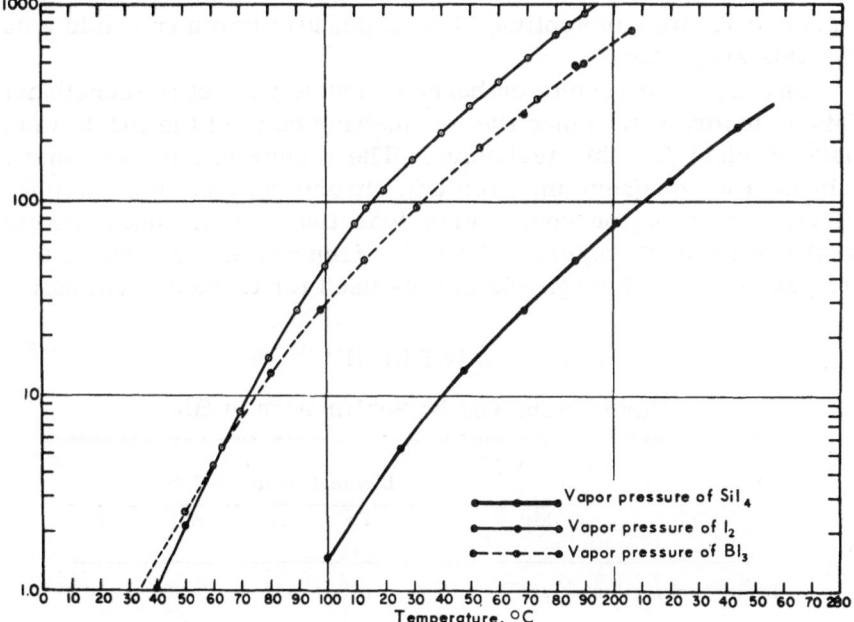

Fig. 3. Vapor–pressure curves.

the bromide is limited to distillation and possibly zone-refining. Sublimation of iodide should produce fairly good purity since only silicon tetraiodide, a major detectable impurity, has an appreciable vapor pressure in the region where boron triiodide will sublime easily. Figure 3, extrapolated from the data of Andersen and Belz [11], indicates the vapor pressures of these compounds, as well as the vapor pressure of iodine, as functions of temperature. Two deductions may be made from these curves. First, unreacted iodine, collected with boron triiodide from the synthesis unit, would be difficult to separate by sublimation. Second, the maximum theoretical purity (in regard to silicon) can be deduced from the vapor pressure using Raoult's law, if the initial purity is known. From data at 60°C (the temperature of our sublimation experiments) and an initial silicon tetraiodide concentration of 1%, the calculated maximum purity for a single sublimation is 50 ppm $SiI_4$. Table II shows that while purification is achieved by sublimation, the theoretical value is not approached. Several sublimers were used in this work, the most satisfactory being the one shown in Fig. 4. The advantage of this apparatus is that the condensed material can be collected without being removed from the apparatus; as a result, a fairly large amount of material can be collected in a single experiment. We can sublime about a pound of boron triiodide a day with this apparatus.

Zone-refining results on boron triiodide are not too conclusive. This is unfortunate, since the low melting point of the iodide would make it ideal for this technique. The results of our experiments indicate that magnesium, titanium, chromium, and aluminum have effective segregation coefficients less than 0.5 for zones traveled at the rate of 2 in./hr. Those for iron and silicon appear to be greater than 1, although the curves used for these determinations

## TABLE II

### Purity Achieved by Sublimation of $BI_3$

|  | Element, ppm | | | | | |
|---|---|---|---|---|---|---|
|  | Mg | Si | Ca | Ti | Al | Fe |
| Before sublimation | 155 | > 5000 | 275 | 310 | > 5000 | 86 |
| After sublimation | < 3 | 2400 | < 50 | < 6 | 68 | < 10 |

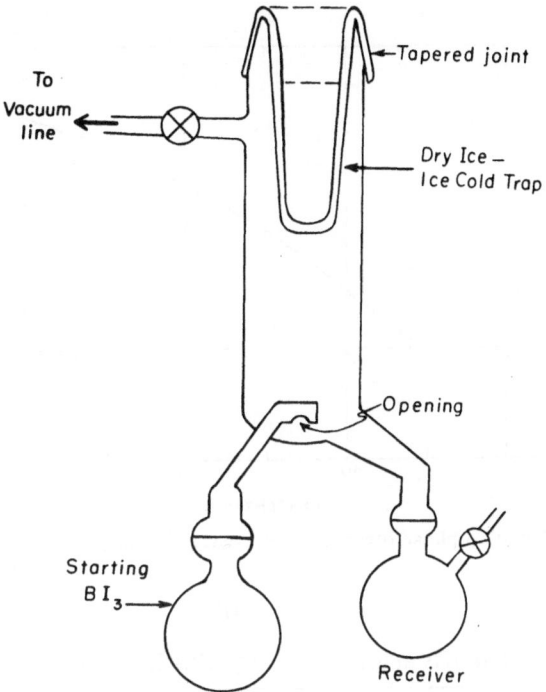

Fig. 4. Sublimation apparatus.

were exceedingly poor. An attempt was made to determine the phase diagram of the system $BI_3$–$SiI_4$ in order to estimate the equilibrium segregation coefficient of silicon tetraiodide, since silicon is the major impurity (with the possible exception of carbon) in boron. The results indicate, as shown in Fig. 5, that the segregation coefficient is actually considerably less than 1, rather than greater than 1, as zone-refining experiments seem to suggest. It is felt that the varying values from zone-refining may be due to the low-temperature eutectic (at about 36°) sucked back into the capillaries in the already-frozen zone material. This would be quite predominant in our experiments, since no interface cooling was employed and a fairly high concentration of silicon tetraiodide was present.

Zone-refining of boron tribromide, using cooling coils to maintain the bromide solid, has been reported [12], but the purity achieved is not known. The authors were not successful with this type of purification.

Distillation can be effectively applied to both the iodide and the

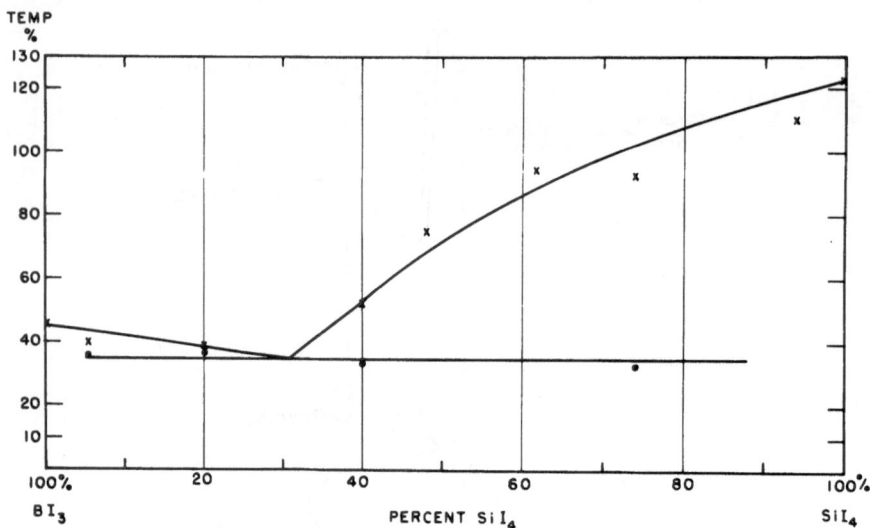

Fig. 5. Tentative phase diagram of $BI_3$–$SiI_4$ system (in weight percent).

## TABLE III

### Purification of $BI_3$ by Distillation

| Method | | Si | Mg | Sn | Fe | Cu | Pt | Cr | Purification factor for silicon |
|---|---|---|---|---|---|---|---|---|---|
| | | | | Impurities, ppm | | | | | |
| 40 - Plate Young column, no reflux | Starting | 1000 | 0.5 | 5 | < 1 | < 0.3 | 10 | ND* | 3.3 |
| | Final | 300 | 1.1 | < 2 | < 1 | < 0.5 | 20 | ND | |
| Glass-bead column 1X, reflux 10-1 | Starting | 300 | < 2 | < 2 | < 1 | < 0.5 | 20 | ND | 3 |
| | Final | 100 | < 2 | < 2 | < 1 | < 0.5 | 40 | ND | |
| Glass-bead column 2X, reflux 10-1 | Starting | 40–80 | 1 | < 2 | ND | ND | – | ND | 5–10 |
| | Final | 7 | 0.5 | < 2 | ND | ND | – | 7 | |
| Glass-bead column 1X, no reflux | Starting | 300 | < 2 | 2 | ND | ND | – | ND | 1 |
| | Final | 300 | < 1 | 2 | < 1 | 0.5 | 30 | ND | |
| Oldershaw column, reflux 20-1 | Starting | 40 | 0.5 | ND | ND | ND | 20 | ND | 10 |
| | Final | 4 | 0.5 | ND | ND | ND | 40 | ND | |

*ND indicates not detected.

bromide. The authors have used several types of distillation equipment for the iodide and are of the opinion that the Oldershaw column is the most satisfactory in terms of throughput and purity. The same type still has been used by Russian workers [10], although the efficiency of their purification is not known. Some of our distillation results on boron triiodide are shown in Table III. It is also apparent from this figure that a second distillation through the Oldershaw column did not significantly improve the purity of boron triiodide. Distillation of boron tribromide has also been extensively investigated by the authors, although, in general, unpacked columns or columns packed with quartz chips have been used, rather than more elaborate assemblies. In general, it has been found that after three distillations through a 60-in. column, the only impurity found spectroscopically is silicon (1–5 ppm).

Decomposition of the halide intermediate has been performed in a wide variety of ways, but decomposition on a resistance-heated filament, in vacuum or in hydrogen, is the most used. This is the simplest method and appears to give the best yields. Several filament materials have been used, but a high-purity boron filament is the most satisfactory. It has the advantage that the deposited material does not have to be separated from the filament prior to further treatment. Contamination from the substrate is eliminated also. The heating of the boron filament can be a problem, however, since it is necessary to increase markedly the power going into the filament during an experiment, in order to keep the outside surface at the decomposition temperature. This can produce strains that will break the filament, if care is not used.

At this point, we would like to compare the bromide and iodide processes in order to emphasize certain advantages and disadvantages of these intermediates. Both processes have a low boron-to-intermediate weight ratio, which means that a large amount of intermediate must be prepared in order to produce a small amount of boron. Boron triiodide contains about 2.5% boron, while boron tribromide contains about 4% boron. Thus, the bromide must be favored from this point of view. In addition, unless the iodine by-product can be effectively recycled in the iodide process, the cost of this process is considerably higher than that of the bromide process. The bromide, being a liquid, is easier to collect and transfer, but the low melting point of the iodide can be easily maintained during transferring and collecting operations. Taking all these factors into consideration, one would expect the bromide process to be considerably less expensive than the iodide process.

An advantage of the iodide process is that it is less limited in the number of purification techniques to which it can be subjected, and, hence, a purer intermediate possibly could be produced via the iodide. Thus, while the bromide process relies primarily on distillation, the iodide process could involve distillation followed by zone-refining prior to decomposition. A somewhat lower decomposition temperature is possible with the iodide, which also favors a purer product. The iodide also can be decomposed easily in vacuum, eliminating the necessity for hydrogen reduction.

The principal advantage of the iodide, however, is the possibility of producing a boron with a carbon content lower than that obtainable from the bromide. This is important because carbon is the main nongaseous component in boron. The advantage lies in the fact that, at present, no carbon–iodine compounds are known that are stable above 200°C, and, hence, any carbon-containing compound should decompose during the distillation process. Carbon tetrabromide and bromoform, on the other hand, are stable at the boiling point of the bromide.

Finally, we would like to present some of our results on the boron produced by these processes. Some of these results, which will be so labeled, are on boron rods that have been zoned in an electron-beam zone-refiner. In general, we have found that this reduces the gas content of the boron rod significantly, but the effect on carbon content is uncertain. Table IV shows the emission

## TABLE IV

### Emission Spectrographic Zoning Results (in ppm)

| Element | Starting boron | Iodide boron | Bromide boron |
|---------|----------------|--------------|---------------|
| Al      | 99             | ND*          | ND            |
| Si      | 1.6%           | 3            | 1–5           |
| Cu      | 358            | ND           | ND            |
| Ni      | 96             | ND           | ND            |
| Mg      | 3000           | ND           | ND            |
| Fe      | 2000           | ND           | ND            |
| Mn      | 2000           | ND           | ND            |
| Sn      | 9              | ND           | ND            |
| Cr      | 14             | ND           | ND            |
| Ti      | 12             | ND           | ND            |
| Co      | 11             | ND           | ND            |

*ND indicates not detected.

## TABLE V

Mass Spectrographic and Emission Spectroscopic Results (in ppm)

| Element | Mass spectrographic | | Emission spectroscopic | |
|---|---|---|---|---|
| | Bromide-process boron | Iodide-process boron | Starting boron, this study | Iodide-process boron |
| Al | 600 | 1 | | |
| Si | 100 | 11 | 1.6% | 4 |
| Ca | 10 | 100 | ND | |
| Co | 100 | ND | ND | |
| Mo | 4 | — | ND | |
| Cu | 1.5 | — | 358 | 2 |
| W | 30 | 13 | ND | |
| Ni | 1 | — | 96 | |
| Mg | — | — | 3000 | |
| Fe | 130 | 3.6 | 2000 | |
| Mn | — | — | 2000 | |
| Sn | — | — | 9 | |
| Cr | 200 | — | 14 | |
| Ti | — | — | 12 | |
| Co | 100 | — | 11 | |
| C | >500 | 500 | | |

spectrographic results of impurities detected in typical boron samples, together with an analysis of the starting boron. It can be seen that the emission spectrographic results indicate no significant variation in impurities between the two processes. Mass spectrographic results, shown in Table V, however, do indicate changes in the concentration of impurities.

The carbon analysis of the samples was done by Professor J. M. Walker using a technique discussed elsewhere.* His results indicate that boron from the bromide process contains about three times as much carbon as boron from the iodide process.

Table VI gives data from an analysis of a boron sample (made by the bromide process) before and after zone-refining. As mentioned previously, the gaseous components appear to be more significantly reduced than the metallic impurities and carbon. This particular sample had four zone passes. Thus, it appears that,

*See Walker and Starks, "Carbon Determination in Hyperpure Elemental Boron Using Gas Chromatography," this volume, p. 63.

## TABLE VI

### Analysis of Bromide-Process Boron

| Element | Unzoned crystal, ppm | Zoned crystal, ppm |
|---------|---------------------|-------------------|
| C | 400 | 910 |
| N | 550 | 900 |
| O | 5000 | 790 |
| Mg | 25 | — |
| Al | 67 | 7.6 |
| Si | 1300 | 35 |
| P | 4.2 | — |
| K | — | 12,600 |
| Ca | — | 4.3 |
| Sc | 0.2 | — |
| Ti | 0.9 | — |
| V | 0.6 | — |
| Mn | 1.2 | — |
| Fe | 140 | 2.5 |
| Co | 0.2 | — |
| Ni | 2.9 | — |
| Cu | 2.2 | — |
| Ga | 0.2 | — |
| Ge | 6.5—65 | 10—50 |
| Rb | 50 | trace |
| Sr | 0.15 | — |
| Mo | 7.5 | trace |
| Ru | 4—25 | — |
| Rh | 0.14 | — |
| Pd | 0.5 | — |
| Pt | 30 | — |

while the zone-refining of boron may not be a significant purification technique for some impurities, it is of definite use in ridding the final boron product of volatile impurities.

## REFERENCES

1. Nies, N. P., Jr., J. Electrochem. Soc. 107: 817 (1960).
2. Stern, S. R., and L. Lynds, J. Electrochem. Soc. 105: 676 (1958).
3. Armington, A. F., W. D. Potter, and L. Tanner, J. Appl. Phys. 35: 730 (1964).
4. Starks, R. J., and J. T. Buford, Electrochem. Tech. 1: 108 (1963).
5. Armington, A. F., and G. F. Dillon, Trans. AIME 224: 631 (1963).
6. Schumb, W. L., F. L. Gamble, and M. D. Banus, J. Am. Chem. Soc. 71:3225 (1949).
7. Renner, T., German Patent 1,056,592 (1959).
8. McCarty, L. V., and D. R. Carpenter, J. Electrochem. Soc. 107: 38 (1960).

9. Armington, A. F., G. F. Dillon, and R. F. Mitchell, Trans. AIME 230: 350 (1964).
10. Ivanov-Emin, B. N., L. A. Nisel'son, and I. V. Petrosevich, Zhur. Priklad. Khim. 34:2378 (1961).
11. Andersen, H. C., and L. M. Belz, J. Am. Chem. Soc. 75: 4828 (1953).
12. DeLazlo, H., L. Light and Co., Bucks, England, private communication.

# Studies on Crystallization of Pure Boron

T. Niemyski, I. Pracka, R. Szczerbiński,
and Z. Frukacz

*Institute of Physics, Polish Academy of Sciences*
*Warsaw, Poland*

The optimum conditions for the growth of $\beta$-rhombohedral boron crystals were studied. Boron powder was prepared by the decomposition of boron trichloride in a hydrogen atmosphere. The powder was then melted in an argon atmosphere by induction heating to form rods. The melt was contained either in quartz boats lined with pure boron nitride or in internally water-cooled silver boats. Single-crystal growth was accomplished without using crucible methods, by passing a floating liquid zone along the vertically mounted rod [1], or by using a pedestal method, with a boron rod as support for a molten drop, and pulling from this drop. The crystal growth was performed in a vacuum, using either induction or electron-beam heating. Single crystals of several centimeters in length were obtained. It was also attempted to crystallize highly doped boron. Structural and electronic phenomena attributed to the influence of impurities are discussed.

## PREPARATION OF POLYCRYSTALLINE BORON RODS

Boron powder was obtained from the thermal decomposition of $BCl_3$ in a hydrogen atmosphere [1,2]. The product was deposited inside a quartz tube or on the surface of a boron rod. The boron deposit was then melted by high-frequency induction heating. The rods, which are the most convenient form of boron to use for crucible-free crystal growth, were prepared by melting boron powder in quartz or silver boats. The quartz boats were lined with pure, pressed, boron nitride powder. The silver boats were internally cooled with a fast flow of water. The conversion of the boron powder into solid polycrystalline rods was accomplished by moving a molten zone along the crucible. It was found that rapid cooling of the liquid zone to a temperature somewhat below the melting point of boron and subsequent slow cooling yielded rods with the best mechanical properties. For this reason, preparation of the boron rods by the stepwise passage of a broad molten zone is better than continuous shifting of the zone. Of the two kinds of boats used, the quartz boats lined with boron nitride were more resistant to local overheating of the liquid boron, which condition

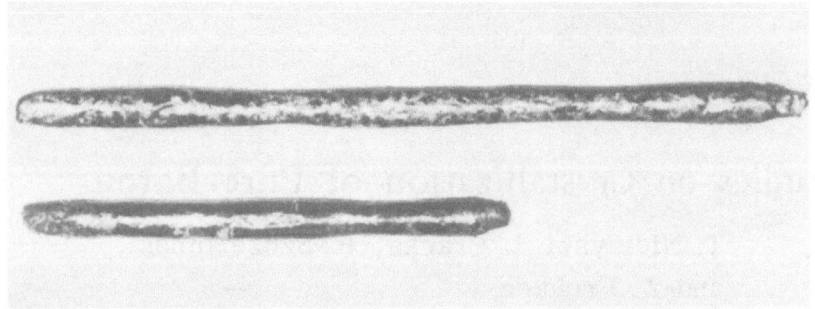

Fig. 1. Boron rod obtained in the boat by the melting of boron powder by means of induction heating.

allowed the formation of a broad liquid zone that consequently produced a small temperature gradient during the cooling process. The main disadvantage of the boron nitride-lined quartz boat was the contamination of the boron with boron nitride, which hampered the subsequent crystal growth.

Boron rods with a clean surface could be prepared in the water-cooled silver boats. The temperature gradient on both sides of the molten zone is great, however. It was thus necessary to precisely stabilize the power from the heating generator and the pressure of the cooling water. These conditions and the fact that only a narrow molten zone could be obtained must be considered as disadvantages. Also, any deviations from the established conditions caused reactions between the molten boron and the boat.

Boron rods, 1 cm in diameter and up to 30 cm in length, were obtained in both types of boats (Fig. 1).

## SINGLE-CRYSTAL GROWTH

Three types of apparatus for crucible-free growth of $\beta$-rhombohedral boron single crystals were used. The apparatus for crystal growth by multiple passages of a floating molten zone was contained in a vacuum tight chamber. An induction coil was placed inside the chamber and connected to a 4-Mcps generator. The power of the generator could be precisely regulated in the range 2–15 kW. The boron rod was vertically mounted between two tantalum holders which were connected with the driving mechanisms. The holders could be rotated and moved upwards or downwards, separately or synchronously. Up to 800°C, the rod was heated indirectly through the tantalum holder; above 800°C, it was heated directly by the high-frequency field. The liquid zone had a length approximately equal to the diameter of the rod. The geometry of the induction coil was adjusted to the diameter of the rod and to the impedance of the generator. When the coil was properly chosen, the length of the molten zone could be regulated to a great extent by the power of the generator. The crystal growth was carried out at different velocities of the liquid zone, with or without rotation of the entire rod, and with rotation of the upper part of the rod only. The liquid zone was passed several times. The crystal growth was performed in argon, in hydrogen, in a mixture of these two gases, and in a vacuum.

Similar experiments were made in an apparatus with electron-beam heating. This apparatus was equipped with mechanisms similar to those described above. However, the crystallization conditions were different from those of the induction-heating apparatus: The heating power per unit of heated surface was very high, the accelerated electrons heated the surface of the rod only, and, as a consequence, the temperature distribution was somewhat varied. Also, ionic bombardment took place.

A third type of apparatus that was designed for crystal growth employed the pulling of a crystal from a molten drop. Induction heating and an ambient gas or vacuum were used, as above, for the protection of the melt. The seed and the rod with the molten drop could be rotated and moved up and down independently. The diameter of the molten drop was approximately twice the diameter of the rod. The diameter of the pulled crystal was approximately four times smaller than the diameter of the drop.

These three types of apparatus allowed the evaluation of the influences of the ambient gas and the temperature gradient on the crystallization process.

## RESULTS AND DISCUSSION

A close relationship was found between the quality of the boron rods and the crystallization process. The crystallization of the rods obtained in the boats lined with boron nitride was difficult because of the contamination. Boron nitride formed crystallization centers which caused the simultaneous growth of a large number of crystals on the surface of the solidifying zone. The quantitative removal of boron nitride was very difficult. Also, the rods had a tendency to crack during the crystal growth.

The diameters of the rods prepared in silver boats usually were not constant throughout the length of the rods, and these rods broke easily. The rods prepared in a vacuum had better mechanical

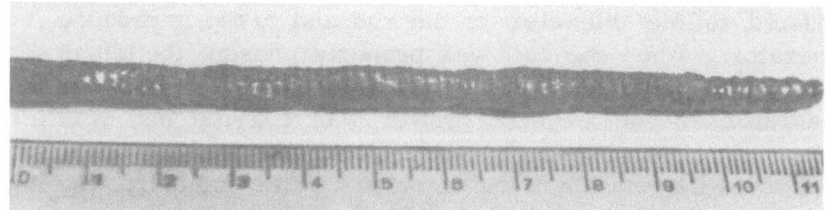

Fig. 2. Boron crystal with a distinct spiral line on the walls.

properties than those made in ambient gases. Also, the surfaces of the vacuum-prepared rods were cleaner.

Liquid boron has a relatively high surface tension, which is undoubtedly a great advantage for crucible-free crystallization. The high viscosity of liquid boron at temperatures close to its melting point introduces certain difficulties. High viscosity makes a quick rotation of the seed unfeasible; a seed rotation of only 2 rpm (at a pulling rate of about 15 cm/hr) caused distinct marks on the surface of the growing crystal (Fig. 2). On the other hand, an increase of the speed of the floating zone or of the pulling speed from the molten drop in the range 1–15 cm/hr markedly improved the conditions for the growth of single crystals.

When the surface of the rods (even those of irregular diameter) was sufficiently clean, one zone pass resulted in a noticeable directional growth of crystals. Monocrystalline zones of a length of several centimeters were obtained after several passes without the use of a seed (Fig. 3). When a seed crystal is used, it must be carefully fused with the melt. The surface of the melt must be clean, of course. Even so, a polycrystalline layer always appears

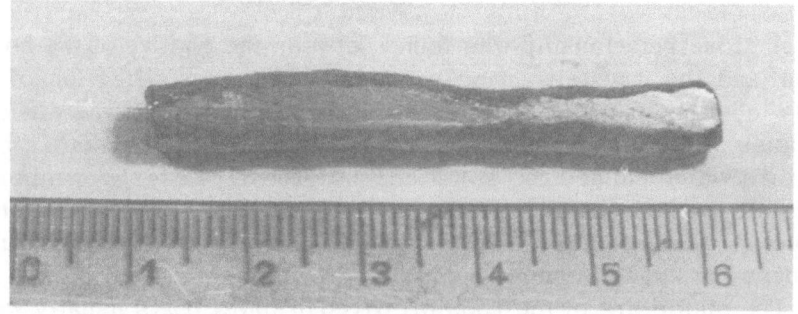

Fig. 3. Monocrystalline areas growing spontaneously in the rod crystallized without seed.

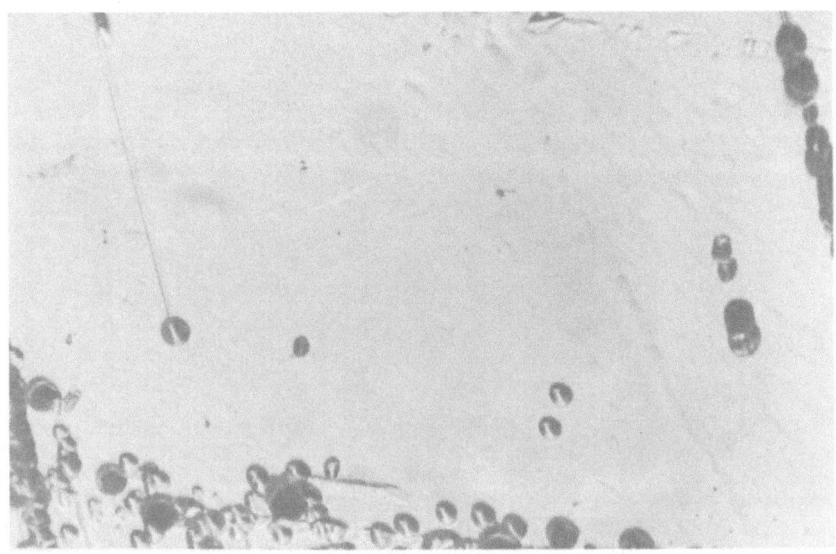

Fig. 4. Characteristic (for boron crystals) increase of dislocation density from center to perimeter of the rod.

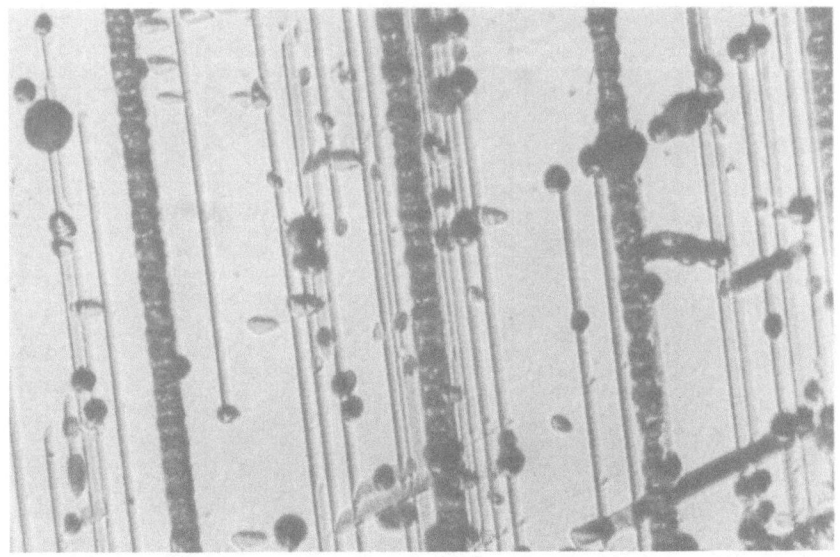

Fig. 5. Etched crystal fracture surface.

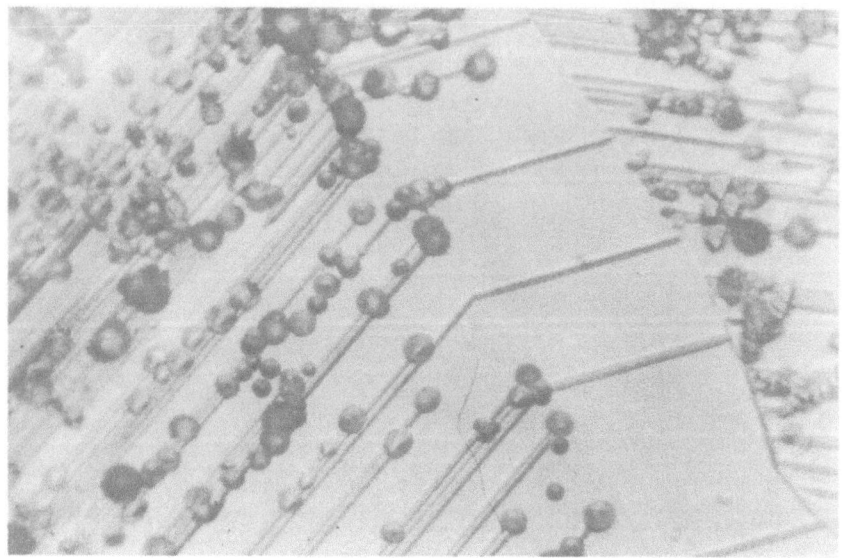

Fig. 6. Etched grain boundaries in boron rod.

Fig. 7. Spiral dislocations on the etched surface.

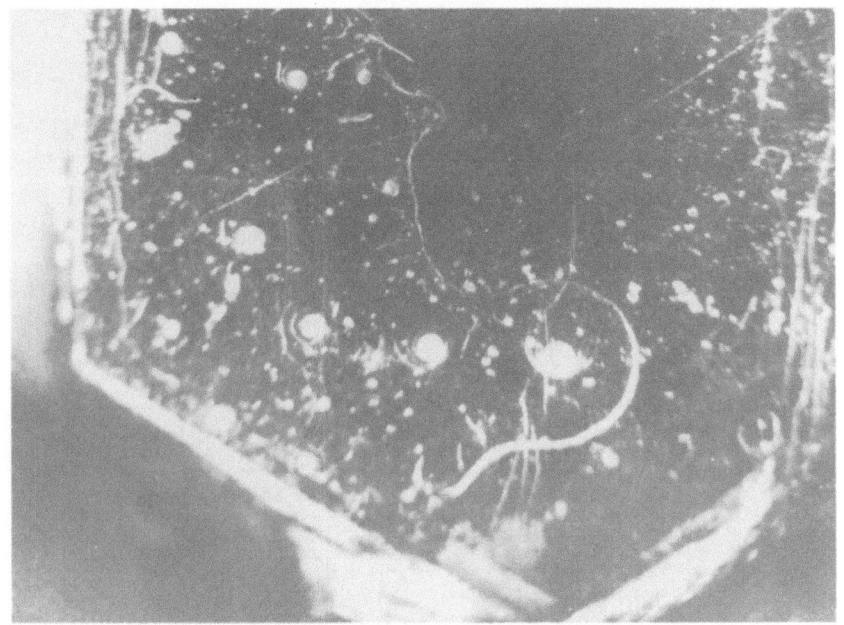

Fig. 8. Aluminum in spiral dislocations.

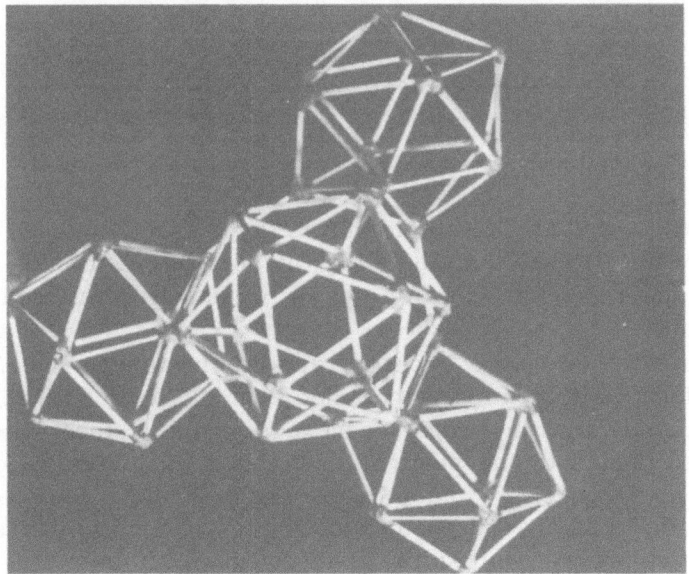

Fig. 9. Fragment of boron unit-cell model.

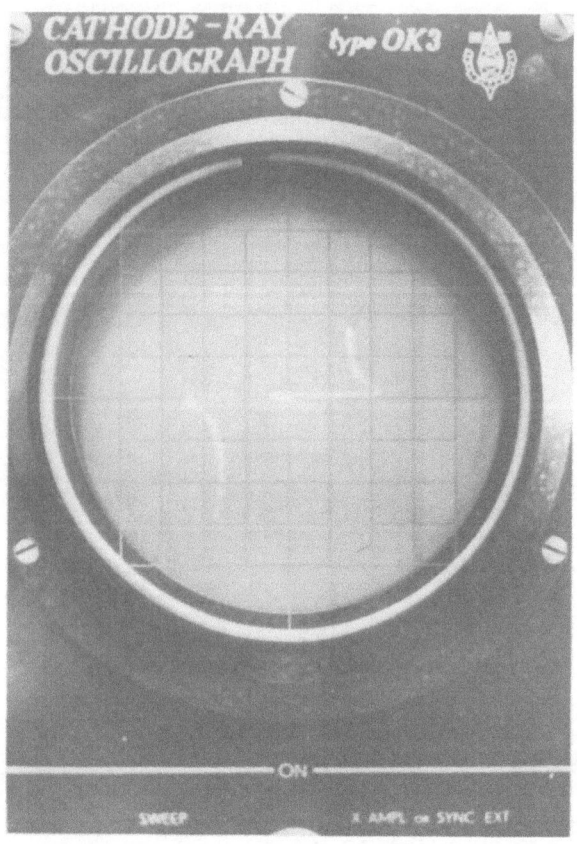

Fig. 10. Current – voltage characteristic for doped boron crystal.

around the crystal. The origin of this layer is not clear as yet. It is interesting to note that generally the peripheral crystals did not grow inward; the interior of a pulled rod usually remained mono-crystalline.

Boron crystals have a characteristic fracture surface. When the surface was chemically polished with 20% $K_3Fe$ $(CN)_6$ plus 20% KOH and etched in aqua regia, an increase of the dislocation density from the center to the perimeter of the rod could be recognized (Fig. 4). Figures 5 and 6 show the polished and etched grain boundaries. The crystal photograph of Fig. 7 seems to indicate that growth originating from spiral dislocations is an important factor.

The three methods of crystal growth that were applied provided different temperature distributions during crystallization. An analysis of the temperature distribution for the growth of crystals in a vacuum shows that thermal strains may occur, causing cracking of the rods. However, other factors could be responsible for the cracking. In this connection, it is noted that exceptionally brittle crystals of boron were formed in argon, especially when that gas contained small amounts of nitrogen. Boron crystals prepared in vacuum and then heated in hydrogen were also very brittle. The $\alpha$-rhombohedral modification of boron was observed on fracture surfaces in some instances.

Peculiar behavior of some impurities during the boron-crystal growth was observed. Boron highly doped with aluminum ejected aluminum atoms into the hollow spiral dislocations (Fig. 8). It seems that certain modifications of boron occur locally, together with high, local concentrations of certain impurities, indicating the enhancement of these crystal modifications by specific impurities. Certain impurity atoms probably come to rest in the screened spaces between icosahedrons (Fig. 9), and lead to phenomena nontypical for semiconductors with impurities, e.g., avalanche currents. The avalanche current occurs at room temperature at voltages exceeding a certain threshold value. The increase of the avalanche current with voltage is steep. It is followed by a branch of the current-voltage characteristic where the resistance is large and negative (Fig. 10). It seems that the phenomenon of avalanche current is connected with the ionization of impurity atoms which otherwise do not participate in the electrical conductivity at room temperature. These phenomena have new, potential, practical applications. Preliminary experiments on crystallization from the gaseous phase of boron have shown that, depending on the temperature range, different, stable, crystalline modifications of boron are formed.

## REFERENCES

1. Niemyski, T., and Z. Olempska, J. Less-Common Metals 4:235 (1962).
2. Niemyski, T., Z. Olempska, and I. Pracka, Proc. Intern. Conf. Phys. Semicond. (Exeter), 1962, p. 722.

# Preparation and Characterization
# of Single-Crystal Boron

Irving R. King, Frank E. Wawner, Jr.,

Gerald R. Taylor, Jr., and Claude P. Talley

*Texaco Experiment Incorporated*
*Richmond, Virginia*

Although boron was first isolated over 150 years ago, little information concerning the properties of this material is available even today. This is probably due, in large part, to the relative difficulty of obtaining pure material and the fact that it exists in amorphous form and several crystalline modifications. Renewed interest in the properties of this material has been stimulated in recent years by the potential use of boron in semiconductor devices, high-energy fuels, and structural materials. To learn more about these properties a program was initiated in this laboratory to produce and characterize high-purity, single crystals of boron. A vapor deposition technique was used to form polycrystalline boron rods on fine tungsten filaments. The polycrystalline rods were then zone-refined to produce single crystals of boron. Single crystals thus produced were characterized in several ways. The techniques used and the results obtained are reported below.

## BORON PREPARATION

### Production of Polycrystalline Rods

Chemical vapor plating is considered the best method for producing many elements of high purity. This is particularly true in the case of elemental boron [1]. In this investigation, high-purity boron was obtained by reducing boron tribromide vapor with hydrogen in the vicinity of a small-diameter, electrically heated filament. A schematic diagram of the deposition apparatus is shown in Fig. 1. The general techniques involved are covered by Powell et al. [2].

To assure that the polycrystalline rods thus formed were of high purity prior to zone-refining, all hydrogen used was purified by diffusion through palladium, and the boron tribromide was triply distilled in a special glass still operated without the use of lubricant on any of the joints. Boron tribromide was stored in quartz containers, and a quartz chamber was used for the boron deposition. To keep contaminants at a minimum, the tungsten substrate filaments

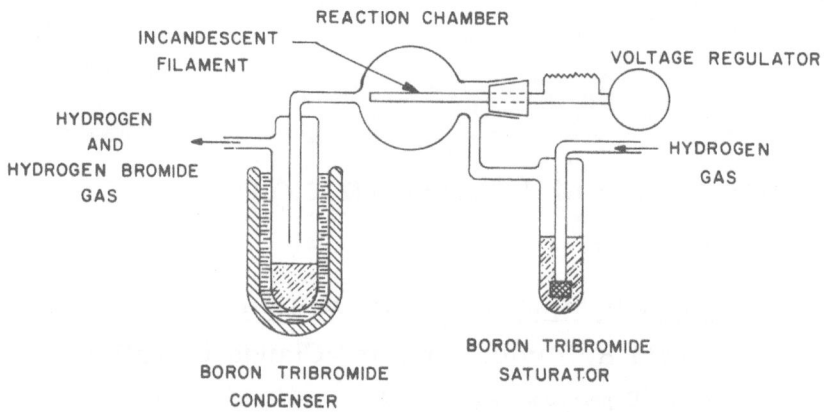

Fig. 1. Schematic diagram of apparatus for depositing boron.

were held to 6 and 12 $\mu$ in diameter. Following these precautions, no impurities were detectable by infrared analysis in either the hydrogen or the boron tribromide. Mass spectrometric analysis of the boron tribromide did reveal some mass peaks that apparently corresponded to small amounts of iron, carbon, and silicon. No impurities were detected in the hydrogen.

A number of rods, 3–4 mm in diameter and 10 cm long, were prepared with both size substrates mentioned above to fulfill the task of single-crystal specimen preparation and chemical analysis before and after zone-melting. The purity of these polycrystalline samples was investigated by wet-chemical analysis for total boron and by emission-spectrographic analysis. Nitrogen, oxygen, hydrogen, and carbon analyses were obtained from Dr. Manley W. Mallet of the Battelle Memorial Institute. Vacuum fusion and carbon dioxide combustion techniques were used.

The average boron content by titration was 99.9%, with an average deviation of ± 0.1%. The accuracy of the method was estimated at 0.2%. The calculated tungsten content was near 20 ppm for the rods prepared on 6-$\mu$-diameter tungsten substrates. Emission-spectrographic analysis indicated the presence of approximately 40 ppm iron, 20 ppm silicon, 3 ppm copper, and 1 ppm magnesium. The gas analysis showed that the hydrogen content was 39 ppm; however, no oxygen or nitrogen was detected. Their detection limits were 7 and 10 ppm, respectively. The combustion analysis showed 840 ppm carbon present. Therefore, it is apparent that, at this stage, carbon constitutes the major impurity in the boron prior to zone-refining. It is interesting to observe that for the rods prepared on

$6-\mu$-diameter tungsten substrates, the concentrations of iron and carbon are greater than the starting impurity of the substrate wire.

### Preparation of Single-Crystal Boron

The floating-zone melting method has been used quite success-fully for producing large single crystals of a number of materials, including silicon, germanium, and tungsten [3,4]. This technique enables one to grow the crystals without fear of contamination from a crucible material. An electron-bombardment, floating-zone melting apparatus for producing large single crystals of boron was designed and constructed. The apparatus was similar to that described by Calverly et al. [5]. To provide an exceptionally clean vacuum for the zone-refining, an ion pump, capable of working

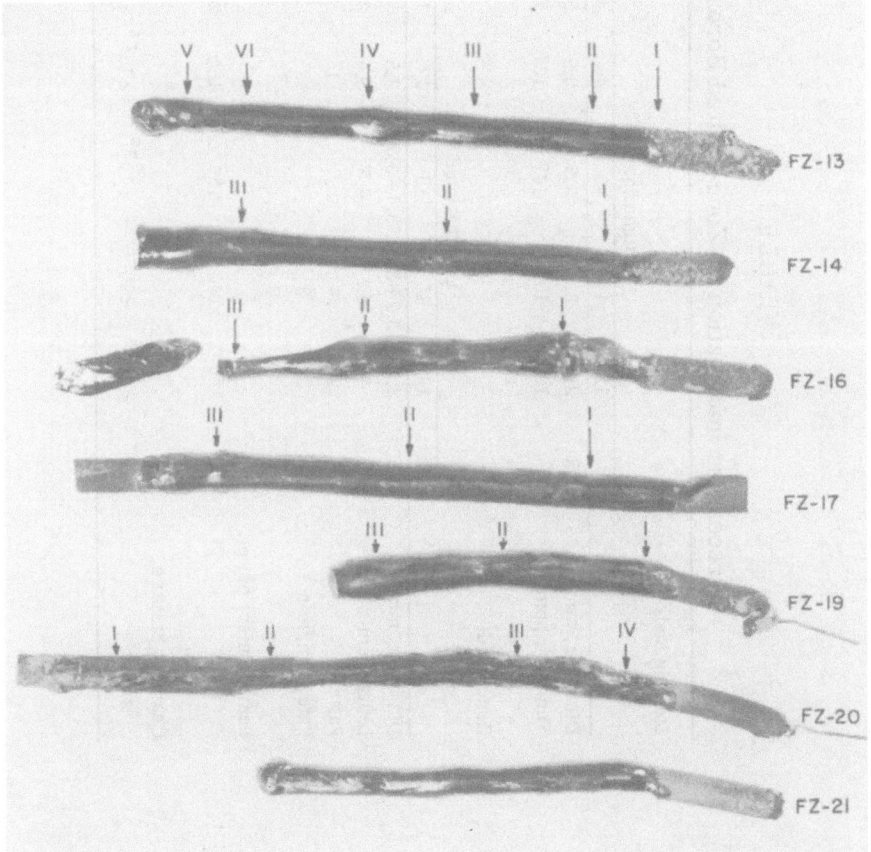

Fig. 2. Zone-melted boron rods showing areas where X-ray diffraction photographs were taken.

## TABLE I

### Processing Information on Zone-Melted Boron Rods

**Starting Rod:**

| | Designation Z | | | | | | | |
|---|---|---|---|---|---|---|---|---|
| | 28 | 19 | 44 | 45 | 44 | 46 | 48 | 52 |
| Diameter, mm | 3,5 | 3,5 | 3,5 | 3,5 | 3,5 | 3,5 | 3,3 | 3,2 |
| Tungsten, ppm | 17 | 107 | 107 | 107 | 107 | 107 | 118 | 21 |

**Zoned Rod:**

| | Designation FZ | | | | | | | |
|---|---|---|---|---|---|---|---|---|
| | 13 | 14 | 15 | 16 | 17 | 19 | 20 | 21 |
| Diameter, mm | 3,5 | 3,5 | 3,2$-$3,7 | 3,4$-$3,7 | 3,5 | 3,8 | 3,2$-$3,6 | 3,2$-$3,4 |
| Length, cm | 4,8 | 4 | 4,6 | 4,5 | 5,0 | 3,0 | 6,0 | 3,6 |
| Passes | 1 | 2 | 4 | 2 | 1 | 3 | 3 | 4 |
| Rate, mm/min | 3 | 3 | 3 | 3 | 3 | 3 | 3 | 3 |
| Bombardment power, watts | 135 | 150 | 140 | 160 | 160 | 200 | 160 | 100 |
| Chamber pressure, mm Hg | $<10^{-6}$ | $<10^{-6}$ | $<10^{-7}$ | $<10^{-6}$ | $<10^{-6}$ | $<10^{-6}$ | $<10^{-6}$ | $<10^{-7}$ |

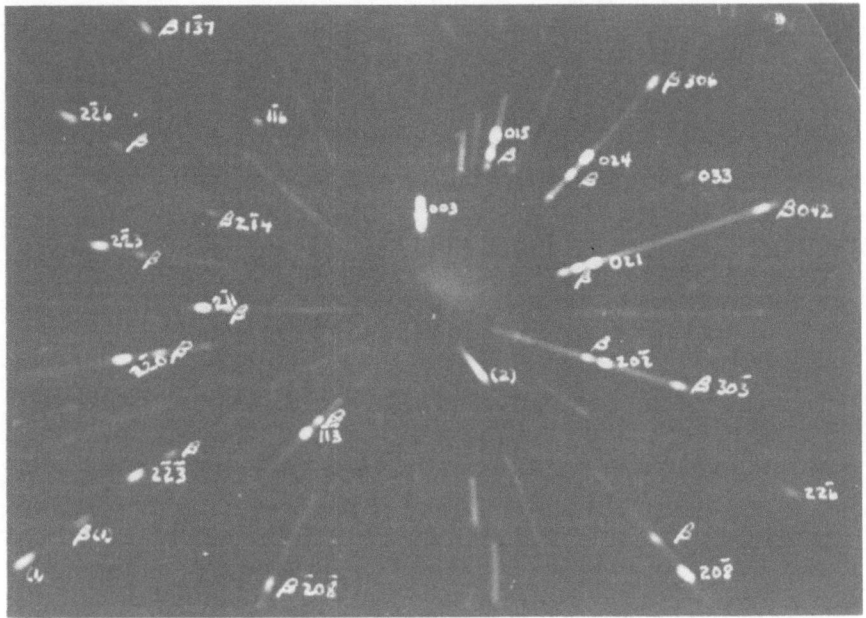

Fig. 3. Typical oscillation photograph of boron single crystals. The pattern has been indexed according to the beta-rhombohedral structure. The continuous radiation streaks can readily be seen. Note: (1) An unindexed reflection corresponding to a $d$-value of 2.038 A. (2) A continuous streak from 104 planes.

pressures in the $10^{-7}$ mm Hg range, was incorporated into the system. A number of polycrystalline boron rods were zone-melted to provide single-crystal samples for chemical analysis and electrical measurements.

A summary of the processing information on some of the runs is given in Table I. The usual procedure was to make the second zoning pass in a direction opposite the first pass in order to allow the large grains or single crystals that developed during the first pass to continue to grow by starting with this section for the second pass. Subsequent passes were made in the same direction as the second pass to carry out zone-refining. Figure 2 is a photograph of the rods.

## BORON CHARACTERIZATION

### X-Ray Diffraction

Oscillation and rotation X-ray diffraction photographs were taken of the zone-melted samples to verify their structure and single crystallinity. Figure 3 is typical of the oscillation photographs

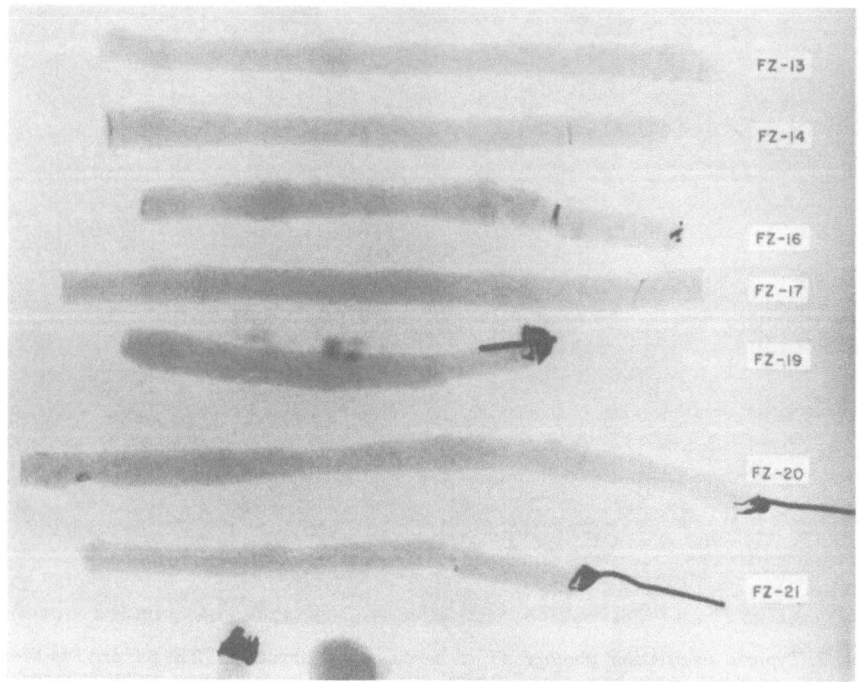

Fig. 4. X-ray radiograph of boron rods. Tungsten accumulations, tungsten substrate filaments, and platinum wire supports can be seen near ends of rods. Dark spot in center is due to reflection from plate holder.

obtained from the rods. These photographs show that the rods are single-crystal over a considerable portion of their lengths. The main reflections on this photograph have been indexed with data from the literature [6] showing that the crystal is of the high-temperature beta-rhombohedral phase. An unfiltered copper target was used as a radiation source. The beta reflections and the continuous radiation streaks stand out on the photograph; however, they cause no confusion in interpreting the pattern. The oscillation angle was 10° with a sample-to-film distance of 10 cm. The X-ray equipment was operated at 35 kV and 15 mA for each 15-min exposure.

Three to four photographs were taken of each zone-melted rod to test for variations in orientation or phase changes along the length of the rods. The majority of the rods X-rayed were single-crystal throughout the major portion of their bodies, and all were of the beta-rhombohedral phase. A photograph of the zoned rods, indicating where the X-ray diffraction photographs were taken, is shown in Fig. 2. In all of the crystals, the C axis was roughly along the length of the zoned rod varying from 5 to 15° from the cylindrical axis. An X-ray radiograph (Fig. 4), taken of the rods in approximately the same orientation as in Fig. 2, shows the platinum support wires and accumulation of tungsten and other impurities at the ends of the rods. Density gradients in the radiograph along the length of each rod are probably due to their nonuniform diameter. The dark area in the center of the picture is a reflection from the plate holders and is not in the crystal. Near the zone-melted ends of each rod, there were reflections from two, three, or more crystals. This is to be expected since the zoning passes did not extend to the same point each time, and when the molten zone cooled, the solid interfaces met at the center of the molten area. Hence, there is probably a sharp boundary between single-crystal material of the orientation of the majority of the rod and smaller single-crystal areas produced by variation of the zoning passes. On occasion, small crystallites formed on the surface; these contributed to the diffraction pattern. If there were many, a powder pattern was detected, superimposed on the main single-crystal pattern. If there were few, other single-crystal patterns appeared on the diffraction photograph. These patterns could easily be isolated from that of the main crystal body, since the reflections would be much less intense and usually in a slightly different orientation.

Purity. Wet-chemical analysis of several zone-melted samples gave an average total boron content of 99.9%. The tungsten content for these rods was calculated on a simple dilution basis to be in the ppm range, and approximately 40 ppm iron, 20 ppm silicon, 3 ppm copper, and 1 ppm magnesium were detected by emission-spectrographic analysis. The combustion conductometric method for carbon gave an average value of 830 ppm. The lowest value obtained was 400 ppm. Vacuum-fusion analysis indicated 15 ppm oxygen and 3 ppm hydrogen. No nitrogen was detected; the detection limit was 3 ppm.

The zone-melting operation apparently did not significantly reduce the impurity content except in the case of hydrogen and, possibly, tungsten. Radiograms obtained on zone-melted samples indicated tungsten was concentrated in the advancing molten zone.

Resistivity. Measurements of resistivity as a function of temperature offer a convenient means of determining the energy gap and the intrinsic resistivity of a sample, as well as affording an

## TABLE II

### Electrical and Optical Properties of Single-Crystal Boron

| Designation FZ | Calculated tungsten content,* ppm | Number of zone passes | Resistivity,† Ω-cm | Thermo-electric type † | Thermo-electric power,† μ V/°C | Band gap, eV | Infrared spectra | Photo-conductivity observed | Recti-fication observed |
|---|---|---|---|---|---|---|---|---|---|
| 1 | 2360 | 2 | $1.5 \cdot 10^6$ | p-type | 520 | 1.32 | – | yes | no |
| 2 | 2360 | 2 | $3.0 \cdot 10^6$ | p-type | 400 | 1.39 | – | – | – |
| 4 | 2060 | 2 | $5.2 \cdot 10^5$ | p-type | – | – | – | yes | no |
| 6 | 1470 | 2 | – | – | – | – | $0.9 \sim 8.5 \mu$ "window" | – | – |
| 7 | 85 | 4 | $6.6 \cdot 10^6$ | p-type | – | – | – | yes | no |
| 8 | 13 | 2 | – | p-type | – | – | – | – | – |
| 9 | 16 | 1 | – | p-type | – | – | – | – | – |
| 13 | 17 | 1 | $4.9 \cdot 10^6$ | p-type | – | 1.35 | – | yes | no |
| 14 | 107 | 4 | – | – | – | – | – | – | – |
| 16 | 107 | 2 | – | – | – | – | – | – | – |
| 17 | 107 | 1 | $4.3 \cdot 10^6$ | p-type | – | – | – | yes | no |
| 19 | 107 | 3 | – | – | – | – | – | – | – |
| 20 | 118 | 3 | $3.1 \cdot 10^6$ | p-type | – | – | – | yes | no |
| 21 | 21 | 4 | $2.7 \cdot 10^6$ | p-type | – | – | – | yes | no |

*Before zoning.
†At room temperature.

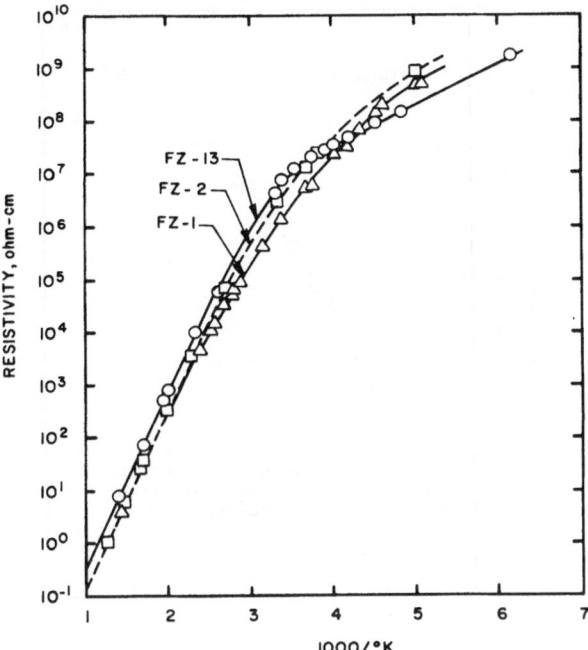

Fig. 5. Resistivity of single-crystal boron as a function of reciprocal temperature.

indication of its purity. Measurements of room-temperature resistivity of a number of crystals are shown in Table II, and complete resistivity-versus-temperature curves for three of the samples are shown in Fig. 5. Both the total-resistance method and the two-probe method were used in these determinations. Agreement between the two methods was quite good (Fig. 6).

In the total-resistance method, electrodes are connected to each end of the crystal, either by fusion or by silver paint, and the potential drop across the entire crystal and the total current through the crystal are determined. Resistivity is then obtained from the relation

$$\rho = (V/I)\,(A/L)$$

where $V$ is the potential drop across the length of the sample, $I$ is the total current through the sample, $A$ is the cross-sectional area of the sample, and $L$ is the length of the sample. This technique eliminates the use of probes and the problems associated with their use, but introduces a secondary problem of determining the effective cross-sectional area over an appreciable length of crystal.

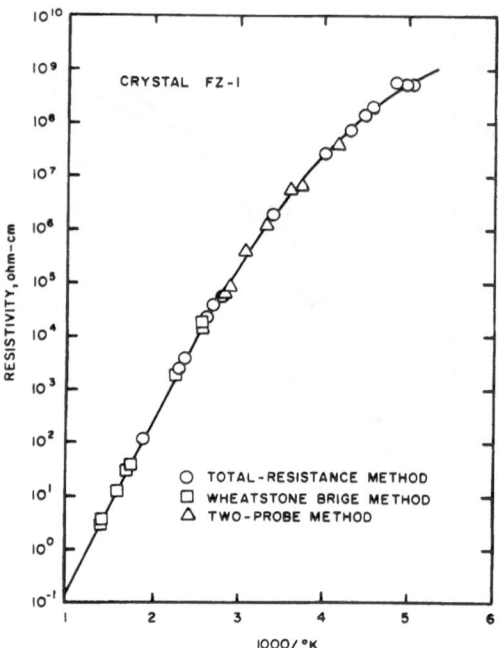

Fig. 6. Comparison of various methods of measuring resistivity of single-crystal boron.

The technique is also lacking in that it gives no information concerning the homogeneity of the sample.

In the two-probe method, the same setup as described above is used, except that the potential drop, instead of being measured across the entire length of the crystal, is measured between two probes in contact with the surface of the crystal. In this technique, careful attention must be given to probe spacing, pressure contact on the probes, and the area of contact between probe and sample. Probe spacing was measured to within ±0.05 mm with a calibrated stereoscopic microscope. Contact pressure was fixed by the apparatus. The area of contact between the probe and the sample was kept to a minimum by using pointed tungsten probes. Potentials were measured with a Keithly Model 200B electrometer having an input impedance of $10^{16}\Omega$. A high-impedance meter was necessary because of the very high resistance of the sample, especially at the lower temperatures. Shielded leads were used when necessary to reduce stray pickup. Since the possibility of temperature rise by Joule heating of the sample was ever present, current flow was held to a small value, and measurements were repeated after current had flowed through the sample for several minutes. A change in resistivity with time indicated Joule heating. In several instances, current through the crystal was plotted as a function of applied voltage. The point at which such a plot deviated from a straight line indicated the current above which Joule heating caused an appreciable temperature rise. Measurements were always conducted well below this point.

A comparison of the results obtained by the various methods of measuring resistivity is shown in Fig. 6. Excellent agreement was obtained in all cases. At the higher temperatures, the sample resistance was also measured on a Wheatstone bridge. The particular sample used in these measurements was 1.5 mm in diameter and 13 mm long. Distance between potential probes was 4.5 mm.

Homogeneity tests were run on a number of the longer crystals. Figure 7 shows a typical curve. Curves were obtained by measuring the potential drop between one end of the crystal and a test probe in contact with the surface of the crystal as the probe was moved along the length of the sample. Inhomogeneities are indicated by changes in the slope of the curve. From a knowledge of the cross-sectional area of the sample and the current flow through the sample, the resistivity at any point along the curve may be obtained. Resistivities determined in this manner agree quite well with those determined by the two-probe method or the total-resistance method. Homogeneity tests were also made on several crystals by the

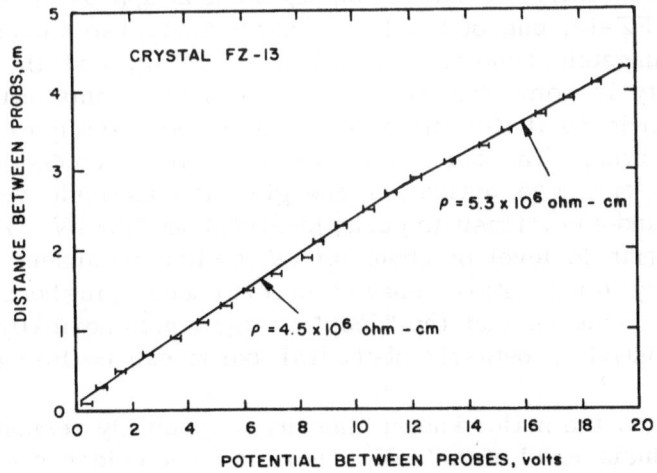

Fig. 7. Typical homogeneity plot for a sample of single crystal boron. Current through crystal, $9.4 \cdot 10^{-8}$ A. Average cross-sectional area of sample, $9.6 \cdot 10^{-2}$ cm$^2$.

resistivity technique. In this technique, the two-probe method was used to determine the actual resistivity at various points along the length of the crystals. Both methods showed the samples tested to be quite homogeneous. Crystal FZ-13, used in Fig. 6, was rotated 120° and the above procedure repeated. Again the results indicated that the sample was homogeneous over the major portion of its length. In most cases, near the ends of the samples, a decided change in slope or resistivity was noted. The resistivity decreased in these regions, suggesting that the impurity content may be greater near the ends of the sample. This would be reasonable, since the zone-melting process carries some of the impurities to the ends of the rods.

No major differences in room-temperature resistivities were noted in the crystals tested, although the maximum initial tungsten content varied over two orders of magnitude and the number of zone passes varied from 1 to 4. In all cases, room-temperature resistivities were of the order of $10^6$ $\Omega$-cm. These are tabulated along with information on other measurements in Table II.

Figure 5 shows the effect of temperature on the resistivity of several single-crystal samples. Temperatures were varied from 163 to 780°K. Resistivities ranged from 1 to $10^9$ $\Omega$-cm. The high-temperature end of the curve represents the intrinsic conduction region and permits an evaluation of the energy gap. Values obtained from the three curves shown in Fig. 5 give an average value of 1.35 eV with intrinsic conduction beginning at approximately 150°C. Crystal FZ-13, one of the last crystals made, had a considerably lower tungsten content than FZ-1 or FZ-2, and the intrinsic resistivity is somewhat higher. At the lower temperatures, this curve tends to flatten off more rapidly and actually crosses the other curves. The curve for FZ-13 can be accounted for on the basis of only two activation energies, the intrinsic band gap of 1.35 eV and an extrinsic impurity level of about 0.32 eV. The often-found extrinsic level of about 0.6 eV is virtually absent [7]. This resistivity–temperature behavior is very encouraging because it has provided evidence that the "deep" energy levels normally found in high-resistivity, beta-rhombohedral boron can be brought under control.

Carbon, the major known impurity, is probably responsible for the extrinsic level of 0.32 eV. There is some evidence in favor of this view. At low temperatures, where extrinsic controlled conductivity predominates, the boron has been found to be $n$-type.

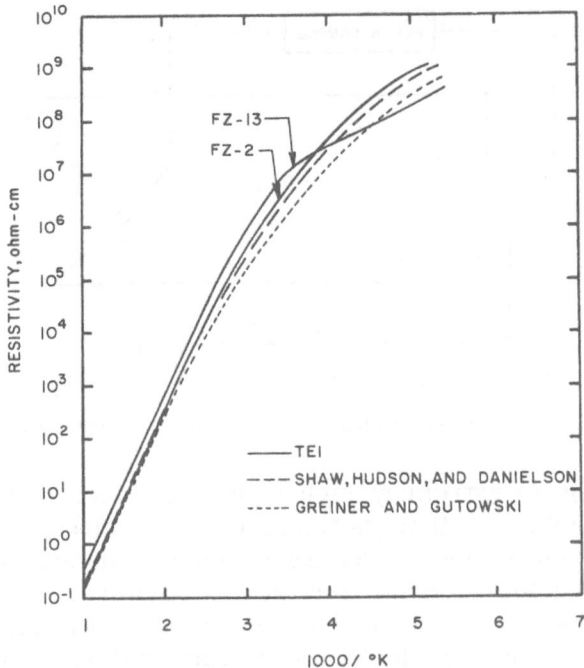

Fig. 8. Resistivities of boron as reported by various laboratories.

This would be the case if carbon were acting as the donor impurity. At higher temperatures, $p$-type behavior is observed. This may be explained on the basis of appreciable intrinsic conductivity combined with a high ratio of hole-to-electron mobility.

Figure 8 shows a comparison of resistivity curves obtained with boron samples produced in this laboratory and similar curves obtained in other laboratories. Energy-gap values determined from such curves range from 1.32 to 1.55 eV. Shaw et al. [8], working with microscopic-sized single crystals of tetragonal boron, obtained a value of 1.55 eV; and Greiner and Gutowski [9] reported a value of 1.39 eV for polycrystalline material.

**Thermoelectric Power.** Thermoelectric power measurements were carried out over a range of temperatures from -20 to 285°C on several boron samples. As in the case with most semiconductors, a high thermoelectric power was observed, several orders of magnitude larger than that of most metals.

Because of the very high resistance of the samples, especially at the lower temperatures, it was necessary to use a microvolt-

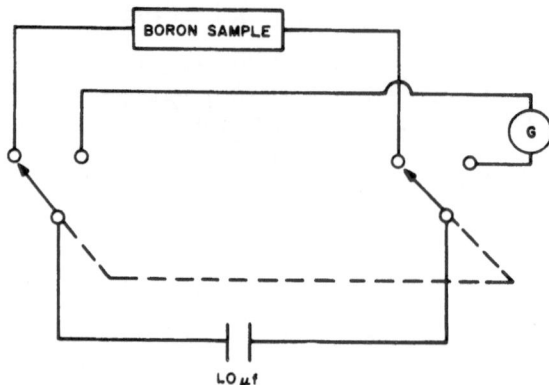

Fig. 9. Electrical setup for thermoelectric measurements by the condenser technique.

meter having an extremely high input resistance. This required careful shielding of all leads to prevent stray pickup. The Keithly Model 200B electrometer, having an input resistance of $10^{16}\Omega$, is ideally suited for such measurements. However, because of its very high impedance and the high resistance of the sample, it was almost impossible to shield the circuit sufficiently to measure reliably the very small voltages involved. Two different techniques were used in an effort to eliminate this problem. Both techniques proved satisfactory. One technique eliminated the need of a high-impedance meter entirely. This was done by connecting a condenser across the ends of the sample, allowing the condenser to charge to the thermoelectric potential established between the ends of the sample, and then discharging the condenser through a ballistic galvanometer and observing the maximum deflection. Then, by charging the condenser to a known potential, using a potentiometer, and adjusting this potential to give the same deflection the boron sample gave, the thermoelectric voltage of the sample could be determined. A knowledge of the temperature gradient across the sample then allowed calculation of the thermoelectric power. A schematic diagram of the setup is shown in Fig. 9.

The other technique simply involved placing the entire measuring system — sample, meter, furnace, etc. — in a doubly-screened cage and measuring the thermoelectric potential directly with the high-impedance voltmeter. The cage was quite effective in eliminating external fields. During actual measurements, the furnace, which operated on alternating current, was turned off to prevent pickup from this source. Because of the large thermal lag of the furnace, no appreciable error in temperature resulted from this procedure.

Both of the above techniques gave similar results. The boron sample was mounted between two platinum electrodes, which, in turn, were mounted between two blocks of aluminum. One of these blocks served as heater, the other as a heat sink, in establishing a temperature gradient across the length of the sample. The temperature of the heater block was adjusted to give a temperature difference of approximately 20°C between the ends of the sample. Junction temperatures between the platinum and the boron were measured by means of two very fine thermocouples imbedded in the junctions. The entire assembly was sealed in a glass tube and placed in a long cylindrical oven for measurements at elevated temperatures, and in an appropriate cold bath for measurements at temperatures below ambient. A drawing of the mounting system is shown in Fig. 10.

Results of the study are presented in Fig. 11, where thermoelectric power is plotted as a function of sample temperature. Although there is some scatter in the results, the curve is fairly well-defined. A maximum value of $760 \mu V/deg$ C for the thermoelectric power occurs at a temperature of about 100°C; beyond this point, the power decreases rather slowly. The polarity of the thermoelectric voltage reverses at a temperature of about -20°C. Although no other detailed curves were run, other samples gave similar values at room temperature. These results are in reasonable agreement with results reported by other investigators. Shaw et al. [8] obtained a maximum value of $640 \mu V/deg$ C. Gaulé, Breslin, Pastore, and Shuttleworth [7] give a value of approximately $500 \mu V/deg$ C.

**Photoconductivity and Rectification.** Preliminary results show that boron does exhibit photoconductivity with the greatest effect appearing in the visible range of the spectrum. Current through

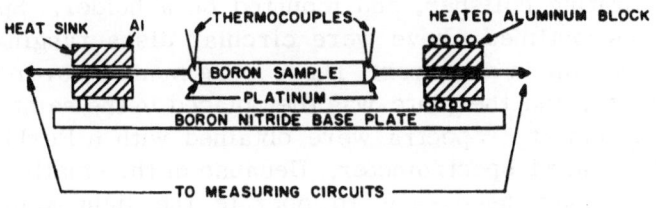

Fig. 10. Sample holder for thermoelectric measurements.

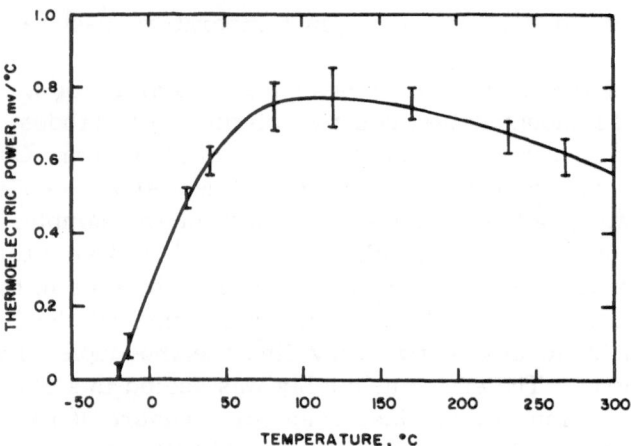

Fig. 11. Thermoelectric power of single-crystal boron crystal FZ-1.

the crystal was more than doubled when the crystal was illuminated with a microscope light. Although semiconductors usually exhibit some rectification effects, no such effects have been observed to date in boron.

Infrared Transmission . Infrared-absorption spectra were obtained from several boron samples. Such spectra are of use in obtaining information concerning band gap, impurity content, absorption coefficients, and transmission characteristics. In the case of boron samples, because of their high optical density and small cross-sectional area, it was necessary to use very thin, highly polished specimens. Specimens were prepared by cutting thin sections (down to 0.1 mm) from a previously zone-melted boron rod, using a Gillings–Bronwill thin sectioning machine. Because of the agitation and heat produced in the cutting process, there is some question as to whether or not sections cut from a single-crystal sample are still single-crystal after being cut. After cutting, the samples were polished, first by hand and later with the aid of a Fisher vibrating polisher, and mounted on a holder. Specimens prepared as outlined above were circular disks ranging in size from 2 to 4 mm in diameter. These were mounted over a hole in a metal plate, and the plate was then placed in the sample holder of the instrument. Spectra were obtained with a Perkin-Elmer Model 13 infrared spectrometer. Because of the smallness of the samples, it was necessary to operate the light source at its maximum intensity. A typical spectrum is shown in Fig. 12.

Analysis of several spectra revealed the existence of an "optical window" in the range of 0.9 to 8 $\mu$. Gaulé et al. [7] have reported a similar "window."

Results shown in Fig. 12 were obtained with a NaCl prism and a Nernst glower. The sample consisted of a thin section of single-crystal boron 0.024 cm thick. The window for frequencies in the range 0.9–8 $\mu$ is obvious, although the exact edge on the short wavelength side remains indefinite. However, if this portion of the curve is extrapolated to the point of no transmission, a value of approximately 0.9 $\mu$ is obtained for the cutoff point. This corresponds to an energy gap of 1.35 eV, which is in agreement with the value obtained from analysis of the resistivity-versus-temperature curves. No absolute values of absorption coefficients were obtained.

Flexural Modulus and Modulus of Rupture . A preliminary value was obtained for the flexural modulus and modulus of rupture of a single crystal of beta-rhombohedral boron by loading a rod of the material as a simple beam. The rod was supported by a knife edge at each end and loaded at its midpoint. The deflection was measured for several different loads up to and including that necessary to break it. A flexural modulus of $50 \cdot 10^6$ lb/in.$^2$ and a flexural strength of 173,000 lb/in.$^2$ were obtained.

## SUMMARY

Large single crystals of beta-rhombohedral boron have been produced and some of the electrical, optical, and mechanical properties determined. Mechanical tests show monocrystalline boron to be a brittle material possessing high strength and modulus.

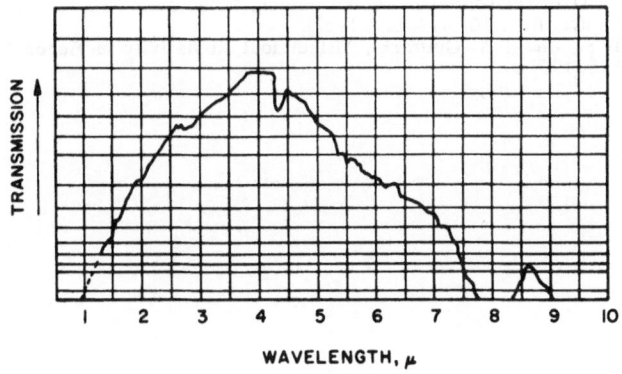

Fig. 12. Infrared spectrum of single-crystal boron showing "optical window."

The electrical and optical properties investigated agree in general with measurements made in previous investigations of primarily polycrystalline material. Beta-rhombohedral boron has a relatively high band gap and thermoelectric power, and an appreciable infrared transmission and photoconductive effect. A very encouraging development was the preparation of single-crystal, beta-rhombohedral boron of high resistivity whose resistance–temperature behavior could be accounted for on the basis of only two activation energies—the intrinsic band gap of 1.35 eV and an extrinsic impurity level of about 0.32 eV. The often-found extrinsic level of about 0.6 eV was virtually absent.

## ACKNOWLEDGMENT

This investigation has been supported by Wright Air Development Division, Air Research and Development Command, United States Air Force.

## REFERENCES

1. Newkirk, A. E., "Preparation and Chemistry of Elementary Boron," 135th Am. Chem. Soc. Meeting, Symposium on Boron, April 1959.
2. Powell, C. F., I. E. Cambell, and B. W. Gonser, Vapor Plating, John Wiley and Sons (New York), 1955.
3. Pfann, W. G., Zone Melting, John Wiley and Sons (New York), 1958.
4. Lawson, W. D., and S. Nielsen, Preparation of Single Crystals, Academic Press (New York), 1958.
5. Calverly, A., M. Davis, and R. F. Lever, "The Floating Zone Melting of Refractory Metals by Electron Bombardment," J. Sci. Instr. 34:142 (1957).
6. Hoard, J. L., and A. E. Newkirk, "An Analysis of the Polymorphism in Boron Based on X-Ray Diffraction Results," J. Am. Chem. Soc. 82:70 (1960).
7. Gaulé, G. K., J. T. Breslin, J. R. Pastore, and R. A. Shuttleworth, "Optical and Electrical Properties of Boron and Potential Application," Boron—Synthesis, Structure, and Properties, Plenum Press (New York), 1960, pp. 159–174.
8. Shaw, W. C., D. E. Hudson, and G. C. Danielson, "Electrical Properties of Boron Single Crystals," Phys. Rev. 107:419–427 (1957).
9. Greiner, E. S., and J. A. Gutowski, "Electrical Resistivity of Boron," J. Appl. Phys. 28:1364–1365 (1957).

# Carbon Determination in Hyperpure Elemental Boron Using Gas Chromatography

Joe M. Walker

*Kansas State College of Pittsburg, Pittsburg, Kansas*

and Ralph J. Starks

*Miami Research Laboratory, Eagle-Picher Company, Miami, Oklahoma*

A convenient, rapid, extremely sensitive, and highly precise method for the determination of carbon in boron has been developed. Powdered boron mixed with tin powder, which serves as an accelerator, is burned under controlled conditions in an induction-heated furnace. The gaseous products are passed through a molecular sieve column at an initial temperature of 100°C. The carbon dioxide remains trapped in the molecular sieve, while the oxygen is swept out by helium after the combustion is completed. The carbon dioxide is then driven off the column by means of a rapid temperature increase. The amount of carbon dioxide is determined by the electroconductometric method or by gas chromatography. (The latter is more sensitive.) National Bureau of Standards steel and iron samples with carbon contents of 0.011–3.28% were used to establish a standard curve. The method permits detection of 0.0005% C under ideal conditions. The time required for a single run is approximately 25 min. The boron samples used in this study contained only a few parts per million of spectrographically detectable impurities. They were prepared by the hydrogen reduction of high-purity $BBr_3$ or by the thermal decomposition of $BI_3$ and deposition on a boron substrate.

## INTRODUCTION

At the first boron conference in 1959, S. Benedict Levin pointed out that progress in the use of boron had been frustrated by the lack of very pure boron. Prior to the first conference, and during the intervening years preceding the 1964 conference, both governmental and industrial organizations have tried to isolate high-purity boron in their laboratories and to develop an analytical method for the determination of the carbon content. Carbon represents the largest impurity in all of the high-purity elemental boron thus far produced.

The first publication on the determination of carbon in elemental boron [1] described the development of a method for oxidizing boron and for converting its carbon content into carbon dioxide. The

63

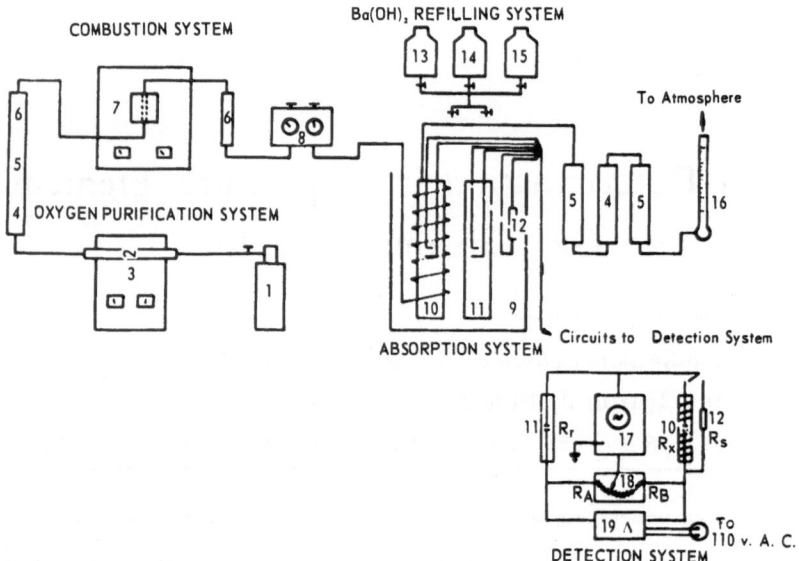

Fig. 1. Schematic diagram of entire system: (1) oxygen cylinder, (2) CuO tube, (3) preheater, (4) Mg $(ClO_4)_2$ scrubber, (5) ascarite scrubber, (6) $MnO_2$ scrubber, (7) high-frequency induction furnace, (8) two-stage pressure regulator, (9) constant-temperature water bath, (10) absorption cell with electrodes, (11) reference cell with electrodes, (12) standard resistor, (13) $Ba(OH)_2$ solution, (14) 0.1 N $HNO_3$, (15) deionized water, (16) soap-bubble flowmeter, (17) oscilloscope, (18) slide-wire, and (19) power supply box.

procedure consisted of firing a powdered boron sample in an oxygen atmosphere with tin as an accelerator under controlled conditions in a high-frequency induction field. At that time, the electroconductometric method was chosen for the carbon dioxide detection. More recently, gas chromatography has been used. A schematic diagram of the original analytical system, which consists of five individual functional units, is shown in Fig. 1.

## APPARATUS AND PROCEDURE FOR THE
## ELECTROCONDUCTOMETRIC METHOD

### Oxygen Purification Unit

A $\frac{3}{4}$-in. copper tube, filled with Mallinckrodt wireform analytical reagent grade copper(ic) oxide, was heated by a Hevy Duty Electric Co. multiple unit tube furnace at a temperature of 600–800°C. Linde hospital-grade oxygen was run through the preheater, which oxidized any carbonaceous matter present in the oxygen stream. The oxygen was then passed through a copper tube (24 in. long and $\frac{3}{4}$ in.

in diameter) which was packed with G. F. Smith Chemical Co. reagent grade magnesium perchlorate, Arthur H. Thomas Co. Ascarite (8- to 20-mesh), and Leco 501-60 specially prepared manganese dioxide. Any possible moisture, carbon dioxide, and sulfur dioxide in the flowline was thereby removed.

### Combustion Unit

The purified oxygen was then introduced into the Leco Model 523 high-frequency induction furnace. To obtain a more uniform combustion, a pressurized system was chosen. Since the original, Leco-designed, silicone rubber connection between the quartz combustion tube and the absorption unit could not hold a pressure in

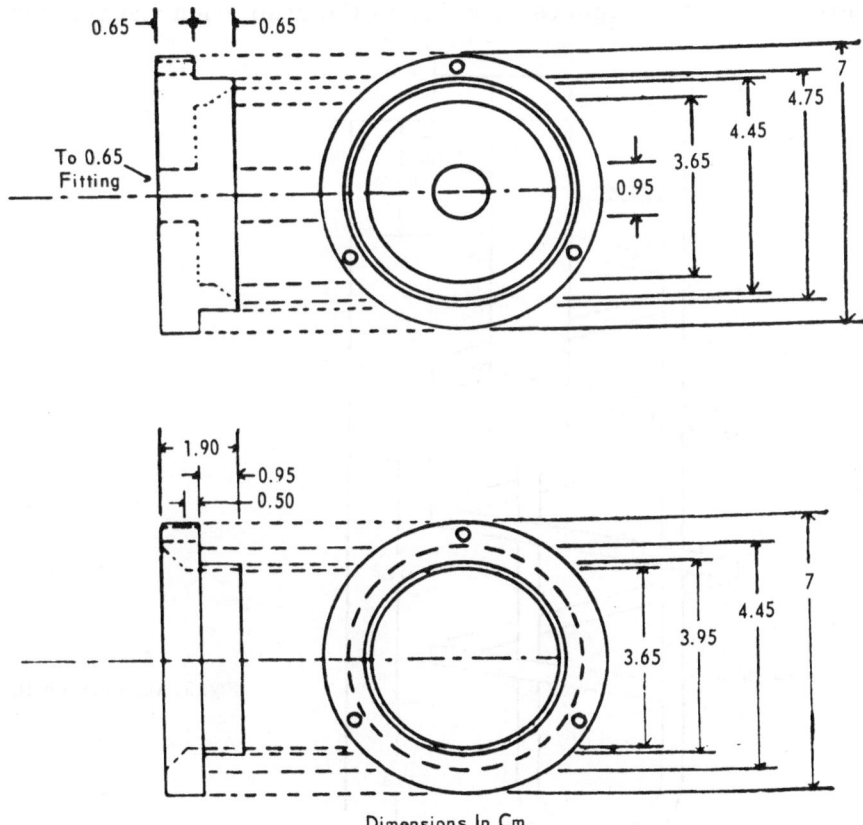

Dimensions In Cm

Fig. 2. Double O-ring seal.

excess of 3 psi, a specially-constructed brass double O-ring seal, which is shown in Fig. 2, was incorporated into the oxidation chamber. Instead of the conventional Leco quartz combustion tube, a quartz tube with a length of $7\frac{1}{2}$ in. and of $1\frac{3}{8}$-in. OD was used. All the flowlines and connections were made of $\frac{1}{4}$-in. copper tubing and fittings. The system has been operated under an oxygen pressure of 25 psi without failure. Leco 501-76 carbon-free tin accelerator (20- to 40-mesh), 528-35 crucibles, and 528-40 covers were used in this study.

### Carbon Dioxide Absorption Unit

The absorption unit consisted of a Beckman Instrument Ltd. Model 9200 two-stage pressure regulator, the spiral-type absorption cell, a plastic cylinder, a Unity Pump Co. BL-131 water-circulating pump, and a Precision Scientific Co. Model 62690 constant-temperature water bath. Figures 3 and 4 show the absorption and reference

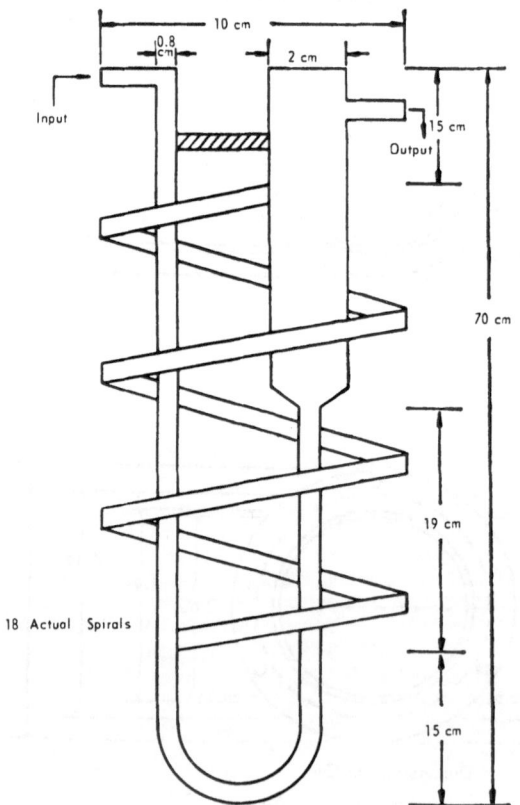

Fig. 3. Absorption cell.

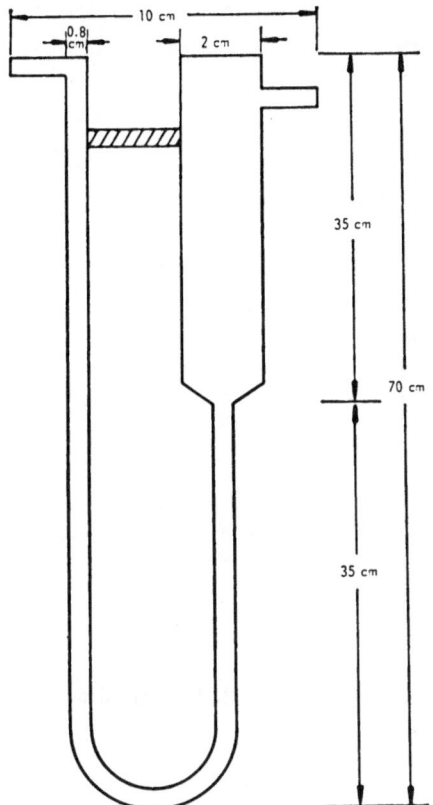

Fig. 4. Reference cell.

cells.    The combustion gas mixture coming from the induction furnace was first run through a manganese dioxide tube and then run via the pressure regulator to the spiral-type absorption cell where carbon dioxide reacted with the barium hydroxide solution.  After the completion of the carbon dioxide absorption, the gaseous mixture was passed through anhydrous magnesium perchlorate and ascarite scrubbers, which prevented the backing-up of carbon dioxide and moisture from the atmosphere.  From the scrubbers, the gas passed through a 50-ml soap-bubble flowmeter into the atmosphere.   The temperature of both cells was controlled by a water bath of 30 ± 0.1°C.

### Detection Unit

The detection unit was essentially a modified Wheatstone bridge as shown in Fig. 1.  It consisted of a specially-constructed power supply, a 400-$\Omega$ standard resistor, a Du Mont Co. Type 241 oscilloscope, a Leeds & Northrup Co. No. 4258 4-digit, 10-turn slide-

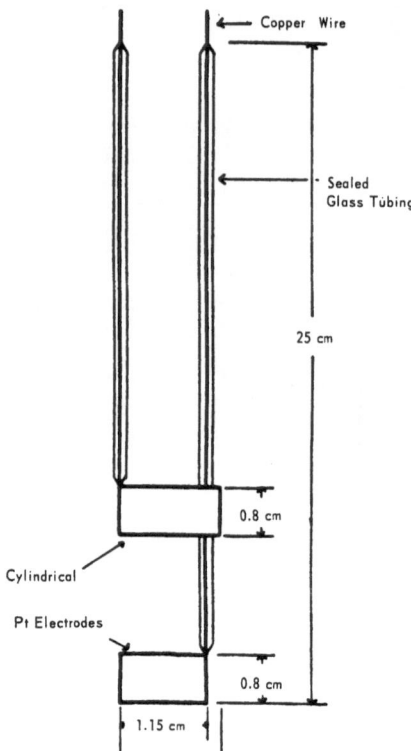

Fig. 5. Electrodes.

wire, and a pair of platinum electrodes (Fig. 5) with cell constants
of 0.270 and 0.280 cm$^{-1}$. When the bridge was balanced, a symmetrical loop was seen on the scope. The decrease of the barium
hydroxide concentration in the absorption cell due to the precipita-
tion of barium ions by the carbon dioxide caused a decrease of
electrical conductivity, which was recognized by a distortion of the
loop on the scope, and which was determined by rebalancing the
slide-wire.

### Barium Hydroxide Refilling Unit

A 0.0065 M Ba(OH)$_2$ solution was chosen in view of the carbon
concentration in the boron samples and the cell constants of the
electrodes. In an open container, such a solution would absorb
carbon dioxide from the air and drastically change its conductivity.
The cells were thus refilled in a closed system. Deionized water
and 0.1 N HNO$_3$ were used for cleaning the cells.

In the original investigation, the oxidation was carried out at
pressures of 2.5–20 psi with flow rates of 50–200 ml/min.

Variations of the flow rate within this range had little effect on the accuracy of measurements.

The carbon content of the Leco 528-35 "carbon-free" crucible varies from 50 to 120 ppm. To eliminate distortions of the results from a varying carbon background, all the crucibles were prefired by weighing 0.4 g of NBS steel and 1.0 g of Leco tin accelerator in the fresh crucible and firing in the induction furnace for 15 min. The unknown sample was then loaded in the prefired crucible while the crucible was still hot to prevent absorption of carbon dioxide from the atmosphere. Different kinds of NBS steel used in the prefiring process caused no observable differences in the results. Table I shows the differences in results obtained with fresh and prefired crucibles.

Boron bars were cracked in a diamond mortar and separated according to size into 100, 50, 20, and 10 mesh (ASTM specification) by means of a set of stainless steel sieves. Samples of boron (0.25 ± 0.001 g) were weighed on a Mettler-Gram-Atic balance and mixed carefully with 1.0 g of Leco tin in the prefired crucible. After the crucible was loaded, the high-frequency induction generator was brought into stand-by operation, and the oxygen flow rate was adjusted. The bridge was balanced and an initial reading was taken. The induction heater was then turned on. While the sample was burning, a second reading was taken. When the oxidation was complete, the induction heater was turned off. The final reading was taken 10 min later.

To obtain high precision, only the middle range of the slide-wire (400–600 units) was utilized. When the reading became larger than 600 units after several runs (due to the decreasing barium hydroxide conductivity), the absorption cell was washed and refilled. The drift of the reference cell was very small, and this cell needed to be refilled only at the beginning of each day.

A calibration curve was established by oxidizing various mixtures of fractional weights of NBS steel samples (0.2–0.8 g) and Leco tin (1.0 g) utilizing the same procedure as above. Steel samples of more than 0.8 g were not fired at higher than 10 psi when prefired crucibles were used, to avoid damage to the quartz tube from a possible overflow.

Trace carbon was found in Leco "carbon-free" tin accelerator, ranging from 2 to 4 ppm. This was recognized by firing plain tin in prefired crucibles and comparing the $\Delta G$ values of 0.6 g of NBS 55e steel plus 1.0 g of tin against 0.6 g of NBS 55e steel. It was

## TABLE I

### Comparison of Results Using Prefired and Fresh Crucibles

| | Prefired crucibles | | | | Fresh crucibles | | |
|---|---|---|---|---|---|---|---|
| Sample number | Weight, Leco tin, g | Weight, 170a steel, g | $\Delta G$, mho $\times 10^{-5}$ | Sample number | Weight, Leco tin, g | Weight, 170a steel, g | $\Delta G$, mho $\times 10^{-5}$ |
| 1 | 0.9990 | 0.3994 | 12.0 | 11 | 1.0009 | 0.3996 | 18.3 |
| 2 | 1.0002 | 0.4002 | 12.4 | 12 | 1.0004 | 0.4010 | 18.0 |
| 3 | 0.9999 | 0.3999 | 12.0 | 13 | 0.9998 | 0.4006 | 15.2 |
| 4 | 1.0008 | 0.3998 | 11.3 | 14 | 1.0000 | 0.3999 | 15.8 |
| 5 | 1.0000 | 0.4006 | 12.2 | 15 | 1.0006 | 0.4005 | 17.8 |
| 6 | 1.0002 | 0.4001 | 11.9 | 16 | 1.0004 | 0.3990 | 16.4 |
| 7 | 1.0001 | 0.4005 | 12.0 | 17 | 1.0007 | 0.4002 | 17.6 |
| 8 | 0.9995 | 0.4002 | 12.2 | 18 | 0.9996 | 0.3992 | 17.5 |
| 9 | 1.0000 | 0.4006 | 12.1 | 19 | 0.9997 | 0.4000 | 15.3 |
| 10 | 1.0006 | 0.4002 | 12.0 | 20 | 1.0006 | 0.4001 | 15.3 |

## TABLE II

### Analysis of Carbon in NBS 170a Steel

| Weight, g | Number of runs | $\Delta G$, mho $\times 10^{-5}$ | Carbon content, ppm |
|---|---|---|---|
| 0.2 | 5 | 5.3 ± 0.089 | 104 ± 1.7 |
| 0.4 | 13 | 10.6 ± 0.324 | 208 ± 6.4 |
| 0.6 | 7 | 16.0 ± 0.582 | 312 ± 11.3 |
| 0.8 | 5 | 21.2 ± 1.073 | 416 ± 21.1 |

Oxygen pressure is 10 psi. Flow rate is 200 ml/min. Bath temperature is 30°C.

also noticed that under 10-psi oxygen pressure, steel samples of less than 0.6 g could not be oxidized without tin as an accelerator. The carbon due to the tin was subtracted each time from the final $\Delta G$ value.

## RESULTS WITH THE ELECTROCONDUCTOMETRIC METHOD

Results for NBS 170a and 55e steels are summarized in Tables II and III. A linear relationship is noted for the range 104–416 ppm C for NBS 170a steel, which has a carbon content of 520 ppm. A deviation from linearity was observed in the range 22–88 ppm C for NBS 55e steel, which has a carbon content of 110 ppm.

Results of carbon analysis for various boron samples are shown in Table IV. It shows excellent agreement for samples having a carbon content lower than 0.2%. A 0.0065M Ba(OH)$_2$ solution was used here. For samples of higher carbon content, a more concentrated solution should be used.

## TABLE III

### Analysis of Carbon in NBS 55e Steel

| Weight, g | Number of runs | $\Delta G$, mho $\times 10^{-5}$ | Carbon content, ppm |
|---|---|---|---|
| 0.2 | 6 | 1.0 ± 0.091 | 22 ± 2.0 |
| 0.4 | 6 | 2.0 ± 0.120 | 44 ± 2.6 |
| 0.6 | 6 | 3.2 ± 0.125 | 66 ± 2.6 |
| 0.8 | 5 | 4.6 ± 0.121 | 88 ± 2.3 |

Oxygen pressure is 10 psi. Flow rate is 200 ml/min. Bath temperature is 30°C.

## TABLE IV

Results of Carbon Determination in Boron by Conductivity Method

| Sample number | Sample size, mesh | Number of runs | $\Delta G$, mho $\times 10^{-5}$ | Estimated carbon content, ppm |
|---|---|---|---|---|
| 1 | 100 | 5 | 9.15 ± 0.215 | 740 ± 17.2 |
| 2 | 100 | 6 | 11.99 ± 0.318 | 970 ± 24.5 |
| 3 | 100 | 3 | 8.29 | 660 |
| 4 | 100 | 6 | 8.3 ± 0.255 | 664 ± 20.3 |
| 5 | 100 | 7 | 4.42 ± 0.071 | 354 ± 5.8 |
| 6 | 100 | 4 | 6.47 ± 0.124 | 538 ± 10 |
| 7 | mixed | 6 | 11.20 ± 0.235 | 905 ± 18.9 |
| 8 | mixed | 6 | 10.7 ± 0.294 | 890 ± 24.5 |
| 9 | mixed | 5 | 13.4 ± 0.637 | 1,070 ± 51.9 |
| 10 | mixed | 4 | 6.8 | 560 |
| 11 | mixed | 2 | 106 | 10,000 |
| 12 | mixed | 3 | 74.4 | 6,000 |
| 13 | mixed | 3 | 31.6 | 2,550 |
| 14 | mixed | 3 | 10.1 | 815 |
| 15 | unknown | 3 | 45.4 | 3,500 |
| 16 | 100 | 2 | 59.5 | 4,800 |

In boron analysis, parameters, such as sample size, sample weight, and tin/boron weight ratio, must be critically controlled to obtain optimum oxidation of the sample. This is not so important for steel, where complete combustion is more easily obtained. Optimum conditions for the following parameters have been investigated.

### Sample Size

The size of boron particles is most important for a complete combustion. Boron samples, consisting of a variety of mesh sizes, were analyzed. More complete oxidation was obtained using fine-powdered boron samples (< 100 mesh). Results for a specific boron sample are shown in Fig. 6. It is believed that for large boron particles, only layers on the surface of each particle can be oxidized. Larger sample sizes have less surface area per unit weight of sample, and thus the combustion is less complete. The excellent analytical reproducibility and the visual appearance of the slag of 100-mesh boron samples indicate that the combustion was probably complete. The slag left in the crucible after firing was a transparent, yellow, glassy material.

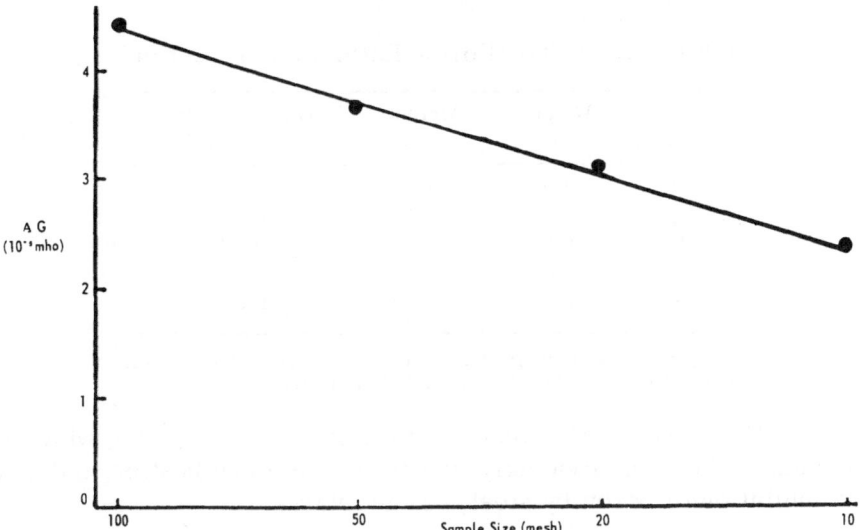

Fig. 6. Plot showing comparative completeness of oxidation for boron samples of various mesh sizes.

### Amount of Sample in Crucible

Boron samples, even when uniformly mixed with tin, burn very poorly in a fresh crucible. On the other hand, the iron oxide envelope of a prefired crucible seems to enhance the boron combustion. Ideally, the amount of sample should be small enough to be surrounded by the iron oxide envelope and large enough to secure the necessary heat input from the induction generator. It was found that $\frac{1}{4}$ g of boron, mixed with 1 g of tin, meets both requirements.

### Tin/Boron Weight Ratio

A minimum weight ratio of tin/boron of 4 was needed for good oxidation. Since the space in the iron oxide crucible is limited, $\frac{1}{4}$ g of boron and 1 g of tin were used in the analysis. Table V shows the influence of the tin/boron ratio on combustion.

### Inhibition by Steel

Steel inhibits the combustion of boron to a certain extent (Table VI). NBS 170a steel (0.4 g) combusted with 1.0 g of tin gave an average $\Delta G$ of $12.0 \cdot 10^{-5}$ mho. Boron (0.25 g) combusted with 1.0 g of tin gave an average $\Delta G$ of $25.0 \cdot 10^{-5}$ mho. Mixtures of 0.4 g of NBS 170a steel, 0.25 g of boron, and 1.0 g of tin gave an

## TABLE V

## Influence of Tin/Boron Ratio in Combustion

| Tin/boron | Weight tin, g | Weight boron, g | Average $\Delta G$ value, mho $\times 10^{-5}$ |
|---|---|---|---|
| 2 | 1.0 | 0.5 | 5 |
| 4 | 2.0 | 0.5 | 12 |
| 2 | 0.5 | 0.25 | 6 |
| 4 | 1.0 | 0.25 | 12 |
| 8 | 2.0 | 0.25 | 12.1 |

Oxygen pressure is 10 psi. Flow rate is 200 ml/min. Bath temperature is 30°C. Size of sample is mixed.

average $\Delta G$ of $26.1 \cdot 10^{-5}$ mho, instead of $37.0 \cdot 10^{-5}$ mho, which is the sum of the two. Evidently, the incomplete combustion is due to the inhibition of boron by steel, or vice versa.

## GAS CHROMATOGRAPHY METHODS

The new work presented below is an extension of previous gas chromatographic work on ferrous metals [2,3]. Previously, we were able to demonstrate a maximum detector sensitivity of approximately 5 ppm [2]. By extending this work, we believe we have reached the same sensitivity for carbon in boron, in contrast to a sensitivity of only approximately 20 ppm from the electroconductometric method described above.

## APPARATUS AND MATERIALS FOR GAS CHROMATOGRAPHY

A diagram of the system is shown in Fig. 7. The apparatus consists of a F & M Scientific Model 500 gas chromatograph, an

## TABLE VI

## Inhibition of Boron Combustion by NBS 170a Steel

| Weight tin, g | Weight steel, g | Weight boron, g | Average $\Delta G$ value, mho $\times 10^{-5}$ |
|---|---|---|---|
| 1.0 | 0.4 | — | 12.0 |
| 1.0 | — | 0.25 | 25.0 |
| 1.0 | 0.4 | 0.25 | 26.1 |

Oxygen pressure is 10 psi. Flow rate is 200 ml/min. Bath temperature is 30°C. Size of samples is mixed. Boron sample is not labeled.

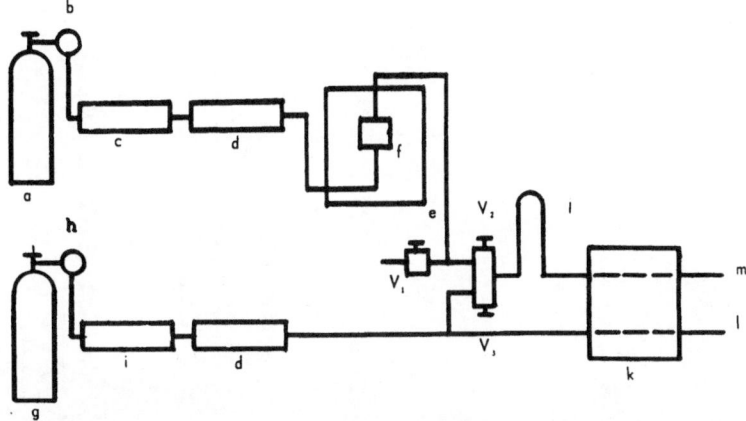

Fig. 7. Schematic diagram of apparatus: (a) oxygen supply, (b) oxygen regulator, (c) Hevy Duty Electric Co. furnace, (d) tube containing magnesium perchlorate, ascarite, and manganese dioxide, (e) Leco induction furnace, (f) quartz combustion tube, (g) helium supply, (h) helium regulator, (i) Burrell furnace, (j) 5-A molecular sieve column, (k) detector block, (l) reference side, (m) sample side, ($V_1$) two-way valve, and ($V_2$ and $V_3$) components of a three-way valve.

8-in. 5-A molecular sieve column, a Leco high-frequency induction furnace, a Hevy Duty Electric Co. tube furnace, and a Burrell high-temperature electric furnace (Fig. 8). All gases were purified as described in the earlier work of Kuo, Bender, and Walker [1].

## PROCEDURE FOR GAS CHROMATOGRAPHY

The same procedure was followed for each determination and also for the preparation of a standard curve using NBS 170a and 55e steels. The column temperature was set at 100°C. The bridge current was set at 130 mA and the system was allowed to attain equilibrium overnight. The helium flow rate was set at 150 ml/min (12 psi) and the oxygen regulator at 11.5 psi, corresponding to an oxygen flow of 100 ml/min. The prefire valve was opened and the crucible loaded with 0.600 ± 0.0009 g of 170a steel and 1.00 ± 0.001 g of Leco tin accelerator. The combustion tube was purged and the furnace turned on. The furnace was turned off after 15 min and the crucible was removed and loaded with the sample after 20 min. While the combustion tube was purged for 1 min, the attenuator was set, the oxygen valve opened, and the helium valve closed simultaneously. With oxygen passing through the column, the sample was combusted for 7 min and swept for 1 min. The helium valve was opened and the oxygen valve closed simultaneously. After 2 min, the

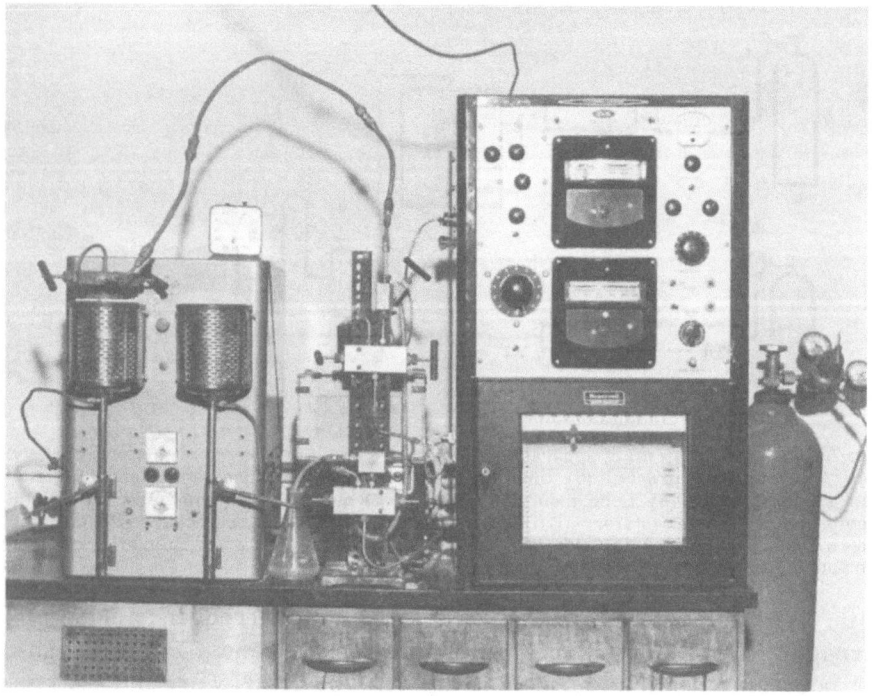

Fig. 8. Gas chromatography apparatus.

attenuation was set properly (depending on carbon content and size of sample), the temperature program started at 42°C/min, and the chart drive started. The carbon dioxide was eluted at 240–260°C, but the column temperature was allowed to reach 500°C after carbon dioxide elution in order to maintain constant column parameters.

The values of the essential parameters were: helium flow rate at the sample side, 150 ml/min at 100°C column temperature; helium flow rate at reference side, 50 ml/min at room temperature; oxygen flow rate at the sample side, 100 ml/min at 100°C column temperature; block temperature, 200°C; temperature limit setting, 500°C; size of boron samples, 100 mesh; weight of boron samples, 0.250 g; and tin/ boron ratio, 4/1.

## DISCUSSION OF THE RESULTS FROM GAS CHROMATOGRAPHY

It was extremely important that parameters, such as sample size, sample weight, and tin/boron ratio, be carefully controlled.

Conditions as found by Kuo, Bender, and Walker [1] were found to give maximum precision and accuracy. It was verified that for the range of 9–16 psi oxygen pressures have no effect on the combustion of boron samples. The oxygen pressure was carefully controlled, however, in order to maintain a constant flow rate.

The length of the column did not affect the peak width as long as the flow rates remained constant. Carbon dioxide peaks obtained from 8-in. and 12-in. columns were nearly identical. In most cases, the carbon dioxide was eluted in the range 240–260°C. The elution temperatures of the 12-in. column were slightly higher.

The boron samples analyzed had carbon contents ranging from 245–22,900 ppm (Table VII). At maximum sensitivity, 5 ppm C gave an integration of 120 counts (Fig. 9).

A major problem in achieving the desired precision and accuracy was the difficulty in attaining complete combustion of the boron samples. This seemed to especially apply to samples with higher carbon content. The precision was improved by mechanically

TABLE VII

Boron Metal Analysis Gas Chromatographic Carbon

| Sample | Number of runs | Carbon, ppm |
| --- | --- | --- |
| M 6312 AN | 4 | 420 ±   8 |
| M 6312 BG | 4 | 1,032 ±  21 |
| M 6401 AO | 4 | 448 ±  27 |
| M 6401 AR | 4 | 760 ±  34 |
| M 6405 CP | 5 | 644 ±  45 |
| M 6405 CJ | 3 | 364 ±   6 |
| M 6405 CR | 5 | 2,476 ±  87 |
| M 6312 BF | 5 | 645 ±  25 |
| M 6404 AP | 3 | 245 ±  29 |
| M 6406 AQ | 5 | 22,900 ± 400 |

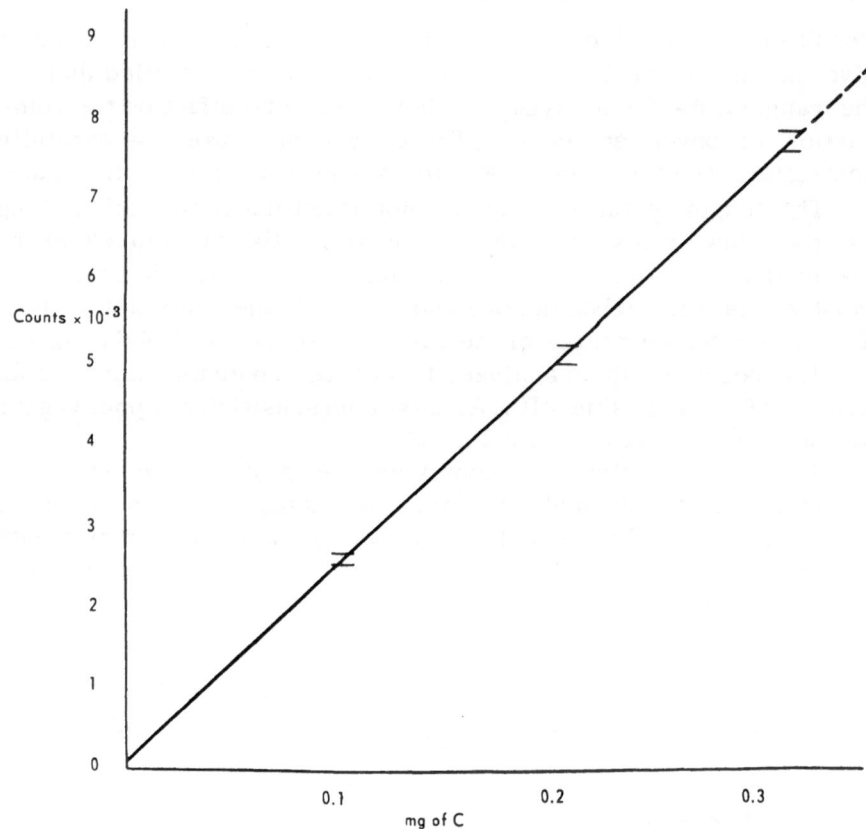

Fig. 9. Plot of carbon content (in mg) vs. counts.

mixing the boron and tin with a stirring rod, rather than by agitation. The nonuniformity of the available boron also affected the precision. Even with seemingly complete combustion, the precision attained was not equal to that attained using NBS steel. Prefiring the crucibles in a bomb, not in a steady stream of oxygen, gave erratic results, probably due to the adsorption of carbon dioxide on the crucibles.

## REFERENCES

1. Kuo, C. W., G. T. Bender, and J. M. Walker, Carbon Determination in Elemental Boron, J. Anal. Chem. 35: 1505 (1963).
2. Walker, J. M., and C. W. Kuo, Carbon Determination in Ferrous Metals by Gas Chromatography, J. Anal. Chem. 35: 2017 (1963).
3. Stuckey, W. K., and J. M. Walker, A. Combustion–Gas Chromatographic Method for the Simultaneous Determination of Carbon and Sulfur in Ferrous Metals, J. Anal. Chem. 35: 2015 (1963).

## DISCUSSION*

Although  the boron bars were cracked in a diamond mortar, we were able to show through two different methods of analysis that little, if any, carbon found in the boron resulted from contamination by the mortar.   We also point out that the minimum carbon content thus far analyzed by our laboratory is of the order of 245 ppm.   It is the authors' contention that the best method for oxidation of boron involves the use of a high-frequency furnace. We feel that unless one uses this method, carbide formation, or at least masking of the carbon content, will occur.  We feel that the temperature during oxidation approaches 2000°C.  As very high purity is approached, the oxidation approaches that of an explosion. The reverse is true as the carbon content increases.

---

*Authors' comments to questions raised during discussion.

# The Beta-Rhombohedral Boron Structure

## J. L. Hoard and R. E. Hughes

*Department of Chemistry, Cornell University*
*Ithaca, New York*

All of the essential features of the structure of beta-rhombohedral boron were described in a recent communication; it is appropriate at this point to elaborate upon some of the general features of this remarkable and elegant structure. To this end, the somewhat succinct report is reproduced here in full to provide background for the subsequent discussion of alternative descriptions.

## THE STRUCTURE OF BETA-RHOMBOHEDRAL BORON [*]

We have determined the structure for beta-rhombohedral boron [1, 2], the phase obtained between 1500° and the melting point, where crystallization of a framework requiring a 105-atom unit becomes kinetically feasible [2]. $B_{12}$ icosahedra are linked in a pattern which apparently is energetically superior to that achieved in the tetragonal [3] or, especially, the alpha-rhombohedral [4] polymorph.

Crystals (approx. 99.5% boron) from the U.S. Borax Research Corporation give $a = 10.145 \pm 0.015$ A and $\alpha = 65°\ 17' \pm 8'$ for the primitive rhombohedral unit. The 105-atom framework derives from the space group $D_{3d}^5 - R\bar{3}m$, and carries (largely substitutional) impurities estimated as 0.11 magnesium, 0.04 aluminum, and 0.01 each of silicon, manganese, and iron atoms. Three-dimensional intensity data from single crystals were counter-recorded using balanced filters with both Cu $K_\alpha$ and Mo $K_\alpha$ radiation. Structure determination through analysis of observed inequalities and statistical relations connecting unitary structure factors, although demonstrably feasible, was short-circuited when one of us (R. E. H.) independently devised a model for the 84-atom subunit described below. The complete structure, free of interpretive ambiguity, then was obtained by Fourier synthesis in successive approximation.

[*]The section under this title by R. E. Hughes, C. H. L. Kennard, D. B. Sullenger, H. A. Weakliem, D. E. Sands, and J. L. Hoard appeared previously in J. Am. Chem. Soc. 85:361 (1963). Copyright 1963 by the American Chemical Society and reprinted by permission of the copyright owner.

Presently, $R$ is 0.15 for all reflections of $(\sin \theta)/\lambda < 0.71$, and bond lengths are 1.65–1.87 A, averaging 1.79 A.

Excepting one atom at $\frac{1}{2}$, $\frac{1}{2}$, $\frac{1}{2}$, the structural unit is contained within a larger aggregate, itself recognizable in the crystal, of 13 $B_{12}$ icosahedra linked in a pattern retaining maximum symmetry. Take $r \cong 1.71$ A and $e \cong 1.80$ A as radius and edge of a regular $B_{12}$ icosahedron, respectively, and $l \cong 1.70$ A for a direct intericosahedral link. Consider a central $B_{12}$ icosahedron radially linked to twelve others, with each secondary icosahedron derived from the central $B_{12}$ through translation of $2r + l$ along the common fivefold axis and rotation of $2\pi/10$ about this axis. (This "eclipsed" configuration for linked $B_{12}$ pairs is observed throughout.) Centers of the secondary icosahedra lie upon the surface of a sphere of diameter $4r + 2l \cong a = 10.145$ A. Within this sphere are 84 atoms which comprise the central $B_{12}$ radially joined to 12 half-icosahedra or pentagonal pyramids; the 60 peripheral atoms occupy vertices of a truncated icosahedron (12 pentagonal, 20 hexagonal faces) in which the 30 edges connecting $B_5$ pentagons in pairs are also links connecting secondary $B_{12}$ icosahedra in pairs nearly along (within 5° of) quasi-fivefold axes. Some compression of the central $B_{12}$ with accompanying expansion of the attached icosahedra is observed. Averaged edge lengths for the respective $B_{12}$ types presently are 1.75 and 1.83 A; intericosahedral links range from 1.65 to 1.75 A with, interestingly, the twelve radial links longest. When centered at a lattice point with symmetry $\overline{3}m$, the 84-atom subunit has six of its quasi-fivefold radii essentially along rhombohedral axes. Lattice translations generate an approximately cubic closest-packed assemblage of these subunits, which are joined in an elegant three-dimensional framework through completion of $B_{12}$ icosahedra centered at the midpoints of all rhombohedral edges. Thus, the 84-atom complex plays a role somewhat analogous to that of a $B_{12}$ icosahedron in the simple alpha-rhombohedral structure [4].

Each subunit still has six pentagonal pyramids with axes inclined alternatively $\pm 10°$ to the equatorial plane; these are capped, using just 20 atoms, to complete $B_{12}$ icosahedra within partially condensed aggregates of symmetry $3m$. Three icosahedra, from three subunits, terminate in a single atom on the threefold axis. This atom has nine neighbors at the vertices of a polyhedron $(3m)$ having three pentagonal and five triangular faces. Six of these nine atoms are shared between icosahedra and display eight-coordination; the remaining three retain six-coordination by each forming a bond with the 105th atom at the cell center $(\overline{3}m)$. This last atom displays octahedral coordination in tying together the two sets of condensed icosahedra centered at $\pm(uuu)$ with $u = 0.385$. Bond distances of 1.72–1.86 A, averaging 1.79 A, characterize the internal crosslinking. (In the alpha structure, three-center bonding provides the crosslinking [4].) Overall, the coordination number for 91 atoms is six, for 12 atoms, eight, for the final pair, nine. Two holes of modest size at $\pm(vvv)$, $v \cong 0.22$, carry no appreciable electron density. Preferential substitution by magnesium and aluminum (probably as $Mg^{++}$, $Al^{+++}$) in the octahedral or nine-coordinate positions seems probable, but is not proven.

Noted for subsequent elaboration are the following: The beta-rhombohedral structure probably is thermodynamically preferred (for ordinary pressures) at every temperature below the melting point. Establishment of the partial framework having 84-atom subunits, but with a different internal crosslinking (probably influenced by foreign atoms), is the obvious, if inconclusive, interpretation of X-ray powder data from some boron samples crystallized too far below the melting point. At still lower temperatures, kinetic mechanism determines the choice of framework [2].

## ACKNOWLEDGMENTS

This work was supported by the National Science Foundation and the Advanced Research Projects Agency. We thank the Cornell Computing Center, Mr. Richard C. Lesser, Director.

\* \* \* \* \*

It is convenient, in discussing rhombohedral framework structures based upon icosahedra, to classify the twelve icosahedral vertices in two sets grouped around the threefold axis. The rhombohedral set, an antiprismatic array of three r vertices at the top and three r̄ vertices at the bottom, subtends at the center a vector triplet which can be used to define a rhombohedral lattice with an angle close to 60°. The equatorial set, a flattened antiprismatic array of three e and three ē vertices, forms a staggered belt around the equator of the icosahedron. In these terms, the important icosahedral framework structure of boron carbide is simply described. Icosahedra are centered at the lattice points of a simple rhombohedral lattice and are directly bonded into a three-dimensional framework structure through r–r̄ links between adjacent icosahedra. No direct intericosahedral bonding exists between equatorial atoms e and ē; rather, such linkages are effected through interposed carbon atoms located on the threefold axis at the termini of a linear chain of three atoms in the center of the cell. A similar description applies to the simple alpha-rhombohedral boron structure where, in the absence of the interposed carbon atoms, the rhombohedral framework structure collapses to the point where intericosahedral bonding is directly effected through relatively labile three-center equatorial bonds.

A slight generalization makes it possible to apply a similar description to the more complex beta-rhombohedral boron structure. The 60 peripheral atoms in the 84-atom unit described above form a truncated icosahedron; alternatively, twelve pentagonal pyramids, or half-icosahedra, occupy the rhombohedral and equatorial vertices of a large icosahedron. Once again, this unit is centered at the lattice points of a simple rhombohedral lattice and, as in the case of boron carbide, a three-dimensional network structure is generated by direct bonding between r and $\bar{r}$ vertices of adjacent units along the rhombohedral directions. This time, rather than forming a simple boron–boron bond, the juxtaposed r and $\bar{r}$ half-icosahedra neatly bond to form full icosahedra. Again, no direct interunit bonding exists between equatorial positions e and $\bar{e}$; these linkages are effected through the intercalation of the condensed, nine-coordinated, boron polyhedra described in detail above. These polyhedra serve the same function in completing the framework as do the terminal carbon atoms in boron carbide, and they occupy equivalent sites in the structure. Thus, the analogy to the boron carbide and related structures is complete, and it is clear that the generic relationships among these seemingly disparate systems arise in a natural way from the underlying icosahedral geometry. A strong case can also be made for a unifying structural principle based upon the existence of a preferred coordination geometry for boron in such systems; structure stability can be associated with the fraction of boron atoms that display pentagonal pyramidal coordination in icosahedra.

It is clear that a rhombohedral angle close to 60° corresponds to an approximately face-centered cubic array of lattice points. In the systems under discussion, the quasi-spherical structure units, $B_{12}$ and $B_{84}$, form the elements of such a cubic closest-packed array, and it is convenient to use the familiar closest-packing nomenclature to identify the gross features of the structures, even to the extent of specifying that the distinctively nonoctahedral, crosslinking, outsized subunits are centered in the "octahedral holes." Nevertheless, to carry the analogy to simple metallic closest-packed systems beyond this coarse representation of structure is to mask with procrustean logic the dominating importance of the directed nature of the bonding in three-dimensional icosahedral frameworks. Thus, for example, to speak of interstitial compounds in the case of derivative structures of these systems not only misapplies the term and ignores the predictive properties of directed pentagonal pyramidal bonding, but also presents an untenable basis for any meaningful discussion of the electronic properties of the structures.

Another X-ray diffraction study of beta-rhombohedral boron was recently reported in a very brief communication [5]; the final set of atomic coordinates, presented without significant discussion, corresponds to a wholly different structure from that described above. Straightforward calculation reveals that the proposed structure involves many boron—boron distances ranging down to 1 A and less— impossibly short bond lengths.

The structure of beta-rhombohedral boron is sufficiently complex so that alternative formulations are of interest for comparison with other icosahedral framework structures. For example, the aforementioned $B_{156}$ unit, in which twelve icosahedra are bonded around one, can be used as the basis for an illuminating description of the structure which involves complete sharing of the six rhombohedrally directed icosahedra between adjacent units on the rhombohedral lattice and partial sharing of the six equatorially directed icosahedra that overlap to form the condensed polyhedra. The same $B_{156}$ unit forms the basic structural element for a complex cubic boride, $YB_{70}$ [6]. Similarly, a formulation of the $B_{84}$ unit in terms of truncated tetrahedra provides new insight into the nature of icosahedral bonding and the mutability of icosahedral packing. There seems to be little doubt that the beta-rhombohedral boron structure will prove to be a keystone in icosahedral crystal chemistry.*

In addition to the well-characterized structures of boron already mentioned [1, 3, 4],† an astonishing number and variety of polymorphs have been reported for the element. The complexity of the system is compounded by an even larger array of borides, some of

---

*An extended discussion of icosahedral structures and polymorphism in boron which elaborates in detail upon all of the points raised here will be found in an article by J. L. Hoard and R. E. Hughes in "The Chemistry of Boron and Its Compounds," E. Muetterties (ed.), John Wiley and Sons (New York), in press. A final report on the details of the analysis of the beta-rhombohedral boron structure is now in preparation by D. Sullenger, C. Kennard, and J. L. Hoard.
†See also reference contained in footnote on p. 81.

which bear direct and obvious relationships to proposed polymorphs. Extended analyses of polymorphism in elementary boron have recently been undertaken [2];[*] it is appropriate here to observe that beta-rhombohedral boron is emerging as the stable thermodynamic form of the element over an extraordinarily wide range of temperatures and pressures. To illustrate the point, we consider a single example, a recent study [7] of boron at pressures ranging up to 150 kb which resulted in the formation of a new polymorph. We remark first that the density, 2.46 − 2.52 g/cc, of the high-pressure product, while considerably higher than that of beta-rhombohedral boron, does not approach that expected were the boron atoms to be forced into a simple closest-packed array, i.e., into a metallic structure, and, indeed, that the new polymorph retains semiconducting behavior. It is further notable that beta-rhombohedral boron remains stable at 80 kb and 1800°C, and that a fragment of the dense boron, when subjected to 30-kb pressure at 1500°C for one minute, reverted to the beta-rhombohedral form. Finally, we note that the experiments, which cover the temperature range 1500 − 2000°C and pressures to 150 kb, provide no evidence for the formation of alpha-rhombohedral boron, the polymorph which has at ordinary pressures a density (2.46 g/cc) 5% greater than that of beta-rhombohedral boron. If, as some investigators hold (we do not), alpha-rhombohedral boron is the stable form at ordinary pressures for temperatures below about 1200°C, the application of 80-kb pressure might be expected to raise the temperature to above 1500°C for the transition, beta-rhombohedral → alpha-rhombohedral; or, failing this, that alpha-rhombohedral boron would be formed at 150 kb and 1500°C. The observed results, in our judgment, enhance the thermodynamic standing of beta-rhombohedral boron, further deprecate that of the alpha-rhombohedral polymorph, and suggest that icosahedral stereochemistry persists in the product obtained above 100-kb pressure.

## ACKNOWLEDGMENTS

The authors gratefully acknowledge the support of the National Science Foundation and the Advanced Research Projects Agency.

---

*See first footnote on p. 85.

# REFERENCES

1. Sands, D. E., and J. L. Hoard, J. Am. Chem. Soc. 79:5582 (1957).
2. Hoard, J. L., and A. E. Newkirk, J. Am. Chem. Soc. 82:70 (1960).
3. Hoard, J. L., R. E. Hughes, and D. E. Sands, J. Am. Chem. Soc. 80:4507 (1958); cf., J. L. Hoard, S. Geller, and R. E. Hughes, J. Am. Chem. Soc. 73:1892 (1951).
4. Decker, B. F., and J. S. Kasper, Acta Cryst. 12:503 (1959).
5. Kolakowski, B., Acta Physiol. Polon. 22:439 (1962).
6. Richards, S. M., and J. S. Kasper, Abstr. Am. Cryst. Assoc. Meeting, Gatlinburg, Tenn., 1965, p. 72.
7. Wentorf, R. H., Jr., Science 147:49 (1965).

# Tetragonal Boron and Borides with Similar Structures

## Hermann J. Becher

*Laboratorium für anorganische Chemie der Technischen Hochschule Stuttgart, Germany*

The structure of tetragonal boron and its relationship to the structure of BeB and to that of the $\beta$-modification of $AlB_{12}$, which according to a new investigation has the composition $Al_3C_2B_{48}$, are discussed. It is possible to prepare tetragonal boron from $BeB_{12}$ by the topochemical substitution reaction

$$BeB_{12} + BCl_3 \rightarrow B_{tetragonal} + BeCl_2$$

A similar reaction of $BeB_{12}$ with $SiCl_4$ yields a boride (Be, Si) $B_{12}$. This boride can be recrystallized from molten aluminum in the form of single crystals with the tetragonal boron structure. The exact stoichiometric composition has not yet been established. Crystals of the $\beta$-$AlB_{12}$ type gave no substitution reaction with $BCl_3$. In the infrared spectrum of powdered $\alpha$-boron, two bands can be assigned to vibrations of boron atoms belonging to the same icosahedron. In $BeB_{12}$ and in tetragonal boron, additional bands are observed which belong to the stretching frequencies of the bonds between neighboring icosahedrons.

## INTRODUCTION

Up to the present, three types of crystalline structure of elementary boron have been studied by X-ray analysis of single crystals [1–3]. Of these, the rhombohedral -boron with 12 atoms in its unit cell is stable at temperatures below 1000°C. The rhombohedral $\beta$-boron with 104 atoms in the unit cell is produced by crystallization from molten boron. The third modification has a tetragonal unit cell with 50 atoms, and can be obtained by decomposing $BBr_3$ in a hydrogen atmosphere at 1300–1450°C. Under these conditions, however, other species of boron are also formed, as can be seen in the Debye–Scherrer patterns [4,5]. The changes in the boron networks may be stabilized by small amounts of impurities [4]. Alpha-boron and $\beta$-boron cannot be transformed into tetragonal boron at temperatures of 1300–1400°C, although the tetragonal boron is formed by the decomposition reaction of $BBr_3$ in this temperature range. Therefore, the three-dimensional connection

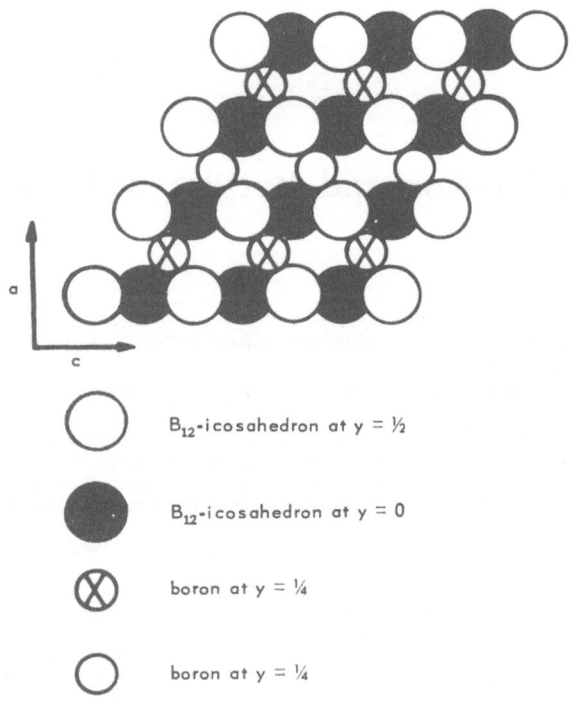

Fig. 1. $B_{12}$-icosahedrons and single boron atoms of tetragonal boron. Projection on (010).

of boron atoms seems to be less stable for tetragonal boron than for either $\alpha$- or $\beta$-boron. The unit cell of the tetragonal modification contains 48 boron atoms in 4 icosahedral groups and 2 additional single boron atoms. The icosahedrons are situated in hexagonal-like layers, as shown in Fig. 1. This arrangement causes a marked deviation of some of the bonds between neighboring icosahedrons from the direction of the line that connects a corner of the icosahedron with its center. The latter direction should be favored by single boron–boron bonds.

According to the theoretical considerations of Longuet-Higgins, there is an electron deficiency in the unit cell of tetragonal boron—26 electrons are needed to fill all bonding orbitals within one $B_{12}$-icosahedron, leaving 10 electrons of the $B_{12}$-unit for the 12 bonds outside the icosahedron [6]. The deficiency could be reduced by polycenter bonds between the additional single boron atoms of the unit cell and the neighboring icosahedrons, but without additional interstitial atoms, the electron deficiency cannot be removed completely [7].

## EXPERIMENTAL

By heating finely powdered $a$-boron with beryllium in a ratio of 12:1, a new phase can be obtained [8]. The Debye–Scherrer pattern of this product agreed with the pattern of tetragonal boron, with the exception of some weak lines. From this observation, from the atomic ratio, and from the density of the phase, the beryllium boride $BeB_{12}$ was established, which has the same network of $B_{12}$-icosahedrons as tetragonal boron, with the exception that there are 4 beryllium atoms in the interstices of the unit cell instead of 2 boron atoms. The formation of $BeB_{12}$ is observed only when $a$-boron is used as the starting material. Beta-boron does not react with small amounts of beryllium below 1000°C. At higher temperatures, the boride $BeB_6$ is formed rather than $BeB_{12}$. This $BeB_6$ has a structure related to the structure of the $a$-modification of $AlB_{12}$ [9], and perhaps also related to the structure of a boron modification with 192 atoms in a tetragonal unit cell, which was derived by Talley and Post from Debye–Scherrer patterns [10], but which still has to be confirmed by a single-crystal examination.

Since it is possible to get larger crystals of $BeB_2$ by the crystallization of powdered $BeB_2$ from molten aluminum above 1200°C [11], the same method was tried in order to obtain larger crystals of $BeB_{12}$. The crystals that were isolated from the melt were found to be mixed crytals of $BeB_6$ and isostructural $a$-$AlB_{12}$. The reaction [11] is

$$2\ BeB_{12} + Al \rightarrow 2\ BeB_6\ AlB_{12}$$

To confirm the similarity of the lattices of $BeB_{12}$ and tetragonal boron, substitution of the beryllium of $BeB_{12}$ by boron atoms was tried [12]. This was made possible by a treatment of the finely powdered $BeB_{12}$ with gaseous $BCl_3$ in a closed tube at 300–360°C. Under these conditions, the beryllium of $BeB_{12}$ is transformed into $BeCl_2$. The remaining boron has the same size and shape particles as the original $BeB_{12}$. By electron-microscopical pictures, the median diameter was found to be about 750 Å. After heating at 850°C in a high vacuum, the product contained 99% boron and was quite stable at room temperature. The product gives the Debye–Scherrer pattern of tetragonal boron. Thus, the removal of beryllium from the $BeB_{12}$ lattice occurs without any change of the icosahedral arrangement. The $BCl_3$ gives a reaction only at the surfaces of the $BeB_{12}$ particles, from where the boron atoms must diffuse into the lattice, displacing the beryllium atoms to the surface where the reaction may continue. The density of the resulting boron agrees with the density of tetragonal boron prepared at 1400°C. The boron obtained by substitution, therefore, must contain 2 additional boron atoms besides the 4 $B_{12}$-groups in the unit cell. These 2 additional boron atoms replaced the 4 beryllium atoms in the original $BeB_{12}$ unit cell.

It is not probable that the beryllium atoms of $BeB_{12}$ occupy the positions of the 2 single boron atoms in the tetragonal boron lattice, since the distances from these points to the next boron atoms are small. Figure 2 shows fourfold positions in the unit cell of tetragonal boron affording more space to the beryllium atoms. Since the single boron atoms shown in Fig. 2 have to be omitted in $BeB_{12}$, there must be a linking of neighboring icosahedrons by four center bonds at these lattice points.

Another boride whose cell dimensions show a relation to tetragonal boron is the $\beta$-modification of $AlB_{12}$. In a recent paper it was shown that $\beta$-$AlB_{12}$ contains stoichiometric amounts of carbon corresponding to the formula 3 $AlB_{12} \cdot C_2B_{12}$ [13]. This phase has a high-temperature form with the same unit cell as tetragonal boron ($a = 8.82$ Å; $c = 5.09$ Å) [13]. Four $B_{12}$-icosahedrons, 3 aluminum atoms, and 2 carbon atoms have to be placed in the cell. Below 850°C, the phase changes into two twinned ortho-rhombic modifications having cell dimensions of $a = 12.34$, $b = 12.63$, $c = 5.08$ Å and $a = 6.17$, $b = 12.63$, $c = 10.16$ Å [13]. Preliminary experiments to substitute the aluminum and carbon atoms in the compound by boron atoms, analogous to the substitution of the beryllium atoms in $BeB_{12}$, did not succeed. At 350°C, finely powdered $\beta$-$AlB_{12}$ is so strongly attacked by $BCl_3$ that the product is completely decomposed. On the other hand, larger crystals of $\beta$-$AlB_{12}$ did not react at all with $BCl_3$ below 375°C [14].

The relationship between the structures of tetragonal boron, $BeB_{12}$, and $\beta$-$AlB_{12}$ is also demonstrated by the possibility of partial substitution of the beryllium atoms of $BeB_{12}$ by silicon by treatment with $SiCl_4$ [14]. The Debye–Scherrer pattern of the reaction product exhibits the spacings of the original $BeB_{12}$, but with some change in the relative intensities, which, after the treatment with $SiCl_4$, show a similarity to the intensities of the Debye–Scherrer pattern

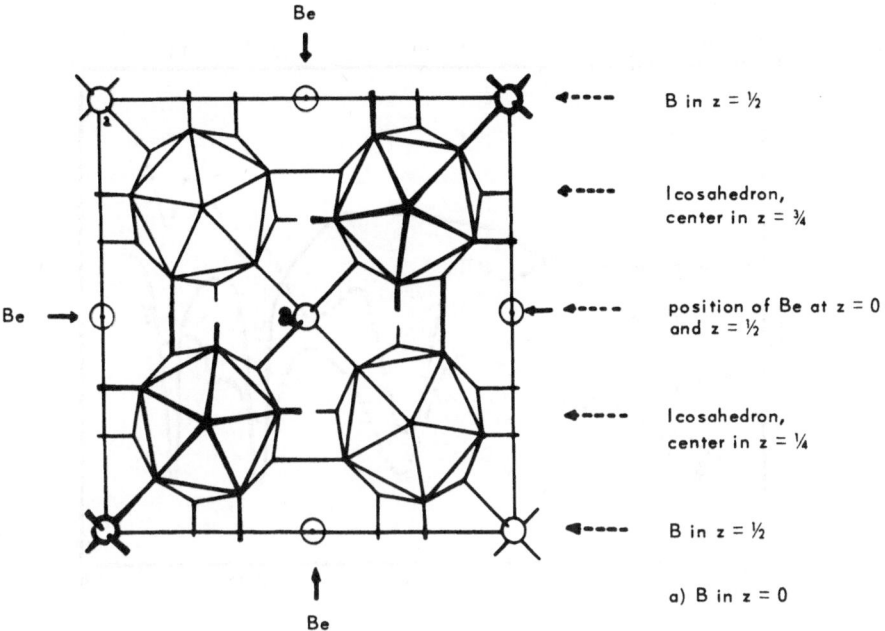

Be

B in z = ½

Icosahedron,
center in z = ¾

Be →

position of Be at z = 0
and z = ½

Icosahedron,
center in z = ¼

B in z = ½

a) B in z = 0

Be

Fig. 2. Unit cell of tetragonal boron. Projection on (001).

of powdered $\beta$-AlB$_{12}$.  The (Be,Si)B$_{12}$ thus obtained can be re-crystallized from molten aluminum as needle- and plate-shaped crystals with the unit cell of BeB$_{12}$ with a small increase of the axis.  Thus, the substitution of silicon for beryllium in the BeB$_{12}$ lattice seems to stabilize the tetragonal arrangement of the ico-sahedrons because the structure is not changed by a reaction between the boride and the molten aluminum, as observed with pure BeB$_{12}$.

The infrared spectra of powdered $a$-boron, tetragonal boron, and BeB$_{12}$ have been examined in the region of the boron–boron vibrations [12].  Figure 3 shows the absorption curves of the products in KBr disks. It is useful to distinguish between vibrations belonging to the bonds within one icosahedron and vibrations of the bonds connecting different icosahedrons.  This is appropriate because the intraicosahedral bonds are tangential and the interico-sahedral bonds are radial to the icosahedrons. Since $a$-boron has only one icosahedron in its unit cell, all those frequencies that concern the bonds between the icosahedrons are inactive in the absorption spectrum.  Only vibrations' along the bonds within one icosahedron are allowed. Selection rules show that an icosahedral

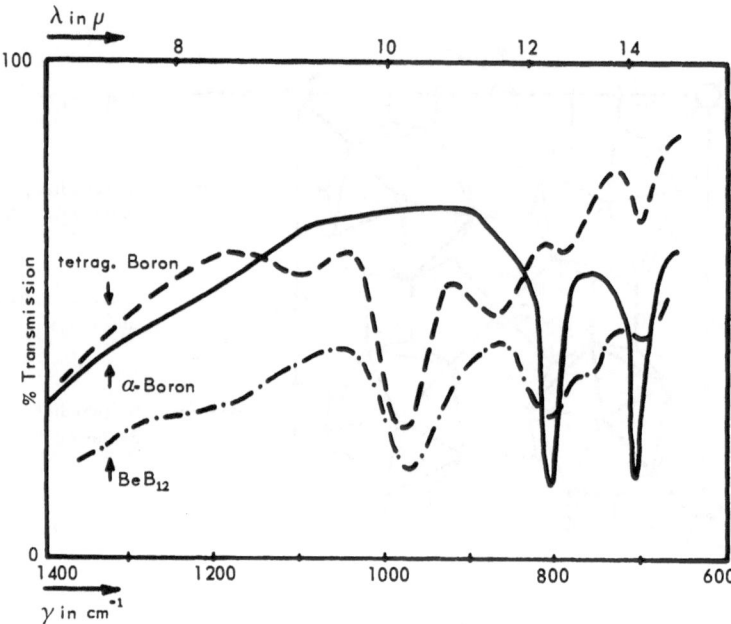

Fig. 3. Infrared absorption spectra of $\alpha$-boron, tetragonal boron, and $BeB_{12}$ from 650–1400 cm⁻¹.

group of 12 boron atoms has only one fundamental infrared absorption band. Since the icosahedron in $\alpha$-boron lies in a rhombohedral cell, it has a site symmetry which is lower than its group symmetry, and a splitting of the allowed vibration may occur. The absorption peaks near 700 and 800 cm⁻¹ in the spectrum of $\alpha$-boron are assigned to this split vibration. This is confirmed by the infrared spectrum of the $[B_{12}H_{12}]^=$ ion, which puts the vibration of a single $B_{12}$-icosahedron at 750 cm⁻¹.

The infrared spectra of tetragonal boron and of $BeB_{12}$ also show bands near 700 and 800 cm⁻¹ which can be assigned to vibrations within the icosahedrons. However, these are broader and not as pronounced as in $\alpha$-boron, because the lattices of tetragonal boron and $BeB_{12}$ allow vibrations of the bonds between the icosahedrons in the infrared spectrum which may couple with the internal vibrations of the $B_{12}$-group. In the spectra of tetragonal boron and of $BeB_{12}$, the two strong absorptions in the region 800–900 and at 975 cm⁻¹ may be reasonably assigned to vibrations of bonds between the icosahedrons. They appear at higher wavenumbers than the internal $B_{12}$-group frequencies, in agreement with the observation that the distances between the adjacent boron atoms of linked icosahedrons are shorter than the distances between neighboring boron atoms of the same $B_{12}$-group.

Pulverized $\beta$-boron exhibited infrared bands still broader than those of tetragonal boron or of $BeB_{12}$ [14]. This may be a question of particle size because sintered $BeB_{12}$ also gave ill-defined absorption bands. The broadening might also be caused by stronger coupling between vibrations of internal and external bonds of the $B_{12}$-groups. The main absorption of $\beta$-boron has been found near $1080$ cm$^{-1}$, and a weaker one has been found near $750$ cm$^{-1}$. This band lies in the same region as the internal $B_{12}$-vibrations of the other samples.

## REFERENCES

1. Hoard, J. L., S. Geller, and R. E. Hughes, J. Am. Chem. Soc. 73: 1892 (1951) and 80: 4507 (1958).
2. Sands, D. E., and J. L. Hoard, J. Am. Chem. Soc. 79: 5582 (1957).
3. McCarty, L. V., J. S. Kasper, F. H. Horn, B. F. Decker, and A. E. Newkirk, J. Am. Chem. Soc. 80: 2592 (1958).
4. Hoard, J. L., and A. E. Newkirk, J. Am. Chem. Soc. 82: 70 (1960).
5. Becher, H. J., and A. Schäfer, Z. anorg. allgem. Chem. 306:260 (1961).
6. Longuet-Higgins, H. C., and M. V. Roberts, Proc. Roy. Soc. (London) 230A:110 (1955).
7. Lipscomb, W. N., and D. Britton, J. Chem. Phys. 33:275 (1960).
8. Becher, H. J., Z. anorg. allgem. Chem. 306: 266 (1961).
9. Sands, D. E., C. F. Cline, A. Zalkin, and C. L. Hoenig, Acta Cryst. 14: 309 (1961).
10. Talley, C. P., S. La Placa, and B. Post, Acta Cryst. 13:271 (1960).
11. Becher, H. J., Acta Cryst. 17: 617 (1964).
12. Becher, H. J., Z. anorg. allgem. Chem. 321: 217 (1963).
13. Matkovich, V. I., J. Economy, and R. F. Giese, J. Am. Chem. Soc. 86: 2337 (1964).
14. Becher, H. J., unpublished work.

## References

# Boron — Another Form

## R. H. Wentorf, Jr.

*General Electric Research Laboratory*
*Schenectady, New York*

A hitherto undescribed form of boron can be prepared by subjecting ordinary forms of boron to pressures exceeding about 100 kb and temperatures between about 1500 to 2000°C. The new form has a density of about 2.52 g/cm³ and yields a characteristic Debye–Scherrer pattern. No large crystals have been prepared.

Elemental boron can be prepared at low pressures in something like four different crystalline modifications [1–3]. The densest of these is the red form described by Decker and Kasper [2], which has a density of 2.46 g/cm³.

A new form of boron has been prepared in this laboratory by exposing low-pressure modifications (beta-rhombohedral and amorphous) of purified boron to high pressures (100–150 kilobars) and moderate temperatures (1500–2000°C) for intervals of a few minutes, cooling the specimen to 25°C, and then reducing the pressure to 1 atm.

This new form of boron was prepared in an apparatus similar to that described by Bundy [4]. The boron was confined in a tube (Fig. 1) and heated by passing a current through the tube (which is made of titanium or tantalum) by way of molybdenum wires at each end. Usually, the boron was placed against the tube wall; in a few experiments, the boron particles were insulated from the tube by a liner made of MgO or they were mixed with MgO or silicon powder. Pressures were estimated by reference to transitions in bismuth at 88 kb, barium at 59 and 140 kb, iron at 131 kb, and lead at 160 kb. The temperatures were estimated from the electric power dissipated in the sample; the error may be as much as 10% due to uncertainties in calibration and gradients in temperature.

The product was a dark, pitch-like solid which was deep red in thin sections; its density ranged between 2.46 and 2.52 g/cm³, as measured by the sink-float technique. Its electrical resistivity

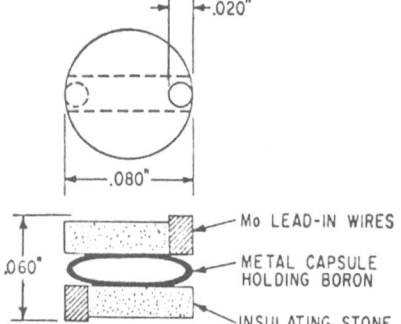

Fig. 1. High-pressure, high-temperature
reaction cell.

Mo LEAD-IN WIRES
METAL CAPSULE HOLDING BORON
INSULATING STONE

was estimated to be about $10^6 \, \Omega$-cm at 25°C, and its resistance fell by about a factor of 3 on heating to about 100°C. Both n- and p-type semiconducting specimens were obtained as determined by measurements of thermoelectric power. (The zone-refined, beta-rhombohedral-boron starting material had a slightly higher resistivity, but about the same relative change of resistance with temperature.)

Debye–Scherrer X-ray diffraction patterns of various preparations showed some weak lines corresponding to some of the known forms of boron, together with a strong pattern which could not be attributed to any combination of the known forms of boron. The d-spacing values of the main lines in typical patterns of the new form are given in Table I. When coarsely crystalline rhombohedral boron was used as starting material, the diffraction spots of these crystals could be distinguished from the smooth lines of the new form.

It is not known whether the product is a single crystalline phase or a mixture of new and previously known forms of boron. Increase of the time of heating at high pressure from 6 to 60 min did not produce any markedly different product; neither did pressures higher than about 100 kb. Temperatures higher than about 2000°C caused the boron to react with its titanium or tantalum container. It is possible that the boron exists in one crystalline phase at high pressures, but then reverts to the observed form as the pressure is reduced. However, during pressure release, the sample did not show any marked resistance changes that would indicate a change in atomic arrangement of the boron.

What is taken to be a new phase of pure boron might possibly be a boron compound with titanium, tantalum, or silicon, for boron can incorporate small amounts of other elements in its lattice [5].

## TABLE I

Debye–Scherrer $d$-Values
for a New Form of Boron

| Å | Relative intensity* |
|---|---|
| 4.4 | m |
| 4.1 | m |
| 3.75 | ms |
| 3.50 | m |
| 2.51–2.54 | s |
| 2.30–2.32 | s |
| 2.15 | w |
| 2.10 | w |
| 2.00 | w |
| 1.98 | w |
| 1.68 | w |
| 1.60 | w |
| 1.50 | mw |
| 1.48 | mw |
| 1.45 | m |
| 1.41 | ms |
| 1.39 | m |
| 1.38 | m |
| 1.35 | m |
| 1.30 | m |
| 1.27 | w |

*m, medium; s, strong; w, weak.

However, this possibility does not seen likely in view of the following observations:

1. The same new features of the Debye–Scherrer pattern are obtained whether the boron is confined in tantalum, titanium, or MgO during preparation.

2. The new form is not observed at conditions of about 1800°C and about 80 kb; the product instead has a density less than 2.39 and shows only the rhombohedral beta pattern reported by Amendola [3].

3. If a fragment of the dense form is buried in hexagonal boron nitride and heated to about 1500°C for about 1 min at a pressure of 30 kb, the product has a density between 2.25 and 2.36 g/cm$^3$, and its Debye–Scherrer pattern agrees well with that of the well-known rhombohedral form of boron [3].

4. The electrical resistivity of the dense form is almost as high as that of the purified boron starting material.

## ACKNOWLEDGMENT

I thank Mrs. D. K. DeCarlo for preparing many Debye–Scherrer X-ray patterns.

## REFERENCES

1. Kohn, J. A., W. F. Nye, and G. K. Gaulé (eds.), Boron—Synthesis, Structure, and Properties, Plenum Press (New York), 1960; J. A. Hoard and A. E. Newkirk, J. Am. Chem. Soc.
   82:70 (1960); S. LaPlaca, ASTM Debye–Scherrer Card No. 11-617.
2. Decker, B. F., and J. S. Kasper, Acta Cryst. 12:503 (1959).
3. Amendola, A., ASTM Debye–Scherrer Card No. 11-618; D. E. Sands and J. L. Hoard,
   J. Am. Chem. Soc. 79:5582 (1957).
4. Bundy, F. P., J. Chem. Phys. 38:631 (1963); Science 37:1057 (1962).
5. Becker, H. J., Z. Anorg. Allgem. Chem. 306:260 (1960).

# Some Results of the Band Structure Calculation of Rhombohedral Boron by the K·p-Method

## Ulrich Rössler

*Institut für Theoretische Physik (II) der Universität Marburg/Lahn*
*Marburg, Germany*

By use of group theory, information on the energy-band structure of lattices whose symmetry is that of the space group $D_{3d}^5$ is presented. (The two rhombohedral modifications of boron have such lattices.) The special arrangement of the atoms forming icosahedra is not considered. The possible energy bands can be classified according to their transformation properties under the symmetry operations of $D_{3d}^5$. The K·p-method allows the conclusion that all the energy bands have zero slope at the points $\Gamma$, $Z$, $A$, and $D$ of the Brillouin zone; i.e., we can expect band extremes at these points. However, it is not possible to determine where the band edges are located, where in the Brillouin zone the transitions between valence and conduction band take place, and whether these transitions will be direct or indirect. Optical measurements together with our group theoretical results could possibly lead to a band model for boron.

The K·p-method is a quantum mechanical perturbation calculation in which the matrix elements are treated by group theoretical methods [1–6]. The task is to solve the Schrödinger equation for one electron moving in the periodic potential $V(r)$ of the lattice

$$\left[ \frac{p^2}{2m} + V(r) \right] \Psi(k, r) = E(k) \Psi(k, r) \tag{1}$$

The wave function $\Psi(k, r)$ and, therefore, the electron are characterized not only by the space vector $r$, but also by the vector $k$ of the reciprocal lattice. It follows that the energy $E$ depends on $k$; we call the function $E(k)$ the band structure of the crystal. As a consequence of the periodicity of the potential, the Hamiltonian of equation (1) is invariant against primitive lattice translations. Therefore, we are allowed to choose the wave function $\Psi(k, r)$ as a Bloch function

$$\Psi(k, r) = e^{ik \cdot r} u(k, r)$$

where the function $u(k, r)$ is periodic throughout the lattice. Use of the Bloch function in equation (1) yields

$$\left[\frac{p^2}{2m} + V(r) + \frac{\hbar}{m} k \cdot p\right] u(k, r) = \left[E(k) + \frac{\hbar^2 k^2}{2m}\right] u(k, r) \tag{2}$$

If one assumes the energy $E(k)$ to be known at the point $k$, one can calculate the energy $E(k + K)$ in the neighborhood $k + K$ of the point $k$. In our calculation, $k$ is always chosen to be a point of high symmetry in the Brillouin zone (BZ). If $k$ in equation (2) is replaced by $k$ $K$, we have

$$\left[\frac{p^2}{2m} + V(r) + \frac{\hbar}{m} k \cdot p + \frac{\hbar}{m} K \cdot p\right] u(k + K, r) = \left[E(k + K) + \frac{\hbar^2 (k + K)^2}{2m}\right] u(k + K, r) \tag{3}$$

In this equation, $(\hbar/m) K \cdot p$ can be treated as a perturbation potential of the Hamiltonian of equation (2), for we have chosen $K$ to be a small wave vector. The energy $E(k + K)$ now follows from a secular determinant whose order is the degree of degeneracy of the eigenvalue treated:

$$\left\| (\Psi_{n\rho} | H' | \Psi_{n\nu}) + \sum_{n \neq l} \frac{(\Psi_{n\rho} | H' | \Psi_l) (\Psi_l | H' | \Psi_{n\nu})}{E_n - E_l} - \epsilon \delta_{\rho\nu} \right\| = 0$$

$E_n, E_l$ are eigenvalues; $\Psi_{n\rho}$, $\Psi_{n\nu}$, $\Psi_l$ are the respective eigenfunctions of the unperturbed problem (2); $\epsilon$ indicates the correction of the energy $E_n$ in the second order. $H'$ is the perturbation potential, i.e., in equation (3), $H' = (\hbar/m) K \cdot p$. If spin–orbit coupling is included, two terms must be added to the Hamiltonian. The first one, $H_{so} \propto (\nabla V \times p) \cdot \sigma$, causes the bands to split; the second one, $\propto K \cdot (\sigma \times \nabla V)$, is an additional perturbation potential following from the introduction of Bloch functions.

Now we demonstrate how the matrix elements of the perturbation potential may be treated by group theoretical methods. The perturbation potential is a scalar product of two vectors, namely, $H' = K \cdot R$, and we can thus write

$$(\Psi_n | H' | \Psi_l) = \sum_{i=1}^{3} K_i (\Psi_n | R_i | \Psi_l)$$

where the matrix elements are scalar products of the vectors $\Psi_n$ and $R_i \Psi_l$, and are therefore invariant against an orthogonal transformation, such as a symmetry operation from the group of the wave vector.    Through this invariance, it is possible to determine which matrix elements vanish by symmetry, and to reduce others to a few constants.

In general, all first-order matrix elements of the perturbation potential will vanish if the inversion is an element of the group of the wave vector.  Let us consider the matrix element $(\Psi_{n\rho} \mid R_i \mid \Psi_{n\nu})$. The functions $\Psi_{n\rho}$, $\Psi_{n\nu}$ are eigenfunctions of the same unperturbed eigenvalue $E_n$; and, therefore, (from group theory) are functions with the same parity; i.e., when transformed by the inversion, they either both change their signs or they both do not.  $R_i$ contains $x$, $y$,

linearly (e.g., as $\partial/\partial x$) and, therefore, $R_i$ always changes sign upon inversion.  Thus, always

$$(\Psi_{n\rho} \mid R_i \mid \Psi_{n\nu}) = -(\Psi_{n\rho} \mid R_i \mid \Psi_{n\nu})$$

When the inversion is applied, the matrix element must vanish.  For such a point of the BZ, we will find a zero slope of $E(K)$, and, therefore, an extreme or a saddle point can be expected.  If the space group of the lattice contains the inversion and if spin–orbit coupling is included, every band throughout the BZ is at least doubly degenerate.  This can be shown with the help of the invariance against time reversal.   The time-reversal operator $T$ contains the complex conjugation operator and the spin operator $\sigma_y = \begin{pmatrix} o & -i \\ i & o \end{pmatrix}$ and commutes with the Hamiltonian

$$TH\Psi\,(k,\,r) = HT\Psi\,(k,\,r) = E(k)T\Psi\,(k,\,r)$$

$T\Psi\,(k,\,r)$ and  $\Psi(k,\,r)$ are eigenfunctions of the same energy value.  It follows from

$$T\Psi\,(k,\,r) = e^{-ik \cdot r}u\,(k,\,r) = \phi(-k,\,r)$$

that $T\Psi\,(k,\,r)$ is an eigenfunction at $-k$.  That means it is always valid that $E(k) = E(-k)$.  Now let the inversion $J$ be an element of the space group.  Then

$$J\,[T\Psi\,(k,\,r)] = J\phi(-k,\,r) = \Phi\,(k,\,r)$$

is an eigenfunction at the point $+k$ of the BZ.  The assumption of linear dependence between $\Psi\,(k,\,r)$ and $\Phi(k,\,r)$, i.e.,

$$\Phi\,(k,\,r) = JT\Psi\,(k,\,r) = a\Psi\,(k,\,r)$$

leads to a contradiction.  This is seen by twice applying the operation $JT$.  We find

$$JTJT\Psi\,(k,\,r) = JT\Phi\,(k,\,r) = JT\,[a\Psi\,(k,\,r)]$$
$$= a^* \, a\Psi\,(k,\,r); \qquad a^* \, a > 0$$

However, $JTJT = T^2 = -1$. Thus, there is no linear dependence between $\Phi\,(k,\,r)$ and $\Psi\,(k,\,r)$. Therefore, all energy values at a general point of the BZ are doubly degenerate.

Let us now discuss the band structure calculation of boron. With the K·p-method, every result follows from the symmetry alone, i.e., from the space group of the lattice. Since $a$- and $\beta$-rhombohedral boron have the same space group $R\bar{3}m$ or $D_{3d}^5$ (in the Schoenflies notation), our results are valid for both rhombohedral modifications of boron. The space group $D_{3d}^5$ is symmorphous; i.e., the rotations, etc., are only associated by primitive lattice translations. The point group $D_{3d}$ of $D_{3d}^5$ contains twelve operations: $E,\,2C_3,\,3C_2,\,J,\,2JC_3,\,3JC_2$. $D_{3d}$ is also the group of the wave vector of the points $\Gamma$ and $Z$ [$Z$ at (001) is equivalent to $Z$ at (00$\bar{1}$)] (Fig. 1). As $D_{3d}$ contains the inversion, we expect at these points zero slopes of the energy. The group of the wave vectors at the points $A$ and $D$ includes the inversion (the group elements are $E,\,C_2,\,J,\,JC_2$); therefore, extremes are possible at $A$ and $D$. At all the other points and lines of symmetry, the gradient of $E(k)$ in k-space, $\nabla_k E$, is different from zero for at least one direction. The $E(k)$ functions in the neighborhood of the points $\Gamma$ and $A$ are as follows:

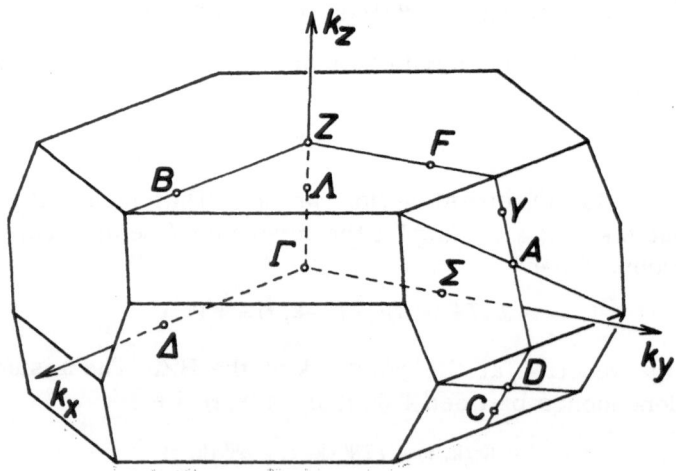

Fig. 1. Brillouin zone of a rhombohedral lattice, such as $a$- and $\beta$-boron.

Without Spin

$$E_{\Gamma_i^{\cdot}\pm}(K) \quad = E(\Gamma_i^{\pm}) + B_iK_{\perp}^2 + C_iK_z^2 \qquad\qquad i = 1, 2$$

$$E_{\Gamma_3^{\cdot}\pm}(K) \quad = E(\Gamma_3^{\pm}) + B_3K_{\perp}^2 + C_3K_z^2 \pm D_3K_{\perp}^2$$

$$E_{A_i\pm}(K) \quad = E(A_i^{\pm}) + F_iK_x^2 + G_iK_y^2 + H_iK_z^2 + L_iK_yK_z \qquad i = 1, 2$$

Spin Included

$$E_{\Gamma_i\pm}(K) = E(\Gamma_i^{\pm}) + B_iK_{\perp}^2 + C_iK_z^2 \qquad\qquad i = (4 + 5), 6$$

$$E_{(A_3 + A_4)}(K) = E(A_3 + A_4) + F K_x^2 + G K_y^2 + HK_z^2 + LK_yK_z$$

These functions are the result of the K·p perturbation approach in second order. The constants $B_i$, $C_i$, etc. are sums over matrix elements of the perturbation potential K·p. At $\Gamma$ and $Z$, we have the four one-dimensional representations $\Gamma_1^{\pm}$, $\Gamma_2^{\pm}$ and the two two-dimensional ones $\Gamma_3^{\pm}$. From spin—orbit coupling we get six additional "extra representations:" $\Gamma_4^{\pm}, \Gamma_5^{\pm}$ are one-dimensional and $\Gamma_6^{\pm}$ is two-dimensional. $\Gamma_4$ and $\Gamma_5$ are degenerate by time reversal. At $A$ and $D$ there are only four one-dimensional representations $A_1^{\pm}$, $A_2^{\pm}$ and four one-dimensional extra representations $A_3^{\pm}, A_4^{\pm}$, which are pairwise degenerate because of time-reversal symmetry. The energy functions have no terms linear in K. $\nabla_k E$ vanishes for all directions at $A$ and $D$. To decide whether there is an extreme or a saddle point, we must know the sequence of the energy bands and their symmetry at the band edges. If there is an extreme, the energy surfaces for $\Gamma_1, \Gamma_2$, ($\Gamma_4 + \Gamma_5$), and $\Gamma_6$ are ellipsoids of revolution with symmetry axes in the $\Lambda$-direction. The representations $\Gamma_3^{\pm}$ are two-dimensional; and the degeneracy is maintained along the $\Lambda$-axis only; in all the other directions, the bands split. If spin is considered, $\Gamma_3^{\pm}$ splits into the time-reversal degenerate terms $(\Gamma_4^{\pm}, \Gamma_5^{\pm})$ and $\Gamma_6^{\pm}$. Because in this case every point is doubly degenerate, no further splitting is possible. The results for $\Gamma$ are also valid for $Z$. At the point $A$ the results for the case without spin qualitatively agree with those for the case with spin. The energy surfaces are general ellipsoids at $A$. The same is valid for $D$. Figure 2 shows the $E(K)$ curves at the points $\Gamma$ and $A$, with and without spin. The bold curves indicate double degeneracy. For the one-dimen-

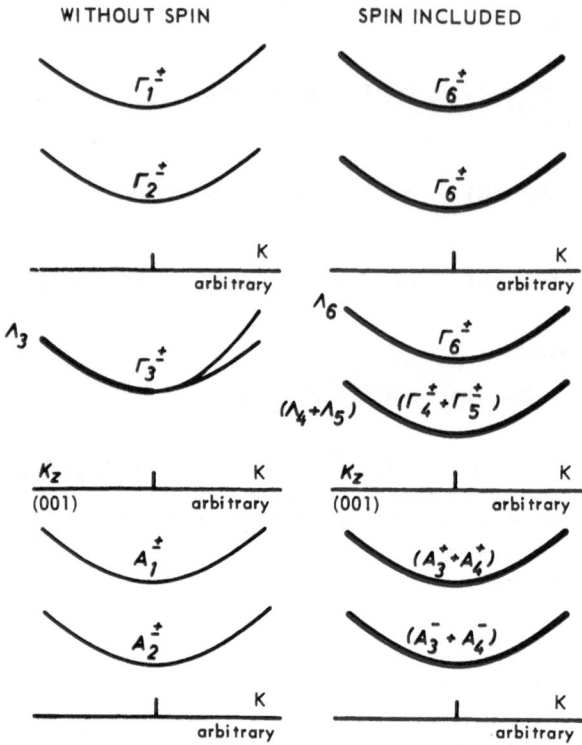

Fig. 2. $E$ vs. K curves in the neighborhood of $\Gamma$ (also valid for $Z$) and $A$ (also valid for $D$) for all possible representations at these points. The bold curves indicate double degeneracy.

sional representations of $\Gamma$ and $A$, the bands are parabolic and do not split. $\Gamma_3$ is doubly degenerate only in the (001) direction and splits if spin–orbit coupling is included.

A quantitative calculation of the band structure of boron will be very laborious because of the great number of atoms in the unit cell. However, it is hoped that this qualitative description of the band structure of boron will help in the interpretation and comprehension of experimental results.

## REFERENCES

1. Lietz, M., and U. Rössler, Z. Naturforsch. 19a:850 (1964).
2. Dresselhaus, G., A. Kip, and C. Kittel, Phys. Rev. 98:368 (1955).
3. Dresselhaus, G., dissertation, University of California (1955).
4. Kane, E. O., J. Phys. Chem. Solids 1:83 (1956).
5. Kane, E. O., J. Phys. Chem. Solids 1:249 (1957).
6. Dresselhaus, G., Phys. Rev. 100:580 (1955).

# Conductivity, Hall Effect, Optical Absorption, and Band Gap of Very Pure Boron

## Wolfgang Dietz and Hans Herrmann

*Consortium für elektrochemische Industrie*
*Munich, Germany*

Semiconductor properties were examined on highly pure and doped, $\beta$-rhombohedral, partly monocrystalline boron. The resistivity was from $7 \cdot 10^{12}$ and $4 \cdot 10^{13}$ $\Omega$-cm at -180°C to 2 and $3.5 \cdot 10^6$ $\Omega$-cm at room temperature. The band gap $\Delta E_0$ ascertained from the intrinsic range was about 1.42 eV, whereas $E_{300°\,K}$ obtained from the optical absorption edge was about 1.53 eV. The temperature dependence of the mobility should be more than -$\frac{3}{2}$ because of the influence of optical phonons. Both the optical and the electrical properties showed the expected anisotropy, which is probably due to the scattering mechanism depending on the direction. All samples examined were $p$-type, though they were in part largely doped with tungsten, titanium, molybdenum, zirconium, vanadium, silicon, and carbon. It was not possible to measure a Hall effect. Because of the sensitivity of the measurement, $\mu$ must be below 1 cm$^2$/V-sec. In the optical transmission spectrum, numerous bands appear both in amorphous and crystalline boron.

Fewer investigations have been made of boron than of any other elemental semiconductor, mainly due to technological difficulties in the production of pure, crystalline boron. Thus, much of the work done to date was carried out with insufficiently pure material. With the exception of diamond, no elemental semiconductor is known with a melting point as high as that of boron (about 2200°C). Because of the varying impurity contents of the materials examined, contradictory values have often been given in the literature. For instance, values quoted for the mobility have been 1 and 55 cm$^2$/V-sec [1,2] and even higher, and values for the band gap, 1.1, 1.55, and 1.41 eV [1,3,4].

The first step of our work was thus to produce highly pure boron in crystalline form. Our material was zone-refined several times. It had a spectrographical purity of 99.9999%; the predominant impurities were: silicon, <0.4; calcium, <0.1; copper, <0.001; and tantalum, <0.01 ppm. The material was of the beta-rhombohedral high-temperature modification of boron. Several large monocrystalline areas could be seen in the rods, and the test specimens

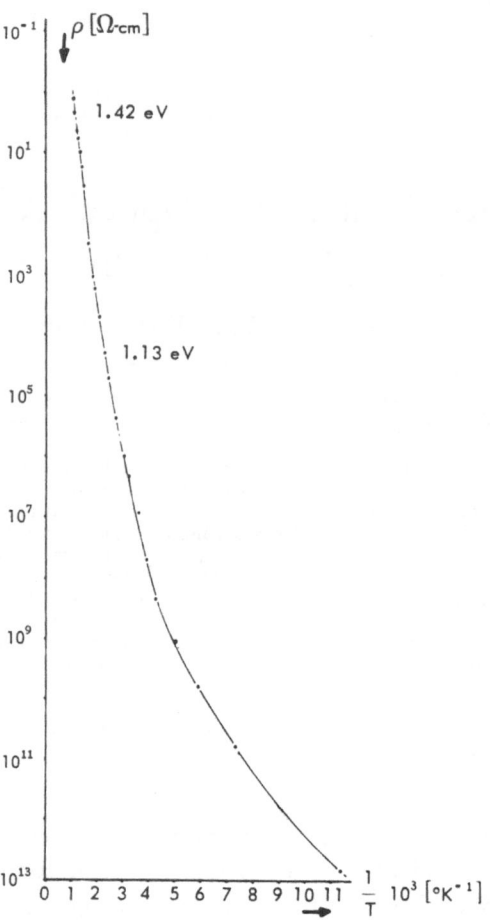

Fig. 1. Resistivity of hyperpure boron as a function of temperature.

were cut out of these. Laue patterns confirmed the monocrystalline character of the specimens. Beta-boron has a very complicated structure, the influence of which on physical properties has not been sufficiently clarified so far.

Resistivity measurements at temperatures from that of liquid air up to about 700°C were carried out with specimens having platinum contacts. The results are shown in Fig. 1. The resistivity extends from $10^{13}$ to 0.1 $\Omega$-cm, i.e., over a range of fourteen powers of ten. Specimens doped with 1% tungsten give a considerably flatter curve, as shown in Fig. 2.

From the slope in the intrinsic range, we obtain an energy gap $\Delta E$ of 1.42 eV. This value indicates the zero-temperature energy gap $\Delta E_0$, if the shift of the gap is assumed to be linear with temperature. From optical absorption measurements, we obtain a $\Delta E$ of 1.54 eV at 25°C by taking the point where the absorption curve is steepest [8] (Fig. 3). The flatter part of the edge may indicate a change from indirect to direct excitation processes. This is confirmed by an evaluation according to the method of MacFarlane et al. [7]: $\lambda^{-1}$ or $h\nu$ is proportional to $a^{1/2}$. This gives a band gap of 1.50 eV at 25°C (Fig. 4). From this we conclude that indirect transitions are probably important in boron.

Since $\Delta E_{300°K} < \Delta E_0$, it must be assumed that in boron the mobilities do not change proportional to $T^{-2/3}$ at elevated temperatures.

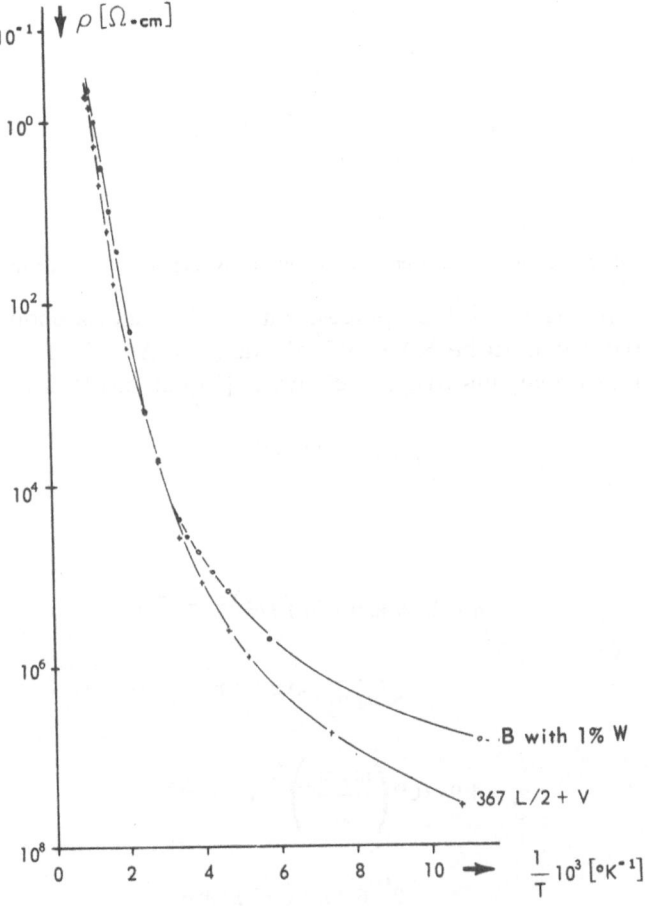

Fig. 2. Resistivity of boron doped with 1% tungsten or vanadium as a function of temperature.

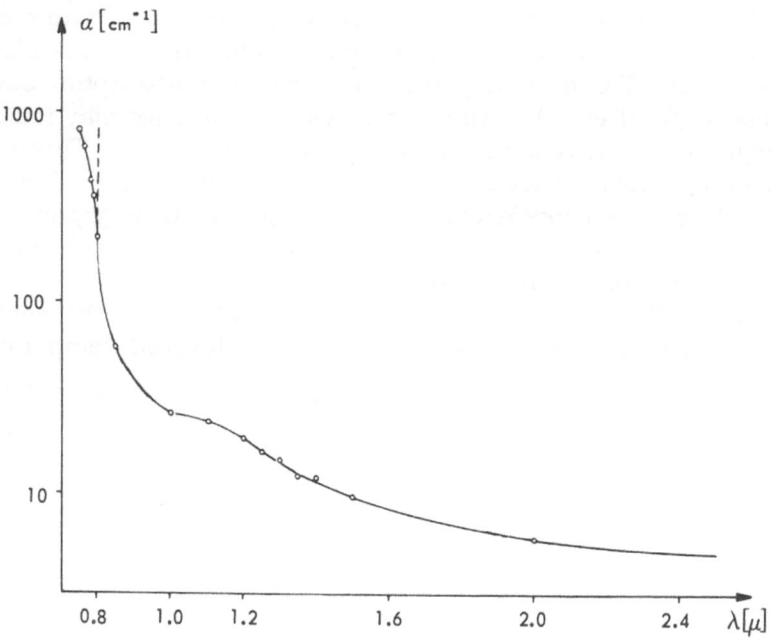

Fig. 3. Dependence of the absorption coefficient of hyperpure boron on the wavelength.

According to Moss [8] displacement of the photoconductivity by temperature $a$ should be $3.4 \cdot 10^{-4}$ eV/deg and $\Delta E_0$ should be 1.64 eV. Pertinent equations are as follows. [Equation (1) is simplified.]

$$\sigma = \sigma_0 \, e^{-\Delta E / 2 k T} \tag{1}$$

$$\sigma = n \, \mu \, e \tag{2}$$

$$n = 2(2 \pi \, m^* k T / h^2)^{3/2} \, e^{-\Delta E / 2 k T} \tag{3}$$

$$n_i = \sqrt{n_0 \, p_0} \; e^{-\Delta E / 2 k T} \tag{4}$$

$$n_i = 4.9 \cdot 10^{15} \left( \frac{m_n \, m_p}{m^2} \right)^{3/4} T^{3/2} \, e^{-\Delta E / 2 k T} \tag{5}$$

$$\mu_{ak} = \frac{2^{1/2} \, 6^{1/3} \, n^{1/3} \, e \, \hbar^2 \, k^2 \, M \theta}{4 \, \pi^{5/6} \, m^{5/2} \, (kT)^{3/2} \, C^2} \tag{6}$$

$$\mu_{opt} = A \ (m/m^*)^{3/2} \ \frac{e^{\theta/T} - 1}{T^{1/2}} \tag{7}$$

$$\frac{1}{\mu} = \frac{1}{\mu_{ak}} + \frac{1}{\mu_{opt}} \tag{8}$$

$$\rho = \rho_0 \ T^{\beta} \ e^{\Delta E / 2kT} \tag{9}$$

When measured values of $n$ and $\mu$ are introduced into equation (2), consistency is obtained only if scattering of the carriers by optical,

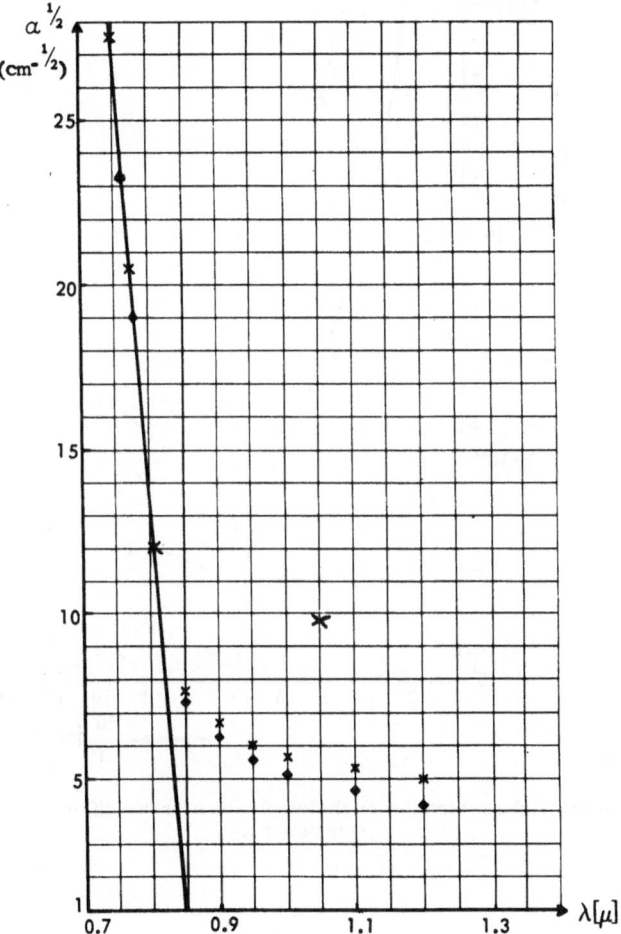

Fig. 4. Dependence of $\alpha^{1/2}$ on wavelength.

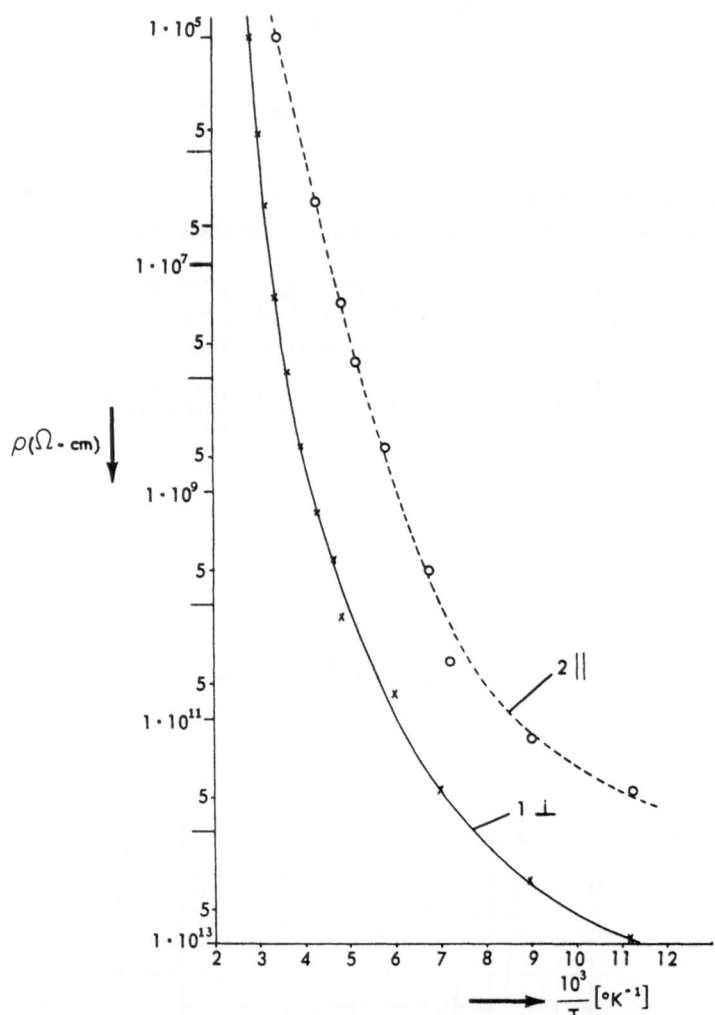

Fig. 5. Dependence of the resistivity on the temperature for two different crystallographic orientations.

as well as acoustical, phonons is assumed. This gives the correct (large) dependence of $\mu$ on $T$. Our results correspond to the values of Hagenlocher [5], who suspects that the mobilities of boron change with temperature as $T^{-7/4}$.*

Another possible explanation for the observed $\sigma$ vs. $T$ behavior may be that polarons are formed in boron. Due to the Frank–Condon principle, the optical excitation energy for these polarons is higher than the thermal one. The very low mobility of boron also seems to indicate the presence of polarons.

The slope in the intrinsic range, and thus the apparent band gap, are somewhat more pronounced when the current is parallel to the (10$\bar{1}$1) rhombohedral plane. On the basis of our optical and electrical measurements, we assume that scattering in boron is relatively anisotropic.

Figure 5 shows the resistivity for two mutually perpendicular directions. At lower temperatures, a distinct difference can be seen. Figure 6 shows the preferred orientation of the rhombohedron

*Regarding scattering on acoustical and optical phonons, a similar case is known from lead sulfide.

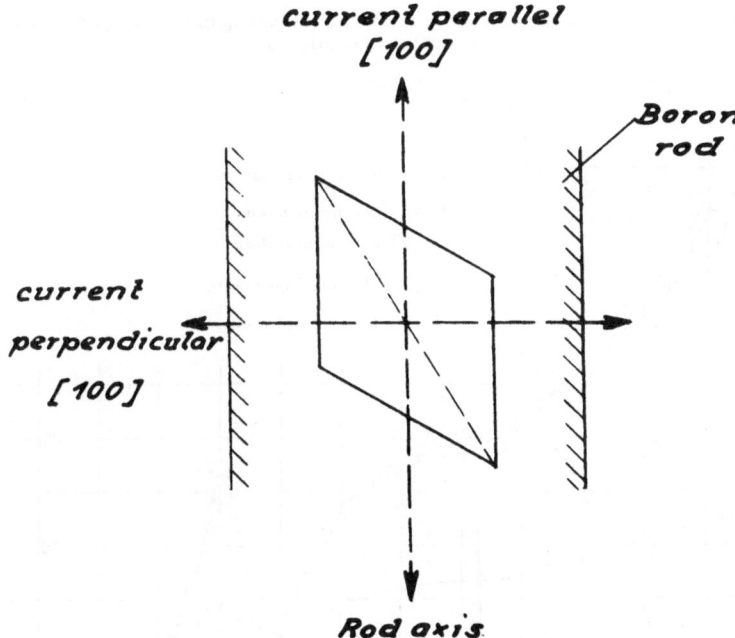

Fig. 6. Schematic drawing showing the two directions in which samples were cut from the boron rod.

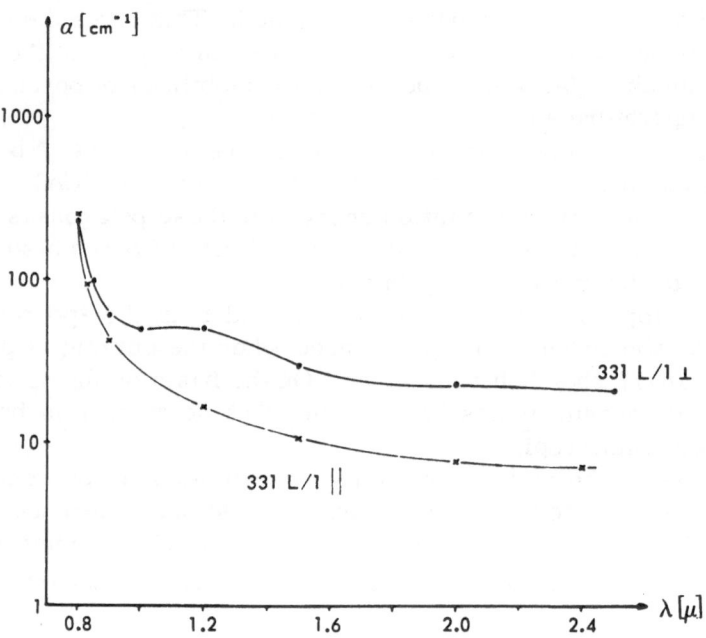

Fig. 7. Dependence of the absorption coefficient on wavelength for two different crystallographic orientations.

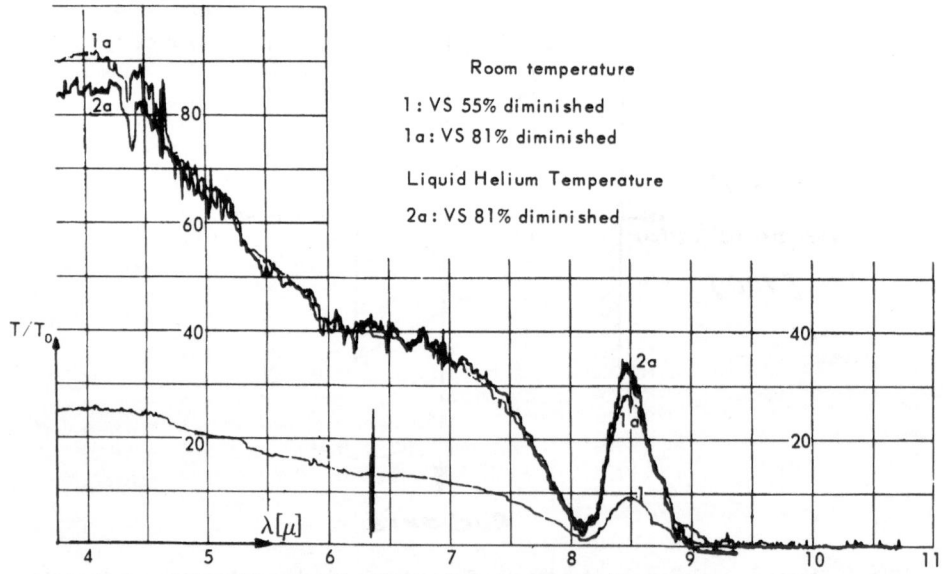

Fig. 8. Transmission curves for compact crystalline boron slices at longer waves for
different temperatures.

in the rods and the two measuring directions. Figure 7 shows the absorption curves for both directions.

Optical transmission curves show the importance of optical lattice vibrations in boron. From about $9-32\,\mu$, no transmission was found in slices with thicknesses of $60-65\,\mu$ (Fig. 8). The measurements were carried out in the long-wave range with a Leitz infrared spectrograph with a cesium bromide prism and micro-mechanism. Using potassium bromide pellets with crystalline boron powder, we obtained a great number of absorption bands ranging from $8.5\,\mu$ onward, as can be seen in Fig. 9. No exact identification has been possible so far. Some of the absorption bands (especially that at $11.8\mu$) also appear in the absorption spectrum of the amorphous boron powder (and in alpha-boron [6]), and thus could belong to the characteristic vibrations of the icosahedrons.

In reference to our electrical measurements again, it is interesting to note that all specimens show an additional activation energy of 1.0 eV. In the optical absorption, we find a wide band exactly at this point in all specimens, which, though not yet classified, seems to indicate a deep level in boron. (See Fig. 3.)

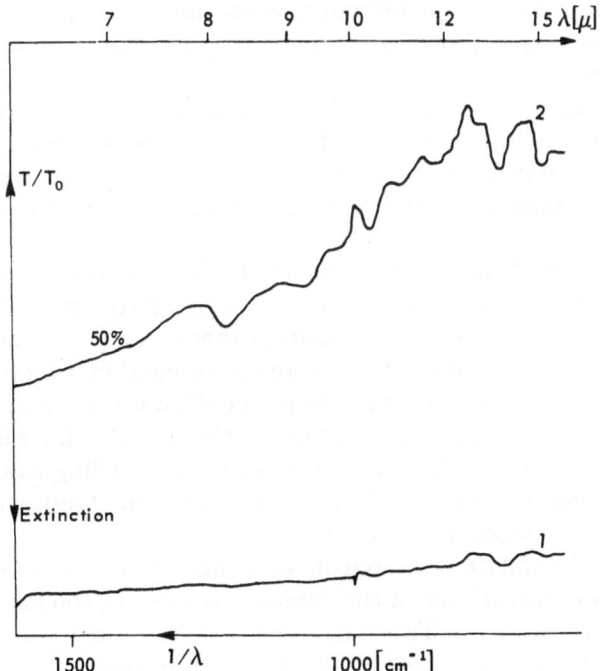

Fig. 9. Transmission curves of crystalline boron powder.

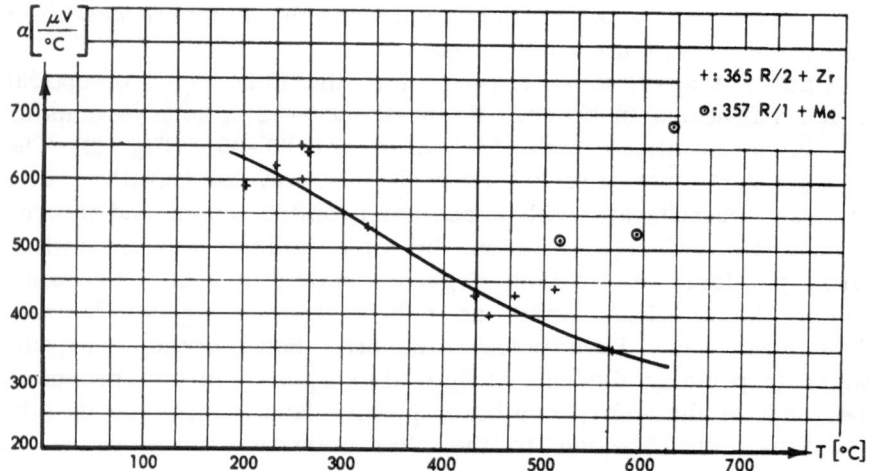

Fig. 10. Dependence of the thermoelectric power of pure boron on the temperature.

It is very surprising that all the specimens examined were
p-type, even when doped with titanium, tungsten, molybdenum,
zirconium, vanadium, carbon, or silicon. The thermoelectric
force was very small for specimens doped with 1% tungsten. Only
one specimen with a high content of vanadium was found to be n-type.
Hagenlocher [5] reports a content of 0.1% carbon and silicon for
his n-boron.

A measuring curve for the thermoelectric force is shown in
Fig. 10. It seems that p-conductivity is a basic characteristic of
boron, which may be connected with the specific bond mechanism in
boron, but vacancies also have to be considered in this complex
crystal.

We also tried to measure the Hall effect on boron, but, in spite
of numerous tests, we did not succeed. From the sensitivity of
our system, it can be estimated that $\mu$ must be smaller than
1 cm$^2$/V-sec. For some of these tests we used completely uniform
monocrystals. The results were confirmed by Mr. Schönwald
(of Siemens & Schuckert, Erlangen, Germany) who kindly tested
some of our polycrystalline specimens. According to comparable
tests by Signals Research Labs, Christchurch, England, it is even
believed that $\mu$ might be below 0.1 cm$^2$/V-sec.

The low mobilities in boron are most likely due to scattering
by the lattice vibrations in the highly complex boron lattice with its
complicated bonds. This also becomes evident in the optical
absorption by free carriers. Since we think that a certain number

of carriers is always present in boron, the absorption will not be greatly influenced by the temperature, as low-temperature measurements prove. (See Fig. 8.)    The dependence of absorption on the wavelength ranges mainly between $\lambda^{1.4}$ and $\lambda^{1.8}$. This plainly indicates scattering of the carriers on lattice vibrations. Only with a highly doped specimen was an exponent of 2.9 obtained.

It appears possible that no really free electrons are involved, but that their motion is due to "hopping." An exact theory for this case has not yet been established.*    It would agree with our assumption of intrinsic carriers or vacancies (or radicals) in boron. One might also speculate that polarons are responsible for the low mobilities.

At longer wavelengths, a broad absorption band at $8.15\mu$ and a small transmission range are seen in the transmission curves. On the basis of measurements on compact boron slices and potassium bromide pellets between 4 and 300°K, this band seems to be independent of temperature. It might represent specific lattice vibration or a $p$-band.

A B–H band and a B–O band were detected in amorphous boron (Fig. 11). These bands show the expected temperature dependence. Of the other impurities, tantalum especially seems to influence the optical absorption. Usually, the absorption constant in the transmission range was $5-8\,\text{cm}^{-1}$. Through transmission measurements, taking layer thickness into account, we found a reflection coefficient

*See Neft, W., and K. Seiler, "Semiconductor Properties of Boron," this volume, p. 143.

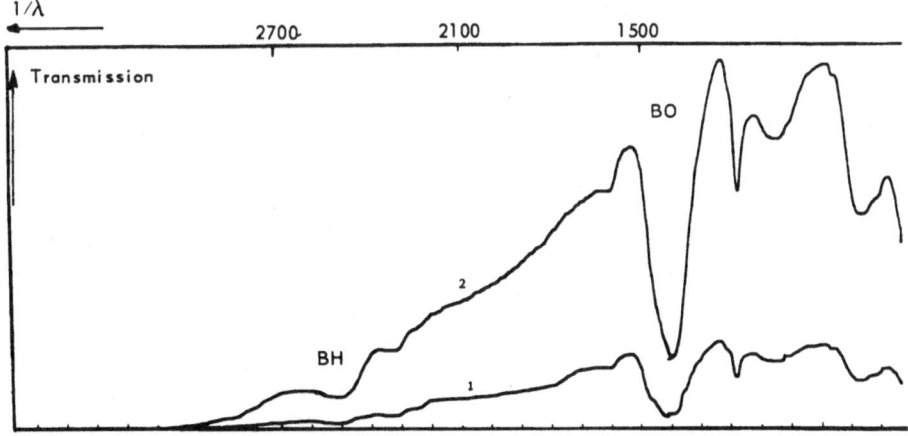

Fig. 11. Transmission of amorphous boron powder.

of 26%, giving an index of refraction of 3.1. The estimate for the dielectric constant thus becomes 9.3.

## ACKNOWLEDGMENTS

We thank Prof. Pick, of the Technological University, Stuttgart, and Dr. Krüger, of Leitz, Wetzlar, Germany, for their kind assistance for the long wavelength and helium-temperature optical measurements. Dr. Köhl, of AEG-Forschungsinstitut Frankfurt, Germany, was kind enough to measure the thermoelectric force at high temperatures. We thank Dr. Schönwald, of Siemens & Schuckert, Erlangen, Germany, for Hall-effect measurements.

## REFERENCES

1. Show, Hudson, and Danielson, Phys. Rev. 107:49 (1957).
2. Hagenlocher, A. K., in: Boron—Synthesis, Structure, and Properties, Plenum Press (New York), 1960, pp. 128-134.
3. Niemyski et al., Proc. Conf. Ultrapure Semicond. Mat., Boston, 1961, p. 67.
4. Geist, D., German Semicond. Conf., Bad Pyrmont (1963).
5. Hagenlocher, A. K., Dissertation, Technische Hochschule, Stuttgart (1957).
6. Wiedemann, Dissertation, Technische Hochschule, Stuttgart (1962).
7. MacFarlane et al., Phys. Rev. 97: 1714 (1955); 98: 1865 (1955); 111: 1245 (1958).
8. Moss, Optical Properties of Semiconductors, 1961.

# Thermal and Optical Band Gaps of Monocrystalline Beta-Rhombohedral Boron*

### Robert A. Brungs, S. J.

*RINS*
*Woodstock College*
*Woodstock, Maryland*

Electrical and optical experiments were performed on relatively large, but impure, single crystals of the beta-rhombohedral form of boron. All samples had a high resistivity (about $7 \cdot 10^5$ $\Omega$-cm at 300°K) which decreased by a factor of $10^6$ between room temperature and 750°K. The thermal energy gap between the valence band and the conduction band was found to be between 1.27 and 1.35 eV for three crystal directions. The optical band gap varied between 1.41 and 1.44 eV for single crystals and was 1.6 eV for a sample containing several crystallites. The maximum error in the thermal and optical values was ± 0.02 and ± 0.08 eV, respectively.

## INTRODUCTION

The present investigation was initiated to study the rather large difference between the thermal and optical band gaps (greater than 0.2 eV) observed by Farabaugh [1] and Krebs [2] using the same samples of polycrystalline beta-rhombohedral boron. While the present results show a difference of less than 0.2 eV between the thermal and optical band gaps (somewhat more in line with expectation), they embody a very unusual result for the thermal band gap. The results of a simple resistivity measurement seem to indicate anisotropy in the band gap.

## SAMPLE PREPARATION

The bulk boron, provided by the United States Army Signal Research and Development Laboratories at Fort Monmouth, was wedge-shaped, $\frac{5}{8}$ in. along a side, and 1 in. thick. The bottom half of this sample consisted of a relatively soft boron "slag." About $\frac{1}{2}$ in. from the top, the sample showed polycrystalline charac-

*This paper originally appeared in the Journal of Physics and Chemistry of Solids 25:701-706 (1964). This work was done at Saint Louis University under V. Jacobsmeyer, S. J., from 1959 to 1962.

teristics; it became more regularly crystalline toward the top. A tenth of an inch from the top, there was a distinct line, apparently a melt line, above which the sample showed monocrystalline characteristics. Spectrochemical analysis showed a purity of only 99.7%, with carbon, silicon, and manganese the major impurities. Back-reflection X-ray photographs of the top surface and the sides of the wedge showed the presence of four single crystals. Laue transmission photographs of the crystals after cutting confirmed their essential monocrystallinity.

Since the crystals were small, it was decided that a diamond saw would not be safe enough for cutting out the crystals. Ultimately, a slab $\frac{1}{10}$ in. thick was removed from the top of the melt. This was cut by hand with 10-mil tungsten wire etched with aqua regia. The final shaping of the crystals was done with a small coping saw, with a 5 by 62-mil tantalum ribbon used as a blade. The cutting technique has since been effectively mechanized by Lonc at Saint Louis University. While the cutting still requires a long time, it has the advantage of being safe enough to use for small crystals.

The five samples which were taken from the top slab were examined with a G. E. single-crystal orienter, Model BR, Type 1. The agreement of data obtained in this study with that reported by Hoard [3] sufficiently confirmed that the samples were beta-rhombohedral boron. The orientation of the samples was as follows: Samples 1, 2, and 4 were single crystals; sample 3, a double crystal; and sample 5 contained many small crystallites which were not randomly oriented. For convenience this latter sample is referred to as quasi-polycrystalline. The orientation of sample 1 was not determined; in sample 2, the width dimension corresponded to the [101] direction, its thickness to the [010] direction, and its length was parallel to the (101) planes. Sample 3 had the same orientation as sample 2. In sample 4, the thickness dimension was parallel to the [010] direction. It should be noted here that the word "direction" is used to designate the normal to a family of planes.

The size of the crystals was approximately 4 by 1 by 0.8 mm with the exception of sample 3, which was 7 by 2 by 0.8 mm.

## PHOTOCONDUCTIVITY

### Apparatus and Procedure

In the photoconductivity measurements, light was focused on the samples and the change of conductivity with wavelength was measured.   A Perkin-Elmer Model 83 universal monochromator and a 500-W CZX tungsten projection lamp provided the incident light.  This light falling on the boron caused an increase of current through the sample; the corresponding voltage increase across a matched series resistor was detected by a DC amplifier connected to a Brown Electronik recorder.  The DC amplifier was a Beckman Model V picoammeter used as a voltmeter.  The samples were mounted in a pyrex tube evacuated to $10^{-5}$ mm Hg.

A G. E. high-pressure mercury arc was used in calibrating the monochromator.  A bridge circuit, two arms of which consisted of matched Kodak Ektron PbS cells, was employed to determine the spectral intensity of the CZX projection lamp.  The use of matched cells acted as a correction for any ambient temperature changes of the apparatus.  The spectral sensitivity of the Ektron cells was very nearly linear for the range of wavelengths used in this experiment.

The spectral variation of the intensity of the CZX source was measured using a 1-mm slit width on the monochromator.  The incident light was chopped at 18 cps.  The intensity values obtained were normalized to take into account the variation in sensitivity of the PbS photodetectors with wavelength.  Since the sensitivity of these cells varied linearly with wavelength over the range of wavelengths used, the measured intensity for a particular wavelength was multiplied by the reciprocal of the sensitivity of the PbS cells for that same wavelength.  The normalized intensity thus obtained was used in calculating the photoconductive sensitivity.  The photosensitivity curves, shown in Figs. 1 and 2, have been normalized to account for variations in the spectral intensity of the source.

It was suggested by Sosnowski and Gaulé* that the spectral intensity in a photoconductivity experiment be kept as small as possible.  In the present experiment, the light intensity was just sufficient to give a workable signal-to-noise ratio for the wavelength range of $1.2-1.6\,\mu$.

Results
    A simple relation between the band gap and the wavelength is given by $E_{g} = hc/\lambda$.  Various criteria have been suggested for choosing wavelength values to be used in calculating the optical

---

*This suggestion was made during the discussion of the present paper at the International Colloquium on Hyperpure Boron, July 1964.

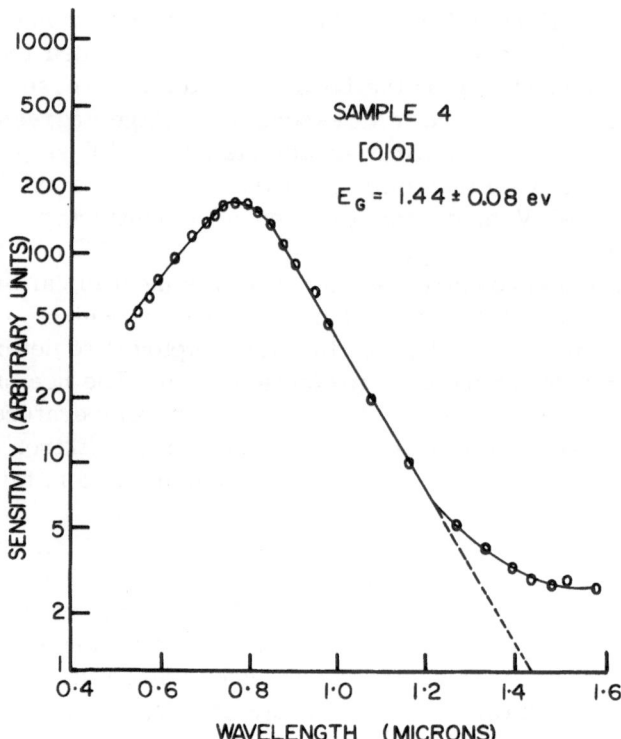

Fig. 1. Sensitivity vs. wavelength for sample 4 with the probes in the [010] direction.

energy gap from data taken from the long wavelength edge [4]. Values have been reported from calculations based on a particular choice of the magnitude of the absorption constant or on the position of maximum slope in the absorption versus wavelength curve [5]. The half-maximum wavelength criterion of Moss [6] was used for the quasi-polycrystalline sample to facilitate comparison with the results of Krebs [2], who used this criterion in his calculations. The values of the optical band gap are given in Table I. The error for optical measurements listed in the table represents the spread of experimental values and is not a statistical error.

Gaulé et al. [7] have reported an absorption edge at approximately $0.8\,\mu$ for beta-rhombohedral boron and their investigation of the photoconductivity revealed a spectral response with a sharp peak at approximately $0.8\,\mu$. This would give an activation energy of about 1.5 eV. Spitzer and Kaiser [8] reported a value of 1.5 eV using absorption coefficient measurements; while Gebhart [9] has obtained, also from absorption measurements, a value of 1.26 eV for samples taken from the same piece of boron as those used by the present author.

The photoconductivity in the present work showed a sharp peak in the range 0.77–0.8 $\mu$ for the single crystals and 0.65 $\mu$ for the quasi-polycrystalline sample. The position of the half-maximum wavelength was 0.8 $\mu$ for the latter sample. The maximum slope position was 0.86–0.88 $\mu$ for the single crystals and 0.76 $\mu$ for the quasi-polycrystal. By using the maximum slope value for the wavelength, the optical band gap is found to be about 1.4 eV for the single crystals, and 1.6 eV for the quasi-polycrystalline sample. The points plotted in Figs. 1 and 2, which are representative photosensitivity curves, are averaged over six experimental runs.

The major source of error in this experiment was the large monochromator slit width needed to give a good signal-to-noise response for the longer wavelengths. The slit width varied from

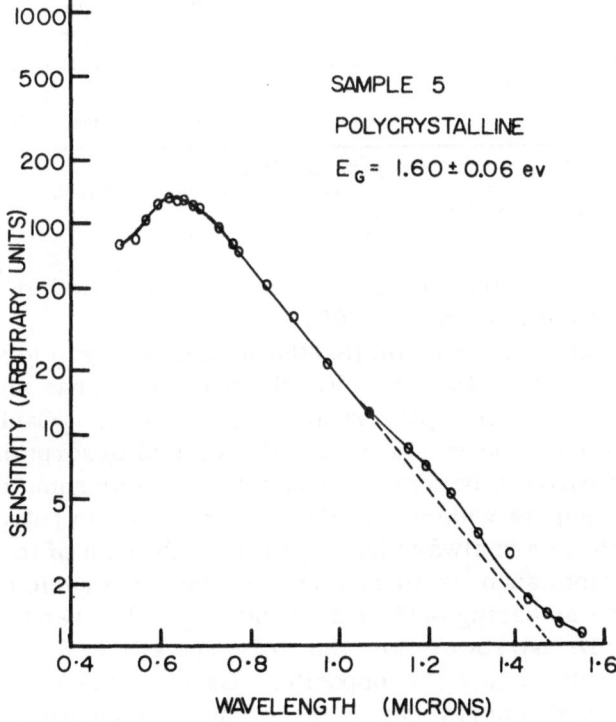

Fig. 2. Sensitivity vs. wavelength for the quasi-polycrystalline sample.

## TABLE I

Comparison of Experimentally-Obtained Activation-
Energy Values for Beta-Rhombohedral Boron

| Investigator | Activation energy (eV) | | Sample |
|---|---|---|---|
| | thermal | optical | |
| Greiner [11] | 1.39 | | polycrystal |
| Farabaugh [1] | 1.33 | | polycrystal |
| Hagenlocher [12] | 1.50 | | polycrystal |
| Shaw et al. [13] | 1.55 | | single crystals, tetragonal boron |
| Krebs [2] | | 1.56* | polycrystal |
| Gaulé [7] | | 1.5 | single crystal |
| Spitzer and Kaiser [8] | | 1.5 | polycrystal |
| Gebhart [9] | | 1.26 | single crystal |
| Dietz and Herrmann [14] | 1.42 | | high-purity single crystal |
| Brungs (present author)† | | | |
| [101] | 1.28 ± 0.01 | 1.41 ± 0.05 | single crystal |
| [010] | 1.27 ± 0.01 | 1.44 ± 0.08 | single crystal |
| (101) | 1.35 ± 0.02 | 1.41 ± 0.05 | single crystal |
| | | 1.60 ± 0.06‡ | quasi-polycrystal |
| | | 1.53 ± 0.06* | quasi-polycrystal |

*Calculated from the half-maximum wavelength.
†The error listed for the thermal data is one standard deviation.
‡Calculated from maximum-slope data.

0.5 to 1 mm for the different samples. At a width of 1 mm, the wavelength band passed was 0.06 $\mu$.

In general, it can be said that the optical band gap for the single crystals is about 1.40–1.45 eV, which is somewhat smaller than that obtained for the polycrystalline material. Gaulé [10] has pointed out that one would expect the optical absorption of single crystals of boron to be somewhat anisotropic. For some directions, the energy gap is wider; for others, narrower. In polycrystalline material, there are always some grains in the path of the light with an orientation such as to absorb the shorter wavelengths with a consequent "smearing out" of the band edge. The band gap should, therefore, be narrower for the polycrystalline material. The present result is just the opposite. Gaulé offers the explanation that the narrow energy gap is the result of impurities which provide energy levels near the band edge. The polycrystalline material has fewer of these impurities because they were allowed to segregate out along the grain boundaries.

## RESISTIVITY MEASUREMENTS

### Apparatus and Procedure

A standard DC bridge circuit was used for these experiments. The voltage $V_S$ was first measured across a precision resistor $R_S$ ($\pm$ 0.5%) in series with the boron sample, and then across the resistivity pressure probes $V_B$ mounted on the boron sample. Since the current was the same in both these circuit elements, the resistance, and hence the resistivity, of the sample could be calculated from the simple ratio $V_S/R_S = V_B/R_B$.

The resistivity probes were made from 12-mil tungsten etched to a point in sodium hydroxide. The current leads were 10-mil platinum wire fused to the boron sample. Rectification, unobservable at room temperature, increased to about 10% at 750°K.

Resistivity measurements above room temperature were made in vacuum ($10^{-3}$ mm Hg) in a chamber constructed from S-band wave guide. A calibrated copper–constantan thermocouple was used up to 540°K; a platinum–platinum + 10% rhodium thermocouple was used for temperatures higher than 540°K. Both thermocouples were accurate to within 1%.

A 2.2-V Willard wet cell was used as the DC source. All voltage measurements were of a null type, and were made with a K-2 potentiometer and a Rubicon galvanometer with a sensitivity of $1.5 \cdot 10^{-3}$ $\mu$A/mm. The dimensions of the samples were measured to an accuracy of 0.01 mm with a traveling microscope. This was considered sufficiently accurate since the chief purpose of the measurements was the value of the slope of the intrinsic curve, and not the absolute magnitude of the resistivity.

### Results

Resistance data were taken for three crystal directions, the [101], [010], and the direction parallel to the (101) planes, the plane of the long diagonal in the rhombohedral system. Figures 3–5 show representative resistivity curves. The temperature was usually raised to approximately 750°K; for sample 2, it was twice raised to 900°K with no observable change in slope. The latter data are not plotted in Fig. 3. The straight-line portion of the curves begins at approximately 500°K, which is in good agreement with Greiner's results [11] for polycrystalline boron. As noted earlier,

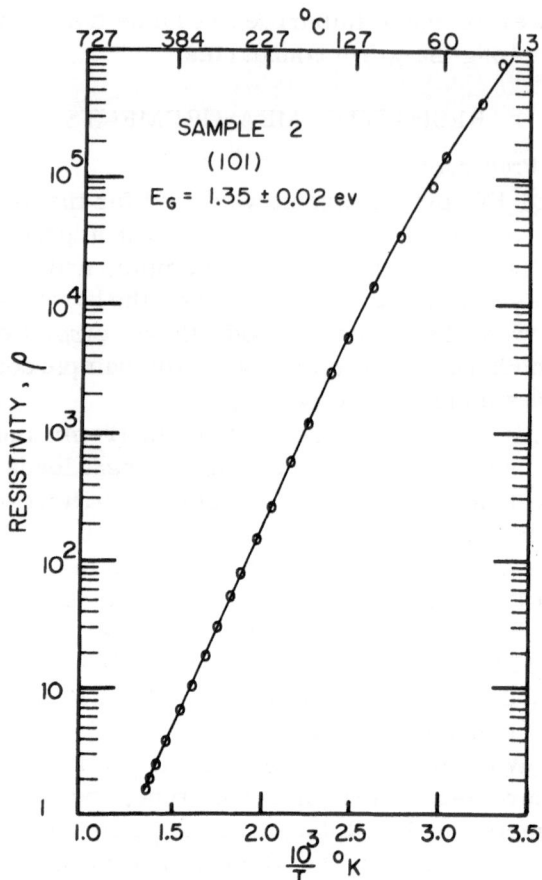

Fig. 3. $\log_{10} \rho$ vs. $10^3 / T$ °K for sample 2, with the current and resistivity probes parallel to the (101) family of planes.

these samples had a high and accidental, rather than doped, impurity content. For this reason, no measurements were made in the extrinsic region below room temperature. The room temperature resistivity was approximately a half order of magnitude less than that of the most reliable data thus far reported [1, 11–13]. This was almost certainly a reflection of the high impurity level. The resistivity of all samples decreased by approximately six orders of magnitude in the temperature range of 300–750°K.

The values of the slopes of the curves obtained by plotting $\log_{10} \rho$ vs. $10^3 / T$ gave energy gap values varying from 1.33 to 1.36 eV, 1.27 to 1.29 eV, and 1.26 to 1.28 eV for samples 2, 3, and 4, re-

spectively.    The values given in the table for the thermal band gap
are the numerical averages over three experimental runs.  These
energy gaps are somewhat smaller than those reported for polycrystalline samples and considerably smaller than those reported
for microscopic single crystals of tetragonal boron.

## INTERPRETATION OF DATA

There is some difficulty involved in interpreting the data obtained from the resistivity experiments.    Data from resistivity
measurements made along the [101] and [010] directions (samples
3 and 4, respectively) give values of the thermal band gap that are

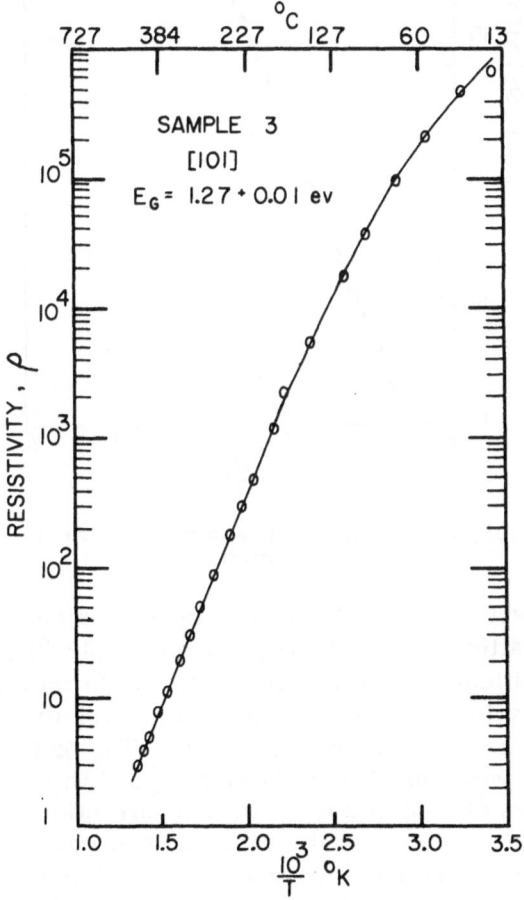

Fig. 4. $\log_{10} \rho$ vs. $10^3/T$ °K for sample 3, with the resistivity probes along the [101] direction.

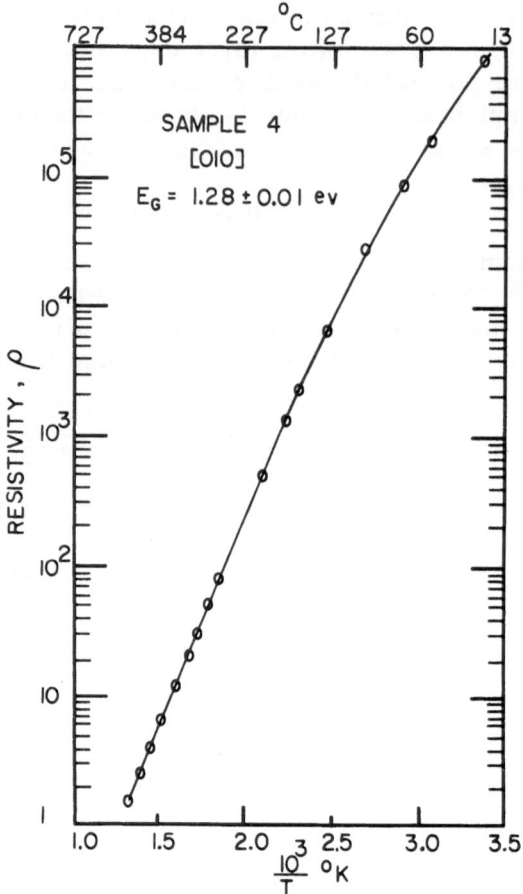

Fig. 5. Log$_{10}\rho$ vs. $10^3/T$ °K for sample 4, with the resistivity probes along the [010] direction.

not significantly different. Those measurements made in the direction parallel to the (101) planes (sample 2) do show a statistically significant difference from the two former results.

The values for the slope of the log$_{10}$ $\rho$ vs. $10^3/T$ curves were calculated by a least-squares analysis. The bottom nine experimental points were used and showed a good fit to a straight line.

In no series of runs in any one crystal direction does the spread of experimental values for the thermal band gap exceed 0.03 eV. The difference of the mean between sample 2 on the one hand, and samples 3 and 4 on the other hand, was 0.074. The equation used to find the error of a measurement was

$$\sigma = \sqrt{\frac{1}{N-1} \sum (X - \overline{X})^2} \tag{1}$$

where $\sigma$ is the error of a measurement; $N$, the number of measurements; $X$, the measurements; $\overline{X}$, the mean. The value of $\sigma$ for sample 2 was 0.015; for samples 3 and 4, 0.01. The error of the difference of the two means ($\sqrt{\sigma_1^2 + \sigma_2^2}$) was 0.018. Since the difference of the mean is 4.1 times greater than the error of the difference, the difference is statistically significant ($P = 0.0001$).

The problem remains as to just what this difference might mean physically. It is true that each set of runs along any crystal direction was made on a different sample, but all the samples were taken from the same layer of the same melt and were contiguous to each other. All the samples were treated in the same way. Nor is the difference to be attributed to impurity differences in the samples, since the calculations were made from data obtained in the intrinsic conduction region. The fact that the optical band gap shows no such variation is to be explained in terms of the inaccuracies incumbent on a lack of detailed analysis of the "tail" of the fundamental absorption edge and on the imperfect quality of the crystals used.

While an anisotropic band gap for boron would not be unexpected, it is surprising that such an unsophisticated experiment should yield a result of this nature. In view of the complexity of the crystal structure of beta-rhombohedral boron, however, and of the present lack of information about some of the basic properties of this element, notably magnetic properties, the results for the band gap along the three crystal directions are included. In the light of the statistical analysis of the data, the divergence in values of the thermal band gap does not seem to be attributable simply to experimental error.

## SUMMARY

Considering the thermal data obtained for the [101] and [010] directions, the value of the thermal band gap is 1.27 ± 0.01 eV, approximately 0.1 eV lower than the data reported by Greiner [11] and 0.05 eV less than that of Farabaugh [1], both of whom used polycrystalline material. It is in good agreement with the band gap of 1.26 eV obtained by Gebhart [9] who used samples obtained from the same piece of boron as those used by the present investigator. The different value obtained along the direction parallel to the (101) planes, 1.35 ± 0.02 eV, has been discussed in the section on data interpretation.

In comparing the values of the thermal band gap with those obtained on high-purity boron by Dietz and Herrmann [14], 1.42 eV, it must be recalled that the samples used in the present work were quite impure. It is most likely that the values found in the present work are too small to represent the forbidden gap. There is most probably a "hopping process" taking place in samples as impure as those used by the present author.

A larger difference is observed between the optical band gaps of the single crystals, 1.41 ± 0.08 eV, and the polycrystalline results. The value of the optical band gap of the quasi-polycrystalline sample, 1.6 ± 0.06 eV, compares well with the results of Krebs [2] for polycrystalline material. The value of the quasi-polycrystalline sample is approximately 0.2 eV greater than that of the single crystals. Until a more detailed analysis of the tail of the fundamental absorption edge has been carried out, there is a necessary inaccuracy involved in the optical band gap; although this inaccuracy should not exceed the error listed in the table. It can be said that the difference of 0.2 eV between the thermal and optical band gaps reported by Farabaugh [1] and Krebs [2] is at least partially due to the polycrystalline nature of the samples.

In general, the result of the present work, evidence of anisotropy in the band gap obtained from a simple resistivity measurement, seems to indicate some strong magnetic peculiarities in beta-rhombohedral boron. This conclusion appears to be supported by the magnetoresistance results of Lonc [15] at Saint Louis University. These show a strong dependence on crystal orientation. Until further data become available, it would seem that this complex rhombohedral form of boron exhibits magnetic anomalies and that measurements may be significantly affected by crystal orientation.

## ACKNOWLEDGMENTS

The spectrochemical analysis was made by W. J. Poehlman of the A. O. Smith Co., Milwaukee. The crystal orientation work was done at the McDonnell Aircraft Corp. of Saint Louis with the cooperation of Drs. William Kester and Gene Carron.

## REFERENCES

1. Farabaugh, E., unpublished M. S. Thesis, Saint Louis University, 1959.
2. Krebs, J., unpublished Ph.D. Dissertation, Saint Louis University, 1959.
3. Hoard, J. L., and A. E. Newkirk, J. Am. Chem. Soc. 82:70 (1960).
4. Moss, T. S., Optical Properties of Semi-Conductors, Academic Press (New York), 1959, p. 41.
5. Burstein, E., and P. H. Egli, The Physics of Semi-Conductor Materials, NRL Rept. 4595, Jan. 1956, pp. 49-50.
6. Moss, T. S., Photoconductivity in the Elements, Academic Press (New York), 1952.
7. Gaulé, G. K., J. T. Breslin, J. R. Pastore, and R. A. Shuttleworth, Boron—Synthesis, Structure, and Properties, Plenum Press (New York), 1960.
8. Spitzer, W. G., and W. Kaiser, Phys. Rev. Letters 1:230 (1958); note corrections, Phys. Rev. Letters 1:382 (1958).
9. Gebhart, F., unpublished Ph.D. Dissertation, Saint Louis University, 1963.
10. Gaulé, G. K., private communication.
11. Greiner, E., and J. Gutowski, J. Appl. Phys. 28:1364 (1957).
12. Hagenlocher, A. K., Boron—Synthesis, Structure, and Properties, Plenum Press (New York), 1960.
13. Shaw, W. C., D. E. Hudson, and G. C. Danielson, Phys. Rev. 107:419 (1957).
14. Dietz, W. H., and H. A. Herrmann, reported at the International Colloquium on Hyperpure Boron, Paris, July 1964.
15. Lonc, W., unpublished Ph.D. Dissertation, Saint Louis University, 1964.

# The Optical and Electrical Constants of Beta-Rhombohedral Boron

Frank L. Gebhart* and

Vincent P. Jacobsmeyer, S. J.

*Saint Louis University*
*Saint Louis, Missouri*

Reflectivity data obtained from two samples of beta-rhombohedral boron were used to calculate the optical properties in the wavelength range 0.45–0.65 $\mu$. One sample was composed of very small crystallites; the other was monocrystalline. The absorption coefficients were approx. $10^5$ cm$^{-1}$ and are attributed to direct optical transitions from the valence band to the conduction band. The absorption threshold energy $E_T$ of the monocrystalline sample was calculated to be 1.26 ± 0.04 eV, which is in agreement with the value for the thermal activation energy $E_a$ of 1.27 ± 0.01 eV obtained by Brungs for the same sample. The index of refraction of both samples was 3 and the dielectric constant 11 in the wavelength range 0.45–0.65 $\mu$. These data are in agreement with the values obtained by Spitzer, Morita, and Lagrenaudie. The extinction coefficient of the polycrystalline sample varies from 1 at $\lambda$ = 0.45 $\mu$ to 0.6 at $\lambda$ = 0.65 $\mu$; for the monocrystalline sample, the extinction coefficient varies from 0.65 to 0.35 over the same wavelength region.

## INTRODUCTION

Research into the optical properties of boron has in the past been largely restricted to wavelengths in the infrared [1,2] above 0.75 $\mu$. The present work presents data taken in the spectral region of 0.45–0.65 $\mu$ on monocrystalline and polycrystalline boron. The data summarized in the figures which follow were obtained from reflection measurements made in air at room temperature.

The choice of the reflectivity method was influenced by several factors. First, boron is particularly well-suited to this method by reason of its hardness and chemical inertness. Secondly, Rizzo [3] has shown that boron does not oxidize to any appreciable extent until the temperature exceeds about 400°C; oxide films, which are a source of serious error for many other materials, should have no great effect on reflectivity experiments on boron. Finally, the reflectivity method yields direct information on four quantities —

*Present address: Lockheed-California Company, Burbank, California.

the absorption coefficient, the dielectric constant, the index of refraction, and the extinction coefficient.

## THE REFLECTIVITY METHOD

In the reflectivity method, a beam of plane polarized, monochromatic light is focused on a polished specimen of the material at an angle of incidence $\theta$. Let the plane of incidence be defined by the $y$-axis, normal to the surface, and the direction of propagation of the incident beam. The $x$-axis is normal to the plane of incidence and lies in the plane of the surface. The reflected light will be elliptically polarized as a result of the phase difference $\delta$ introduced between the intensity components $X$ and $Y$ of the reflected beam. This polarization can be described in terms of these intensities, measured along the $x$- and $y$-axes, and the intensities $A$ and $B$, measured along the major and minor axes of the ellipse. These intensities define the axial ratio, $\tan a = (B/A)^{1/2}$, and the amplitude ratio, $\tan \psi = (X/Y)^{1/2}$. The optical constant of the material can be related to the angles $\theta$, $\delta$, $a$, and $\psi$.

## APPARATUS

The ellipsometer was constructed on the base of an A-O Spenser spectrometer. Monochromatic light was obtained by using a set of five Bausch and Lomb interference filters having half-widths of approximately 20 A. The incident beam is polarized by a Nicol prism. The specimen is placed on the spectrometer table with the axis of the table in the plane of the surface of the sample. Another Nicol prism, which is mounted free to rotate about the direction of the reflected beam, is the analyzer. The angular position of the analyzer can be set to 0.1°.

The intensities $X$, $Y$, $A$, and $B$ are determined by rotating the analyzer to the position corresponding to any one of these quantities, and measuring the output of the Dumont 6467 multiplier phototube with a Leeds and Northrup K-2 potentiometer.

## THEORY

Let the index of refraction be denoted by $n$ and the extinction coefficient by $k$. Ditchburn [4] has shown that the complex index of refraction, $\bar{n} \equiv n - ik$, is related to the phase shift $\delta$, the amplitude ratio $\tan \psi$, and the angle of incidence $\theta$ as follows:

$$\frac{1 + e^{i\delta}\tan\psi}{1 - e^{i\delta}\tan\psi} = \frac{\tan\theta\sin\theta}{(\bar{n}^2 - \sin^2\theta)^{1/2}} \tag{1}$$

When the real and imaginary parts of each side of equation (1) are equated, the expressions for the dielectric constant $\epsilon$ and the conductivity $\sigma$ are obtained.

$$\epsilon = n^2 - k^2 = P(\cos^2 2\psi - \sin^2 2\psi \sin^2\theta) + \sin^2\theta \tag{2}$$

and

$$2\sigma/\nu = 2nk = P\sin 4\psi \sin\delta \tag{3}$$

where $\nu$ is the frequency and

$$P = \left(\frac{\sin\theta\tan\theta}{1 + \cos\delta\sin 2\psi}\right)^2 \tag{4}$$

Equations (2) and (3) are not of immediate use due to the presence of the phase shift $\delta$, which is not directly measurable by the method of the present experiment. It can be shown [5], however, that $\delta$ is related to $\alpha$ and $\psi$ by

$$\sin 2\alpha = \pm \sin 2\psi \sin \delta \tag{5}$$

The sign of $\cos\delta$ appearing in equation (4) is determined from the measurement of the critical angle of incidence $\theta_c$. This angle is such that $\cos\delta < 0$ for $\theta < \theta_c$ and $\cos\delta > 0$ for $\theta > \theta_c$. If $\theta = \theta_c$, then $\delta = \frac{1}{2}\pi$. Since the intensity component of the reflected beam parallel to the plane of incidence assumes its minimum value at $\theta = \theta_c$, $\delta$ may be determined. The critical angle $\theta_c$ of boron at wavelengths employed in the present work is approximately 74°.

The index of refraction $n$ and the extinction coefficient $k$ are obtained from the simultaneous solution of equations (2) and (3). The absorption coefficient $K$ follows from the equation

$$K = (4\pi k)/\lambda \tag{6}$$

## THE SAMPLES

Measurements were made on two samples. The first was a large, triangular specimen approximately $\frac{5}{8}$ in. on a side, obtained from the United States Army Signal Research and Development Laboratory (USASRDL) of Fort Monmouth, New Jersey. It will be referred to as the "RDL sample." Optical inspection and X-ray back-reflection analysis of the original surface of this sample showed it to be composed of a relatively small number of large, roughly hexagonal, crystallites of the beta-rhombohedral modification of boron. The largest dimension of these crystallites was several millimeters in length. The upper 0.1 in. of this specimen was cut away to provide samples for the thermal activation energy studies of Brungs [6]. The surface thus exposed was shown by X-ray back-reflection techniques to be composed of a small number of large crystallites, all of the same approximate orientation [(101) planes parallel to the surface] separated by thin veins of boron of a different structure. The crystallites comprising this second surface were smaller than those of the original surface. Inasmuch as the polishing operation described below obscured the crystallite boundaries, it is not possible to say definitely how many crystallites took part in the reflection. The number is thought to be not more than three. Measurements were made on two areas, designated on the figures as "L" and "R."

Spectrographic analysis of this sample showed carbon to be the principal impurity. The total impurity concentration is about 100 ppm.

The second sample studied was a small specimen of polycrystalline boron obtained from the Norton Abrasive Company. This sample was prepared by the decomposition of $BCl_3$ on an incandescent graphite rod held at a temperature of 1400°C. The total impurity concentration was approximately $10^3$ ppm; carbon, again, is the principal impurity.

Both samples were ground and polished to a high luster with a succession of grades of diamond paste. Immediately before each measurement, the sample was cleaned in concentrated KOH and rinsed thoroughly in distilled water.

The data exhibited in the figures which follow are the averages of four runs apiece for the Norton sample and for each of the two areas of the RDL sample. The spread in the calculated values of the five constants over the four runs varied from 2% in the index of refraction to 12% in the conductivity. There is no significant difference in the measurements made on the two areas of the RDL sample.

## THE OPTICAL PROPERTIES

### Absorption Coefficient

Direct and indirect optical transitions from the valence to the conduction band in semiconductors have been reviewed by Bardeen,

Blatt, and Hall [7] and also by Fan, Shepherd, and Spitzer [8]. In the direct transitions, the wave vector of the electron is not changed. In the indirect transitions, a phonon is emitted or absorbed and the electron wavenumber may change. If we let $\nu_T$ denote the threshold frequency, then for direct transitions the absorption coefficient is

$$K = C'h(\nu - \nu_T)^{1/2} \tag{7}$$

or

$$K = C''h(\nu - \nu_T)^{3/2} \tag{8}$$

where $C'$ and $C''$ are constants. The particular form of $K$ given by equations (7) and (8) depends essentially on the magnitude of the matrix element of the electron's momentum, and this, in turn, depends on the wave functions in the two bands. In germanium, the matrix element is large for electron wavenumbers equal to zero where the wave functions correspond to atomic s-orbitals. In silicon, the matrix element is small for the electron wavenumbers equal to zero due to some wave functions corresponding to atomic p-orbitals. Absorption due to indirect transitions should, in general, increase with a power of $(\nu - \nu_T)$ higher than $3/2$. The data taken on boron indicate that the absorption coefficient $K$ is proportional to $(\nu - \nu_T)^{3/2}$, for which the initial electron wavenumber is not zero. We use these data and equation (8) to calculate the threshold energy $E_T = h\nu_T$.

Let $K_a$ and $K_b$ be any two values of the absorption coefficient, and $E_a$ and $E_b$ the corresponding photon energies. The proportionality constant $C''$ of equation (8) is eliminated when the ratio

$$\frac{K_a}{K_b} = \frac{(E_a - E_T)^{3/2}/E_a}{(E_b - E_T)^{3/2}/E_b} \tag{9}$$

is formed. The threshold energy follows directly from equation (9) —

$$E_T = \frac{RE_b - E_a}{(R - 1)} \tag{10}$$

where

$$R = \left( \frac{K_a E_a}{K_b E_b} \right)^{2/3} \tag{11}$$

The five sets of values of $K$ and $E$ obtained from the successive use of the five interference filters allow the calculation of ten

## TABLE I

### Calculation of Threshold Energy from Absorption Data

| Sample | Threshold energy, $E_T$(eV) | $C''$ $(\text{cm-eV}^{1/2})^{-1}$ |
|--------|-----------------------------|-----------------------------------|
| RDL—R  | 1.28                        | $2.52 \cdot 10^5$                 |
| RDL—L  | 1.24                        | $2.53 \cdot 10^5$                 |
| Norton | 1.24                        | $4.40 \cdot 10^5$                 |

values of $E_T$. The average of these ten values can then be used to determine the constant $C''$ of equation (8). This calculation is summarized for both the RDL and the Norton samples in Table I.

With the values of $E_T$ and $C''$ determined above, values of $K$ corresponding to preassigned values of $E$ can be predicted from the equations:

$$K = 4.40 \cdot 10^5 \, \frac{(E - 1.24)^{3/2}}{E} \qquad \text{for Norton sample} \qquad (12)$$

$$K = 2.52 \cdot 10^5 \, \frac{(E - 1.26)^{3/2}}{E} \qquad \text{for RDL sample} \qquad (13)$$

The results of such a calculation are plotted in Fig. 1. The numerical constants of equation (13) are the averages of those obtained from the right and left sides of the RDL sample.

Brungs [9] found thermal activation energies of 1.35 eV perpendicular to the (101) set of crystal planes, 1.27 eV parallel to the [010] direction, and 1.28 eV parallel to the [101] direction. The terms parallel and perpendicular refer to the orientation of the current probes of the resistivity-measuring apparatus relative to the set of planes or the crystal direction mentioned. The optical threshold energy and the thermal activation energy are the same within the accuracy of the experiments. This is a feature exhibited by other semiconductors [7].

### Index of Refraction

No large-scale variation in the index of refraction was encountered in either the RDL or the Norton samples in the spectral region $0.45-0.65 \, \mu$. This fact is consistent with the view that the observed absorption is due to free electrons making the transition across the energy gap from the valence to the conduction band. Such an absorption process leaves the state of polarization of the

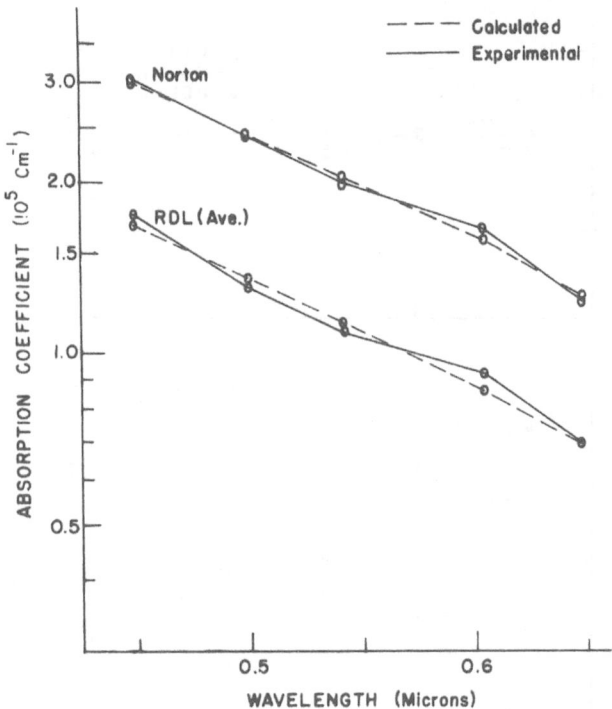

Fig. 1. Absorption coefficient versus wavelength. The dashed curve was calculated from equation (8) using values of constants $C''$ and $E_T$ calculated from points on solid curve.

sample unchanged. The index of refraction of the material consequently remains constant.

The index of refraction of the Norton sample is almost constant in the region of interest, having the value 3.38 at 0.45 and 0.65 $\mu$ and the value 3.42 at the midpoint of the region. The RDL sample yielded a value of 3.44 at 0.45 $\mu$, which decreased linearly to 3.27 at 0.65 $\mu$. Morita [10], working with boron thin films calculated an index of refraction of 1.9 at 0.45 $\mu$, which increased to 3.3 at 1.0 $\mu$. This dispersion is attributed by Morita to a lattice absorption having a threshold at 1.0 $\mu$ ($E_T$= 1.2 eV). Spitzer and Kaiser [2] find an absorption threshold in the vicinity of 1.0 $\mu$, but attribute it to optical transitions across the forbidden energy gap of the same type as those transitions found in the present work.

### Dielectric Constant

The data obtained on the dielectric constant are given in Fig. 2. It is evident that there is no significant difference in the magnitudes of the dielectric constants of the Norton and the RDL samples.

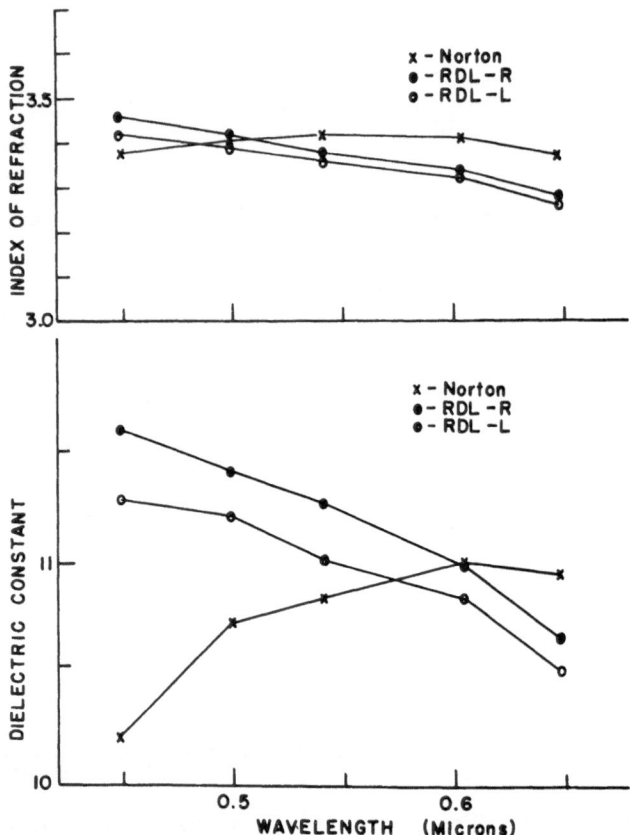

Fig. 2. Index of refraction versus wavelength, and dielectric constant versus wavelength.

The DC dielectric constant is known from capacity measurements [11] to be approximately 14. The value of 10.0 ± 0.5 found by Spitzer and Kaiser [2] for λ > 21 μ for a sample of polycrystalline boron of the beta-rhombohedral modification indicates, therefore, the presence in this sample of an absorption band at some longer wavelength. On the short wavelength side of the 21-μ absorption band, Spitzer obtains the value of 8.4 ± 0.4.

It can be seen from Fig. 3 that the extinction coefficient of the RDL sample is approximately one-half that of the Norton sample at any given wavelength. It follows that the effect of *k* in reducing the dielectric constant is more pronounced in the latter sample.

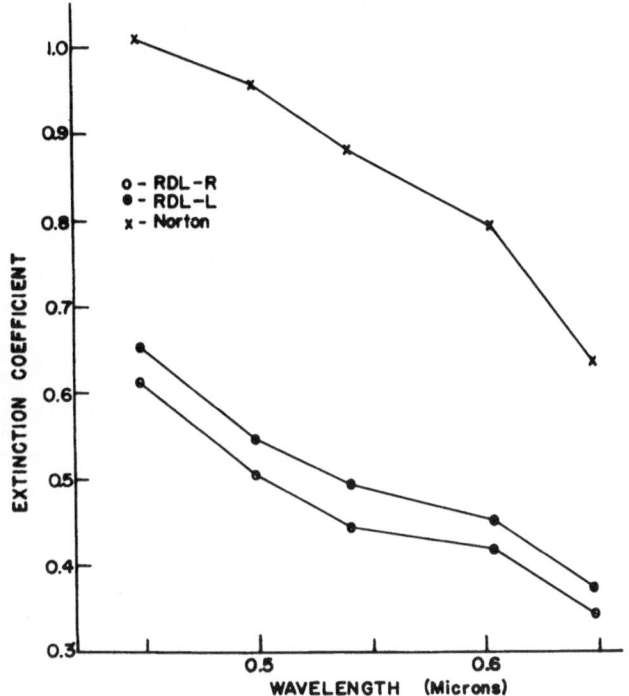

Fig. 3. Extinction coefficient versus wavelength. The three curves of Figs. 2 and 3 are related by $\epsilon = n^2 - k^2$.

## ACKNOWLEDGMENT

The authors are indebted to Drs. William A. Barker and Sook Lee for many informative discussions, and to Gerhart K. Gaulé of USAEL for the boron sample.

## REFERENCES

1. Morita, N., J. Sci. Res. Inst. 48: 8 (1954).
2. Spitzer, W., and W. Kaiser, Phys. Rev. Letters 1: 231 (1960).
3. Rizzo, H., Boron—Synthesis, Structure, and Properties, Plenum Press (New York), 1959, p. 180.
4. Ditchburn, R., J. Opt. Soc. Am. 45: 743 (1955).
5. Partington, J., An Advanced Treatise on Physical Chemistry, Longman's Green and Co. (New York), 1953.
6. Brungs, R. A., and V. P. Jacobsmeyer, J. Phys. Chem. Solids 25: 701 (1964). (See also this volume, p. 119.)
7. Bardeen, J., F. J. Blatt, and L. H. Hall, Proceedings of the Atlantic City Photoconductivity Conference, John Wiley and Sons (New York), 1955, p. 146.
8. Fan, H. Y., M. L. Shepherd, and W. Spitzer, Proceedings of the Atlantic City Photoconductivity Conference, John Wiley and Sons (New York), 1955, p. 184.
9. Brungs, R. A., and V. P. Jacobsmeyer, op. cit., pp. 702, 703.
10. Morita, N., op. cit., p. 11.
11. Lagrenaudie, J., J. Phys. Radium 14:14 (1953).

# Semiconductor Properties of Boron

## W. Neft* and K. Seiler [†]

*Institut für Theoretische und Angewandte Physik*
*Technischen Hochschule Stuttgart*
*Stuttgart, Germany*

Investigations of vacuum-sintered and of zone-refined boron are discussed. The electrical conductivity $\sigma$ at 300°K varied from $4.5 \cdot 10^{-7}$ to $1.2 \cdot 10^{-6}$ $(\Omega\text{-cm})^{-1}$. In the temperature range from -70 to 700°C, $\sigma$ can be described by the sum of two exponential functions

$$\sigma = \sigma_1 \exp(-\Delta E_g/2kT) + \sigma_2 \exp(-\Delta E_s/kT)$$

where $\Delta E_g = 1.39 \pm 0.06$ eV and $\Delta E_s = 0.418 \pm 0.025$ eV. Hall-effect studies yield an upper limit of $10^{-1}$ cm²/V-sec for the carrier mobility. The high-frequency conductivity up to 200 Mcps has been measured between 20 and 100°C. The real part of the complex conductance is always proportional to $\omega_s$, where $0.54 < s < 0.86$. The conductivity at low temperatures is explained by assuming hopping processes of the carriers. The rise and the decay of photoconductivity follow the same time function, which contains two characteristic time constants, $\tau_1 \approx 3$ sec and $\tau_2 \approx 40$ sec. Between 15 and 55°C, $\tau_1$ and $\tau_2$ are independent of temperature and of illumination intensity. Within the investigated temperature and intensity ranges, no saturation of the traps could be observed. The mobility of the carriers reponsible for photocurrent depends very weakly on temperature.

## SAMPLE CHARACTERISTICS

For our investigations we used polycrystalline boron sintered at about 1600°C in a vacuum of $< 10^{-4}$ torr or floating zone-refined polycrystalline boron. The sintered samples contained a series of contaminations, but the total impurity content was below 0.5%. Monocrystalline areas with clearly visible rhombohedral planes were cut out for measurement. We concluded from Debye–Scherrer diagrams that the material was of the beta-rhombohedral form, with serious distortions for some samples. In the following, polycrystalline samples are marked B, and samples with a few monocrystalline areas are marked BK.

The impurity content for the zone-refined material of Eagle-Picher (BZ) amounted to some ppm. By spectral analysis, we found traces of silicon, magnesium, iron, barium, and calcium. For high-

*Present address: Ohm-Polytechnikum, Kesslerstrasse 40, Nürnberg, Germany.
†Present address: W. C. Heraeus, Heraeusstrasse, Hanau, Germany.

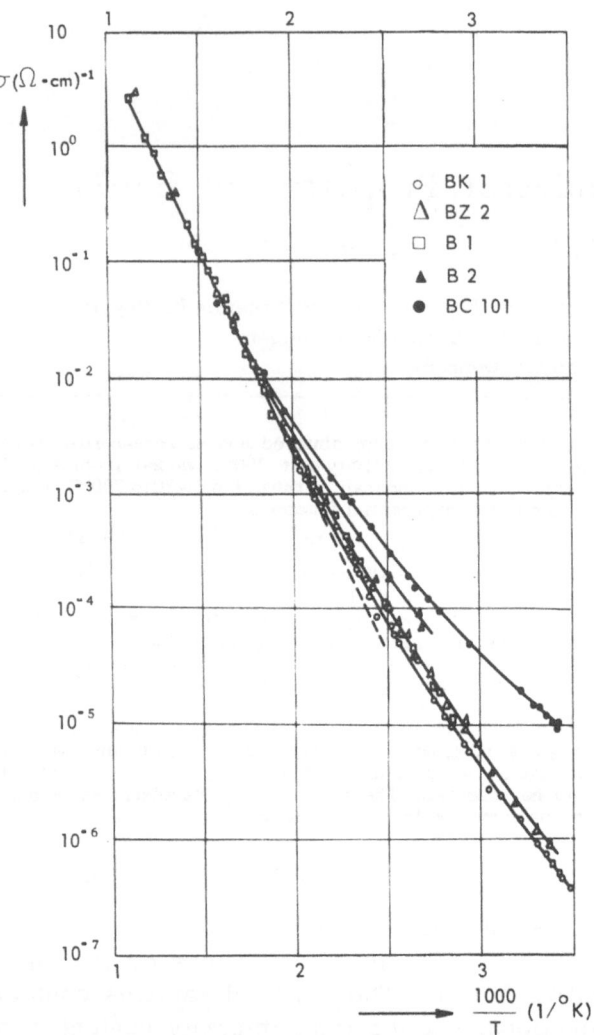

Fig. 1. Intrinsic conductivity of boron.

frequency conductivity measurements, we used additionally zone-refined boron from Wacker–Chemie (BW) containing 2–3 ppm Si, about 1 ppm Cu, 1 ppm Mg, and 0.5 ppm Ca. Both materials had the beta-rhombohedral structure.

## CONDUCTIVITY AND HALL EFFECT

The electrical conductivity above room temperature, up to 730°C, is plotted in Fig. 1. Above about 200°C, the conductivity values of samples with high resistance form a common intrinsic curve. The formula

$$\sigma = \sigma_1 \exp(- \Delta E_g / 2kT)$$

which is correct for a bimolecular process, allows computation of the activation energy $\Delta E_g$ ($\Delta E_g = 1.39 \pm 0.06$ eV), where $\Delta E_g$ corresponds to the band gap at $0°K$, $\sigma_1 \approx 2.7 \cdot 10^4$ $(\Omega\text{-cm})^{-1}$, $k$ is Boltzmann's constant, and $T$ is the absolute temperature. $\sigma_1$ is practically independent of temperature. Figure 2 shows, for comparison, the

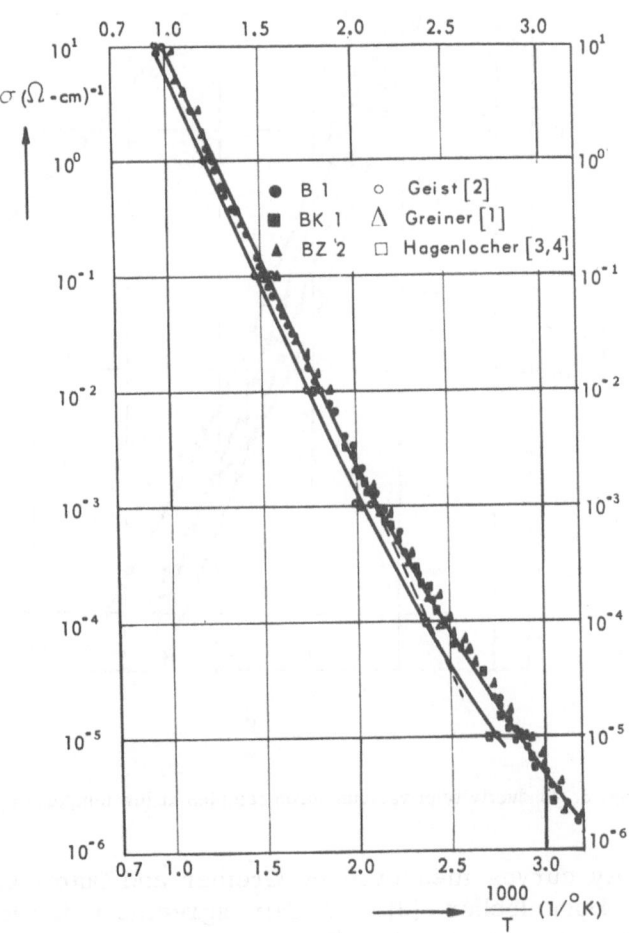

Fig. 2. Intrinsic range of boron (comparison with the results of other authors).

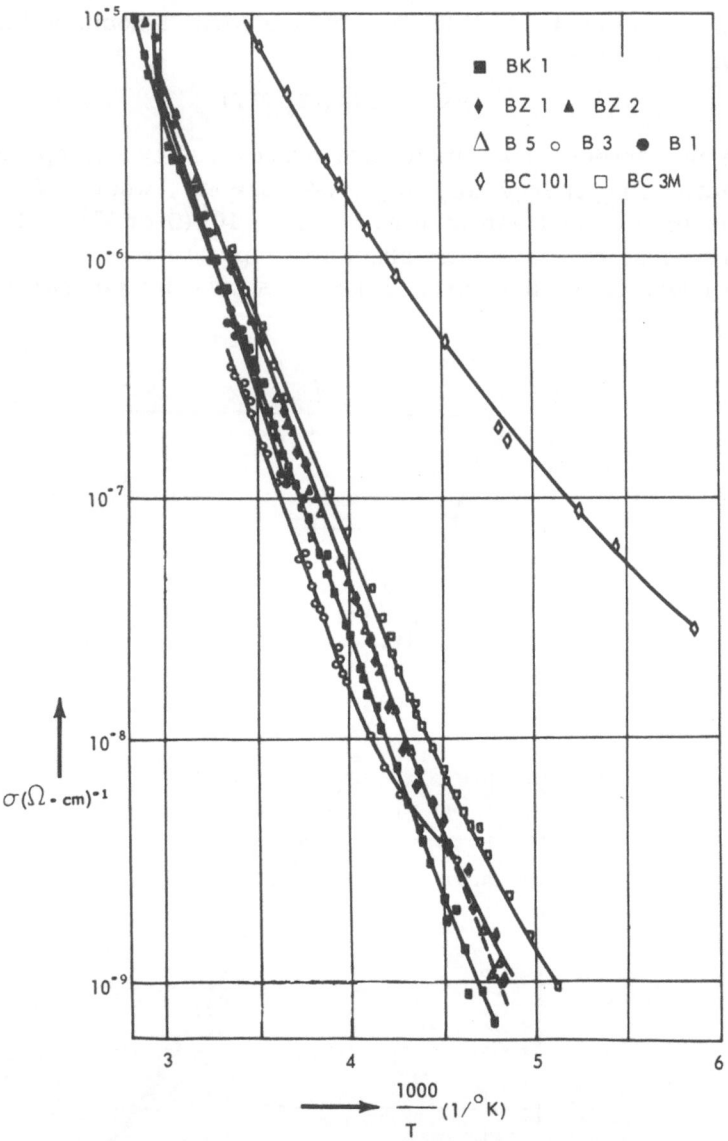

Fig. 3. Conductivity of various boron samples at low temperatures.

conductivity curves measured by Greiner and Gutowski [1], Geist [2], and Hagenlocher [3]. A fair agreement in the activation energy can be observed. Deviations in the absolute value of conductivity amount to a maximum of 20%. It must be borne in mind, however, that some of the points originate from printed figures lacking fine detail.

Below 200°C, impurities in the samples cause the curves to deviate considerably from the straight line of the intrinsic range (Fig. 3; note the enlarged scale). The absolute values of conductivity in this temperature range, however, are no measure for the purity of the boron samples. As confirmed by spectral and optical examination, the zone-refined boron samples surely had the lowest impurity content. In spite of this, the conductivity curves of the sintered sample B 5 and the zone-refined samples BZ 1 and 2, for instance, are almost identical. The samples BCM 3 and BC 101 contained carbon, which diffused into the material during preparation. The carbon content of sample BC 101 was much higher than that of sample BCM 3. Figure 4 shows the conductivity curves of other authors for comparison.

The conductivity of the samples BZ 1, BK 1, B 3, and B 5 at low temperatures may be described by the function

$$\sigma = \sigma_2 \exp\left(- \Delta E_s / kT\right)$$

The activation energy $\Delta E_s$ is determined from the slope of the curves; $\sigma_2$ is almost independent of temperature. Table I lists the results. It is striking that the values for $\Delta E_s$ differ so little. Moreover, the curves measured by other authors also exhibit an activation energy around 0.4 eV in a large temperature range. We conclude an activation energy of $\Delta E_s \approx 0.4$ eV to be typical for the material investigated, apart from the samples doped with carbon.

To obtain predictions about carrier concentrations and mobilities, we tried to measure the Hall effect. These experiments failed. The sensitivity of the apparatus was limited by relatively high noise voltages appearing at the probes. Thermal noise was far below the measured noise level. With the assumption that a potential Hall voltage is equal to the measured noise voltage, it is possible to calculate an upper limit for the mobility. Since the noise level of the sintered samples (B and BK) was about $1\frac{1}{2}$ orders of magnitude higher than the level of the BZ samples, the estimate has been made for a zone-refined sample. Assuming only one kind of carrier responsible for conduction, we get the result $\mu \leq 1.3 \cdot 10^{-1}$ cm$^2$/V-sec. This estimate is valid between approximately - 70 and + 50°C. Above 50°C, the apparatus became appreciably more insensitive.

Because of the obviously very low mobility, it seems doubtful

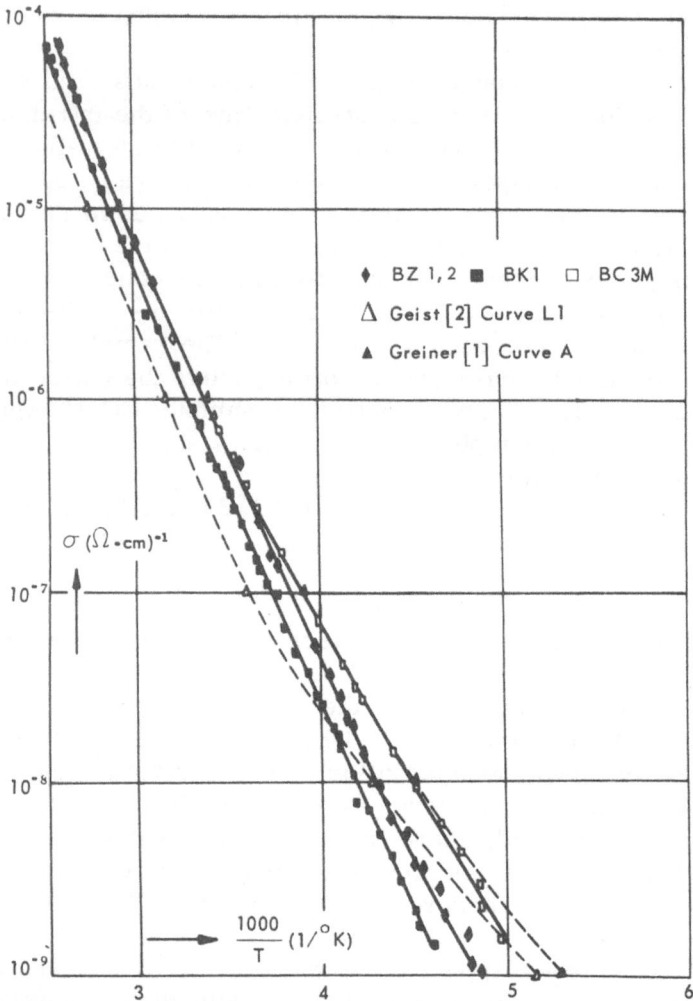

Fig. 4. Conductivity at low temperatures (comparison with the results of other authors).

that the transport of carriers at low temperatures occurs in a conduction band; the effective mass of the carriers would have to be unusually high ($m/m_{eff} > 10^3$). The carriers would move in a very narrow band. The approximation by which the band model has been derived thus does not seem to be correct for our problem. The results of our high-frequency measurements support these considerations.

## TABLE I

| Sample | Temperature range (°C) | $\Delta E_s$(eV) | $\sigma_2\,(\Omega\text{-cm})^{-1}$ |
|--------|-----------------------|------------------|-------------------------------------|
| BZ 1   | + 55 to - 58          | 0.409            | 8.39                                |
| BK 1   | + 72 to - 65          | 0.425            | 9.97                                |
| B 3    | + 25 to - 26          | 0.443            | 13.85                               |
| B 5    | + 21 to - 71          | 0.395            | 3.45                                |

## HIGH-FREQUENCY CONDUCTIVITY

The complex resistance $R_p$ of several boron samples (the vacuum-sintered samples B 10 and B 11 and the zone-refined samples BZ 10 and BW) has been measured from room temperature to 100°C at frequencies up to 200 Mcps. The sample represents a parallel circuit consisting of a resistance $R_p$ and a capacity $C_p$. These are measured separately by a bridge arrangement. We used a Boonton bridge, the zero indicator of which was coupled to the sample by a differential transformer when the frequency was below 200 kcps. Impedances formed by parallel connections of known resistances and capacities were measured in order to check the system.

The samples were disk-shaped and were 0.4–0.8 mm thick with cross sections between 1 and 100 mm$^2$. Layers of silver were fused onto the flat surfaces.

Figure 5 shows the resistance $R_p$ and the capacity $C_p$ versus frequency at 22°C. Above about 1 Mcps, $R_p$ decreases with increasing frequency. The absolute values of $C_p$ vary because of the varying geometry of the samples. In contrast to $R_p$, the capacity $C_p$ depends only slightly on frequency (note the linear scale), yet a small decrease at high frequencies can be stated. The rise of $C_p$ of the samples BW and BZ 10 at 50 Mcps is caused by stray capacities. The absolute value of $C_p$ agrees well with the capacity calculated from geometry and the static dielectric constant.

After a paper of Pollak [5], two fundamental processes must be discussed to find an explanation for the curves for hopping conductivity or a frequency dependence caused by inhomogeneities in a band-conducting material. The frequency dependence of conductivity due to hopping processes has been studied in detail by Pollak and Geballe [6] on compensated silicon between 1 and 20°K.

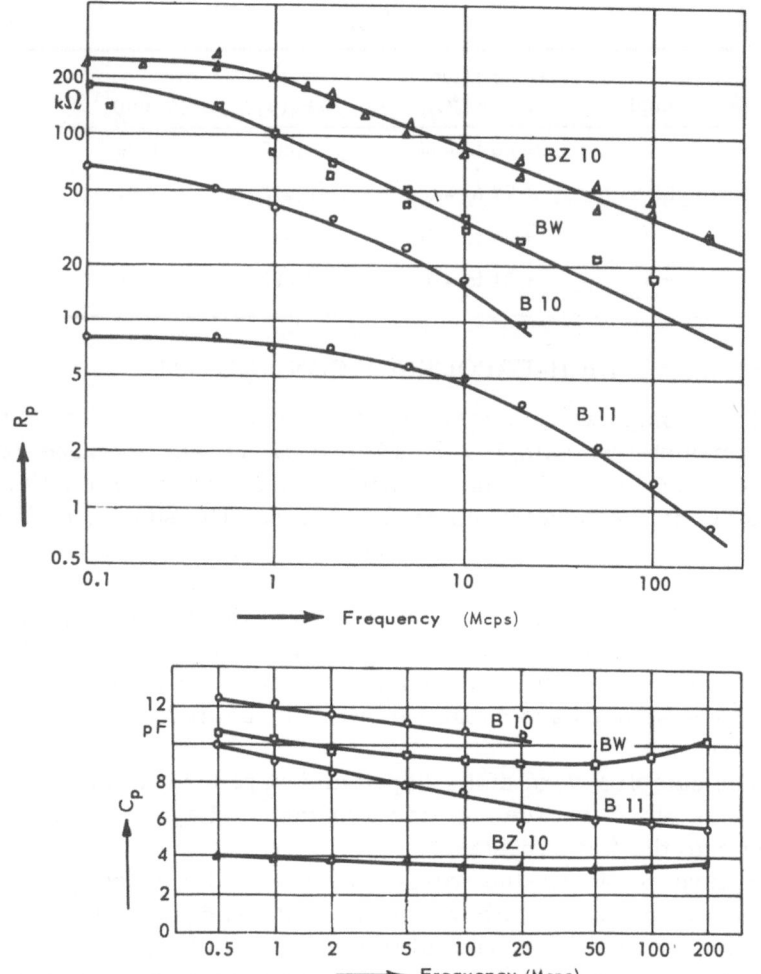

Fig. 5. Frequency dependence of resistance $R_p$ and capacity $C_p$, measured separately by a bridge arrangement for different boron samples. Temperature is 22°C.

Assuming a random distribution of impurities, Pollak and Geballe found for the real part of conductance the relation

$$G_\sim = G_p - G_{po} \quad \propto \quad \omega^s$$

where $G_\sim$ is the AC part of the conductivity, $G_p = 1/R_p$ is the real part of the measured conductivity, and $G_{po}$ is the DC conductivity. In the frequency range $10^5 - 10^7$ cps, $s$ is close to 0.8 for silicon. Figure 6 shows $G_\sim$ versus frequency for the boron samples, where $0.54 < s < 0.86$.   For conductance along boundaries, we would expect $G_\sim \propto \omega^2$.     Moreover, the DC conductivity of the boron samples is independent of the applied electric field up to 400 V/cm between room temperature and -70°C.   Therefore, conductance along boundaries surely can be excluded.

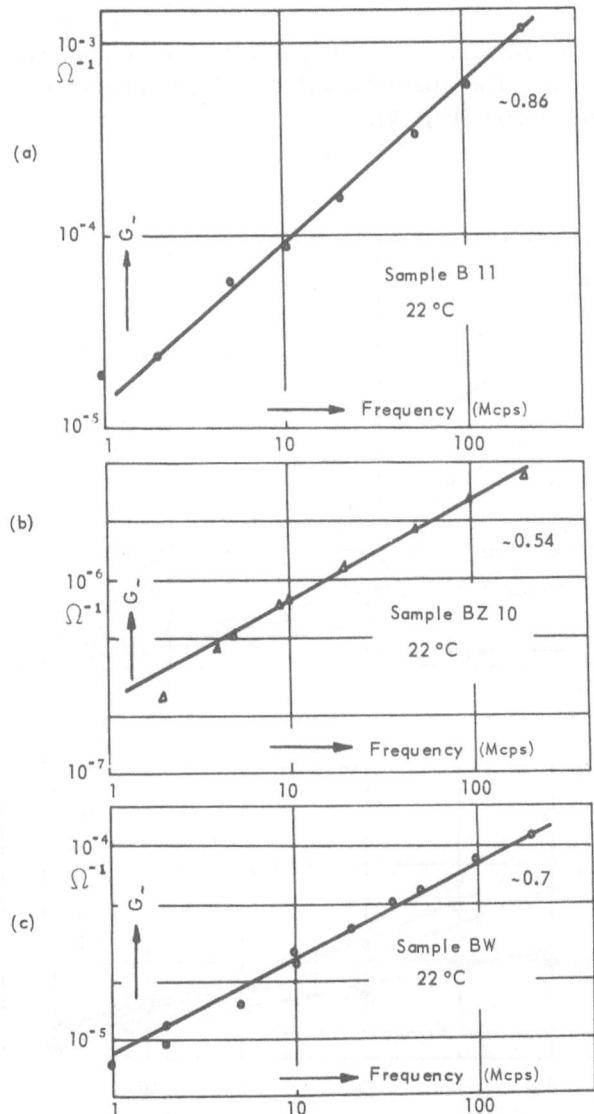

Fig. 6. AC part $G_\sim$ of conductance versus frequency (evaluated from the equation $G_\sim = G_p - G_{po}$, where $G_p = 1/R_p$ is the real part of the measured impedance and $G_{po}$ is the DC conductivity).

After the theory of Pollak, the imaginary part of the conductivity also is proportional to $\omega^s$. The measured capacity $C_p$ contains large, frequency-independent capacities due to the dielectric constant of boron and to the sample holder, which must be subtracted from $C_p$ to obtain the capacity due to the carriers alone. Our experimental results, however, are not sufficiently accurate to determine the latter. The dependence of $C_p$ on frequency and temperature, and the absolute value of $C_p$, correspond to the measurements on silicon (Fig. 7).

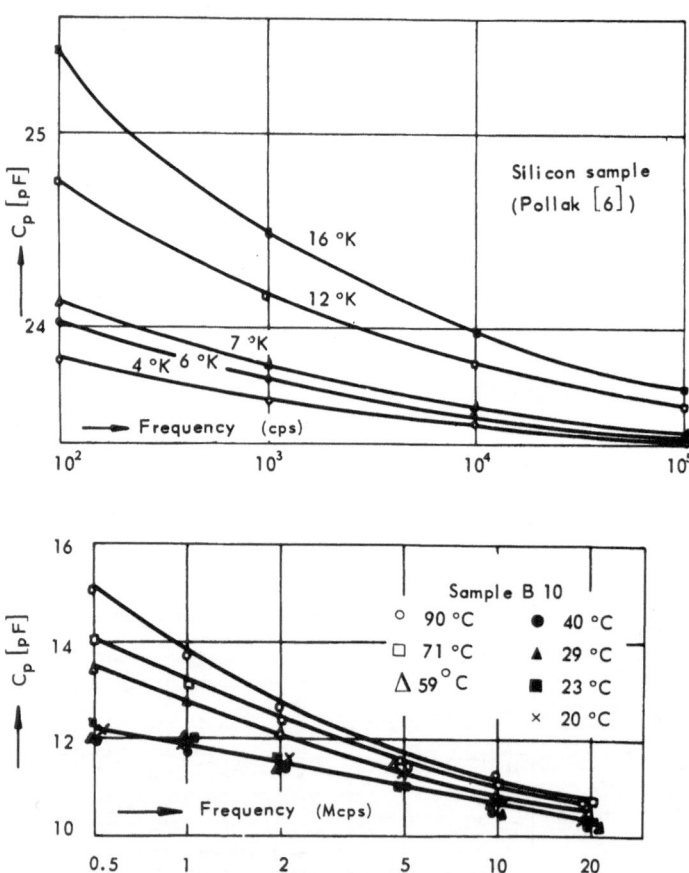

Fig. 7. Capacity $C_p$ for silicon (after Pollak and Geballe [6]) and boron (sample B 10) versus frequency and temperature.

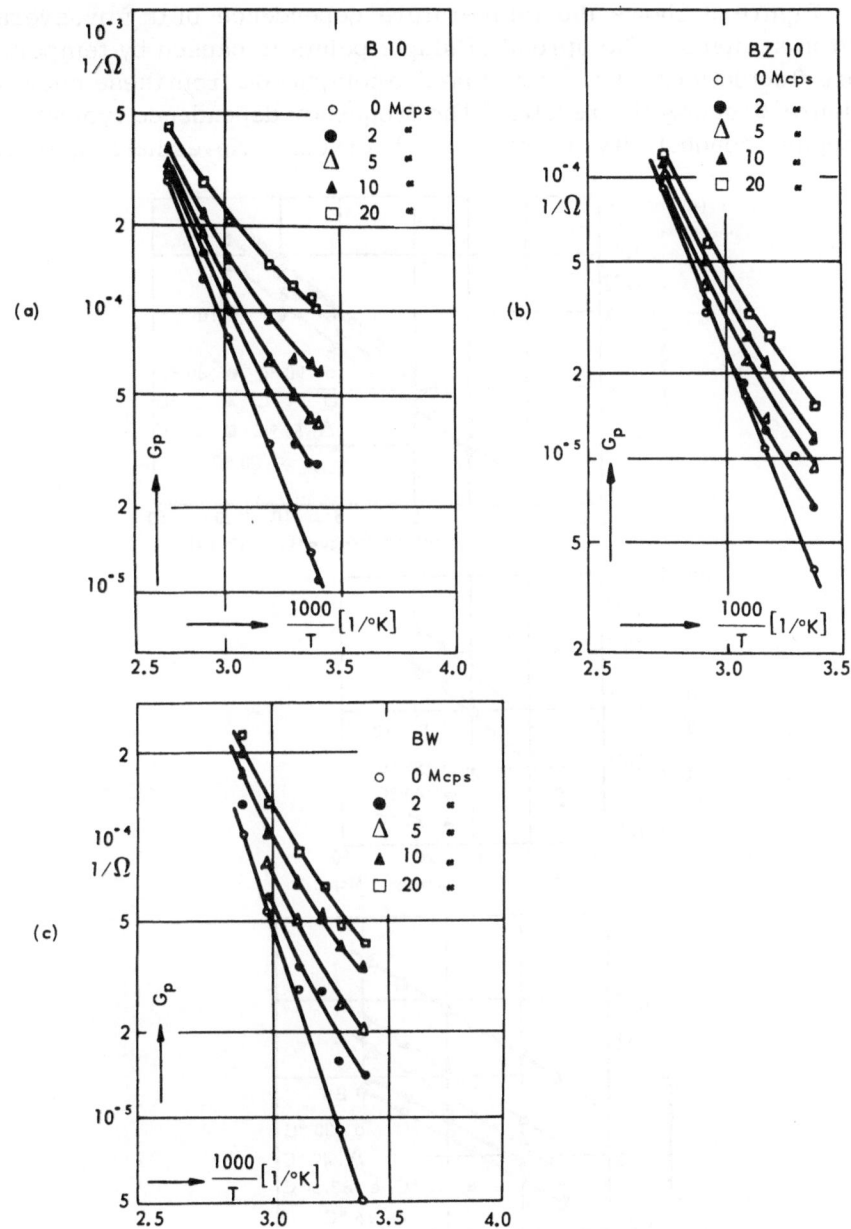

Fig. 8. Temperature dependence of the real part of the impedance $G_p$ for samples: (a) B10, (b) BZ 10, and (c) BW.

Figure 8 shows the temperature dependence of $G_p$ for several boron samples. The spread of single points is caused by temperature fluctuations. It was attempted to compute $G_\sim$ from these curves. Figure 9 shows the results. The frequency dependence typical for hopping conductivity is clearly observed. Nevertheless, these

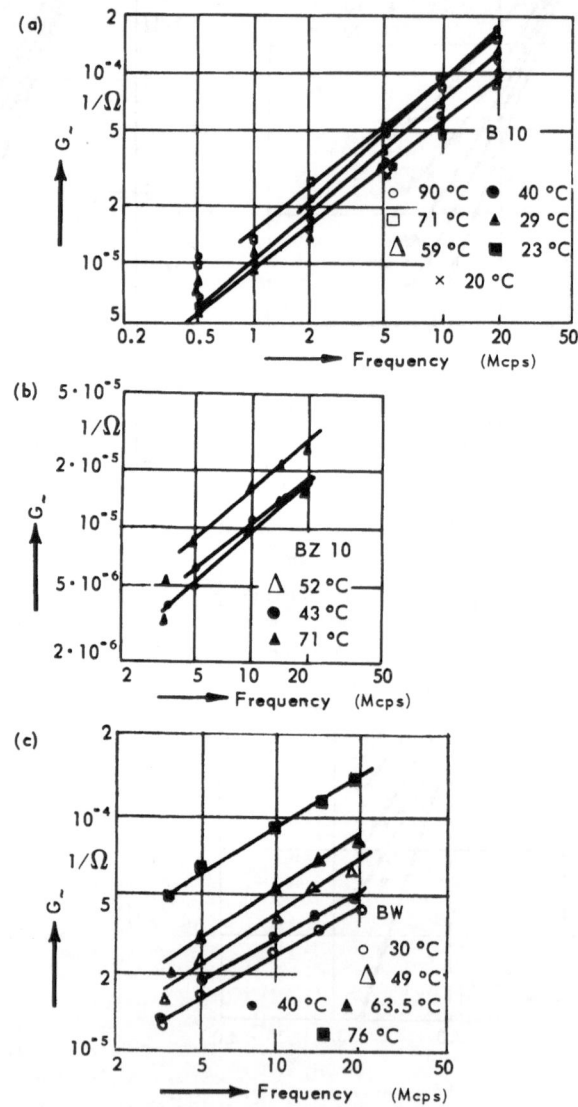

Fig. 9. AC part $G_\sim$ of conductance evaluated from $G_\sim = G_p - G_{po}$ (see Fig. 6) versus frequency and temperature for samples: (a) B 10, (b) BZ 10, and (c) BW.

measurements should be regarded as preliminary. Because of these experimental results, conductivity and Hall-effect conclusions, and the qualitative agreement with the model of Pollak, we believe that the frequency dependence of the conductivity can be explained by a hopping mechanism.

The typical properties of a DC conductivity due to hopping have been discussed by Mott [7], Conwell [8], and Heikes and Johnston [9]. The conductivity of weakly doped, compensated germanium and silicon at low temperatures, as well as the behavior of some oxide semiconductors, have been explained, for instance. If the hopping mechanism is regarded as a diffusion process, the conductivity is given by [8, 10]

$$\sigma = \frac{e^2}{kT} \, a^2 \, \nu_0 \, N_{eff} \, e^{-\,W/kT}$$

where $a$ is the hopping length, $\nu_0$ is the hopping frequency, $N_{eff}$ is the effective concentration of sites free for hopping carriers, and $W$ is the activation energy.

If the picture of hopping conductivity is correct for boron, the temperature dependence of the conductivity at low temperatures is not caused by the change of the number of carriers frozen in impurities, but by the exponential temperature dependence of the hopping frequency and thus the mobility.

The hopping frequencies are obtained from high-frequency measurements. We may consider the frequency at which $G_p$ is reduced to 70% of its DC value as representative for the hopping frequency $\nu_0$. The plot of this frequency versus the inverse temperature gives the exponential behavior which also describes the DC conductivity (Fig. 10, sample BZ 10). The hopping distance was $< 4700$ Å, and the effective concentration $N_{eff}$ was $> 10^{13}$ cm$^{-3}$ for sample BZ 10 [14]. Other samples are similar.

The experimentally found activation energy for this temperature range, $\Delta E_s = W = 0.409$ eV, gives the binding energy linking a hopping carrier to a hopping center. A comparison of these results with theory is not possible, since a quantum mechanical treatment has been given to the case of hydrogen-like impurities only, whereas boron undoubtedly has deep-lying impurity levels. The unusual magnetoresistance of boron probably supports the hopping model for the conductivity. We found the anomalous magnetoresistance in sintered material only. Gaulé and co-workers [4] and Dore [11] described similar effects for zone-refined boron.

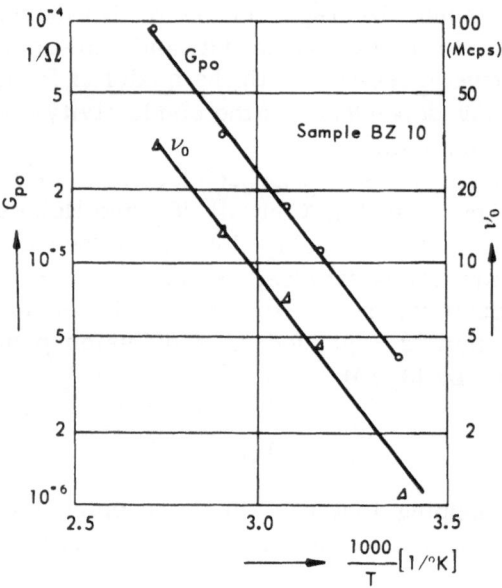

Fig. 10. Hopping frequency $\nu_0$ (defined as the frequency at which $G_p$ amounts to 70% of its DC value) and DC conductivity $G_{po}$ versus temperature.

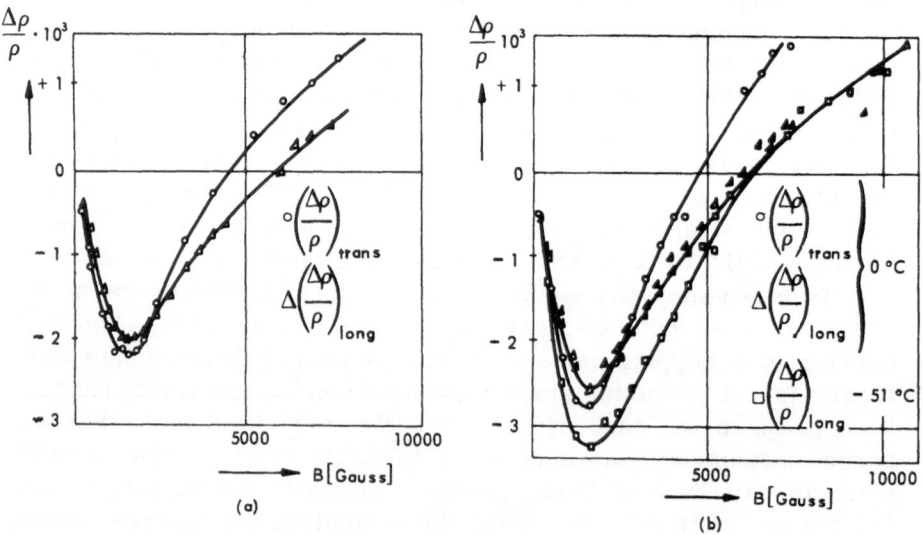

Fig. 11. Transversal and longitudinal magnetoresistance of sample BK 1 at: (a) 22°C, and (b) 0 and − 51°C.

In a magnetic field, the conductivity of sintered boron samples is raised, i.e., the magnetoresistance $\Delta\rho/\rho$ is negative (Fig. 11). The maximum value of $\Delta\rho/\rho \approx 10^{-3}$ depends weakly on temperature between - 60 and 20°C. It is reached at a magnetic field of about 3000 G, is independent of the direction of magnetic field relative to the direction of electric current, and is not influenced by various surface treatments. Boundaries can be excluded with high probability as a reason. $\Delta\rho/\rho < 10^{-5}$ was obtained for zone-refined samples. Recently, Pollak and Watt [12] found a negative magnetoresistance on compensated, boron-doped silicon at 4.2°K in the hopping range. However, $\Delta\rho/\rho$ reaches a saturation value at 15000 G. We are unable to give an explanation for the results.

## PHOTOCONDUCTIVITY

The zone-refined samples showed considerably increased conductivity during illumination with white light. After the light is turned off, the photocurrent decays very slowly, obviously because of trapping centers. (Compare with [4], [11], and [13].) We tried to find the effective trapping mechanism by suitable experiments. The resulting trapping model must be consistent with the results discussed in the two previous sections.

Figure 12 shows a schematic diagram of the measuring circuit. The sample was illuminated by deeply penetrating, filtered light. We used a 0.3-mm thick polished boron plate as a filter, which was prepared from the same material as the measured sample. Basically, the light produces small deviations from thermal equilibrium. The vibrating condenser electrometer senses these as small changes of the voltage $\Delta u$ across R. The voltage drop across R due to the dark current was compensated by an opposing voltage. A pen recorder plotted $\Delta u$ or $\Delta i$ as a function of time.

A typical curve for the rise and decay of photocurrent is shown in Fig. 13. The sample is illuminated from B to C. During the remaining time, the sample is in the dark. The line ABDE represents dark current.

A comparison of the rise curves from white light with those from filtered light gives the first insight into the efficient trapping mechanism. White light (Fig. 14a) evidently creates a high excess concentration of electrons and holes in the conduction and valence bands, respectively. The added charge carriers reach the equilibrium concentration due to incident light intensity in a time that

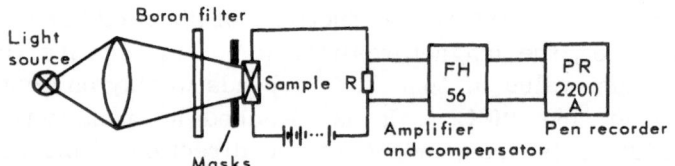

Fig. 12. Schematic diagram of the circuit used to measure photoconductivity.

is small compared to the resolution time of the measuring circuit (approx. 1 sec). Hence, the photocurrent seems to increase immediately to the value at B. The fast rise is followed by a slow increase from B to C, indicating the capture of minority carriers in the traps. Due to the requirement of space charge neutrality, additional majority carriers compensate exactly the charge of the minority carriers in the traps, and, therefore, produce the elevated conductivity. The rise curve BC can be normalized (by multiplication with a constant) to the rise curve of Fig. 14b (measured with filtered light). Therefore, the rise function is the same for white and filtered light. With filtered light, the abrupt rise at $t = 0$ is not observed, since the light intensity is very small. Consequently, the equilibrium concentration due to AB is small and often negligible.

Figure 15 shows the photocurrent versus time for various illumination periods (1–300 sec). The photocurrent does not reach an equilibrium value, even for a long duration of the illumination. Elevated temperature can be excluded as a reason for this, as shown in curve c, which was taken with a light intensity reduced by a

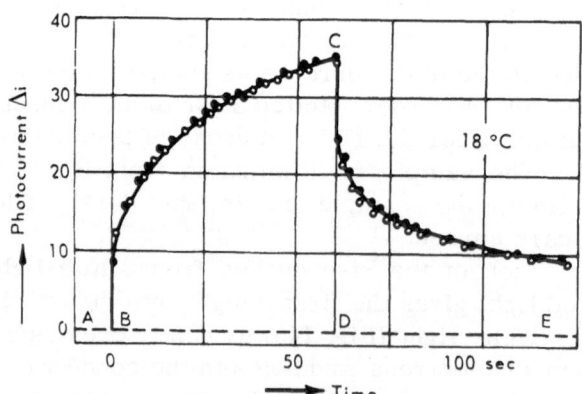

Fig. 13. Typical curve for photocurrent versus time (illumination by filtered light). The sample is illuminated between B and C. The line ABDE represents the dark current.

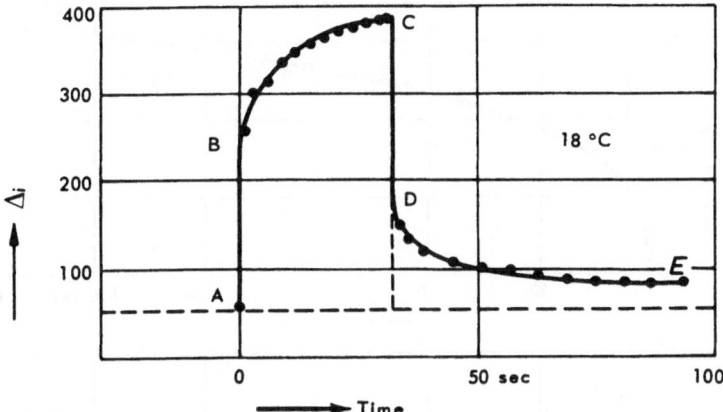

Fig. 14a. Illumination by white light. The unit for photocurrent $\Delta i$ is arbitrary. The rise curve BC has been normalized by multiplication with a constant factor to the rise curve of Fig. 14b.

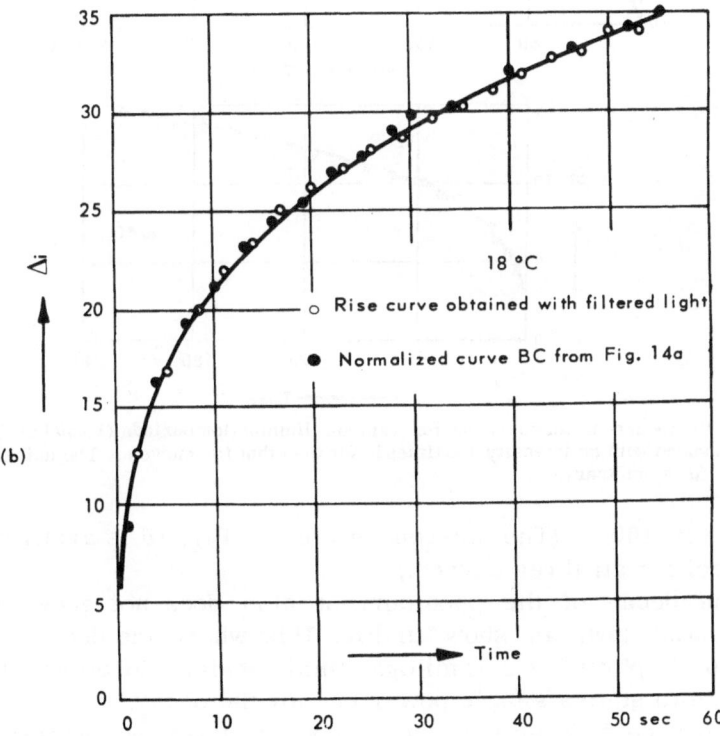

Fig. 14b. Illumination by filtered light. The unit for photocurrent $\Delta i$ is arbitrary.

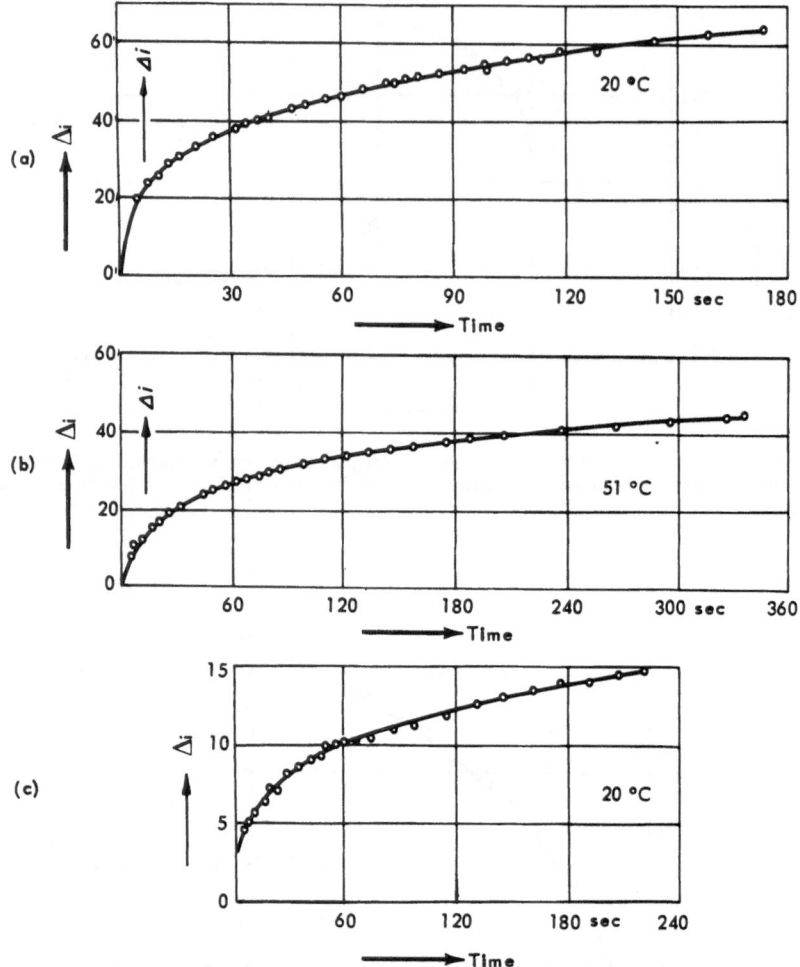

Fig. 15. Photocurrent versus time for various illumination periods (1—300 sec). Curve c is measured with an intensity 100 times lower than that for curve a. The unit for photocurrent $\Delta i$ is arbitrary.

factor of 100.   (The current scale in Fig. 15 is arbitrary, but identical for all three curves.)

The decay of the photocurrent also does not obey a simple exponential law, as show in Fig. 16b, where the decay curve of Fig. 16a is plotted in a semilogarithmic scale. Moreover, it is not possible to state a simple power law for the decay.

From the differentiated rise curve, however, we obtain a result suitable for interpretation.   Assuming the rise function obeys a simple exponential law with a single time constant $\tau$, we can write for the photocurrent

$$\Delta i = \Delta i_0 \, (1 - e^{-t/\tau})$$

and obtain

$$\frac{d\Delta i}{dt} = -\frac{\Delta i_0}{\tau} \, e^{-t/\tau}$$

If the rise curve is graphically differentiated, $\tau$ may be determined

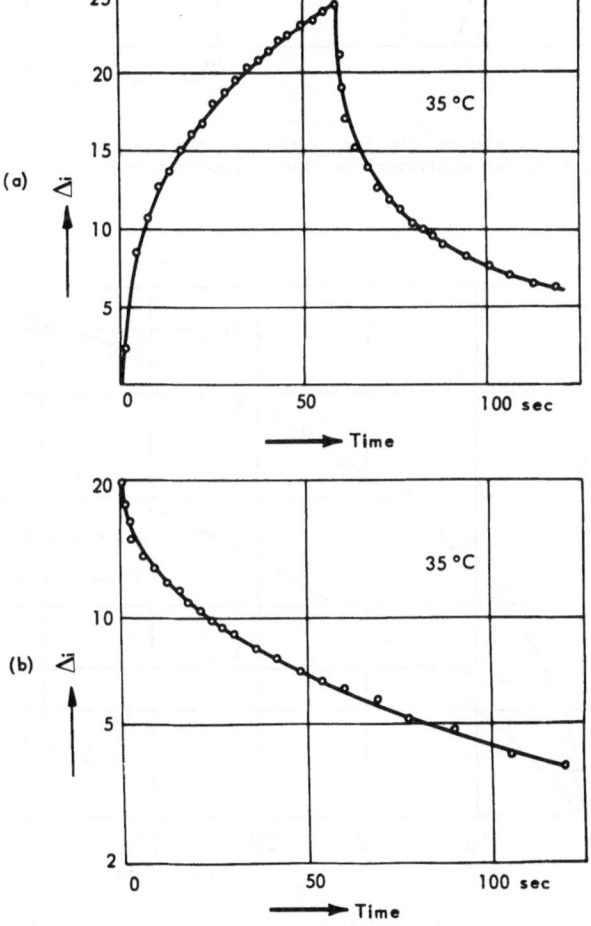

Fig. 16. Rise and decay curves. Note semilogarithmic scale for (b). The unit for photocurrent $\Delta i$ is arbitrary.

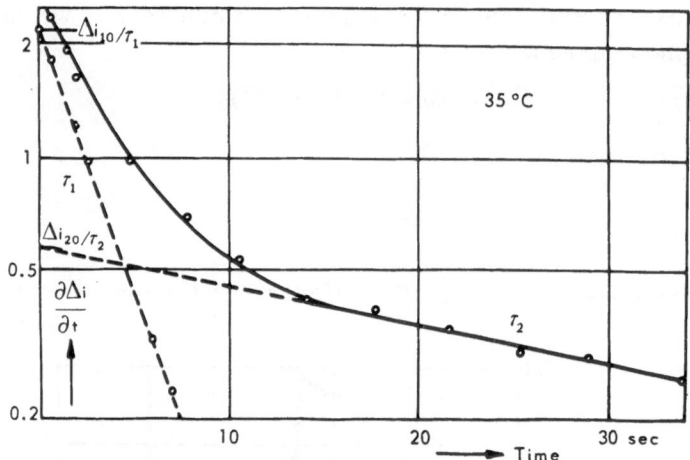

Fig. 17a. The rise curve of Fig. 16a is differentiated and plotted on semilogarithmic scale. In this way, two characteristic time constants $\tau_1 = 3.1$ sec and $\tau_2 = 42.2$ sec are determined.

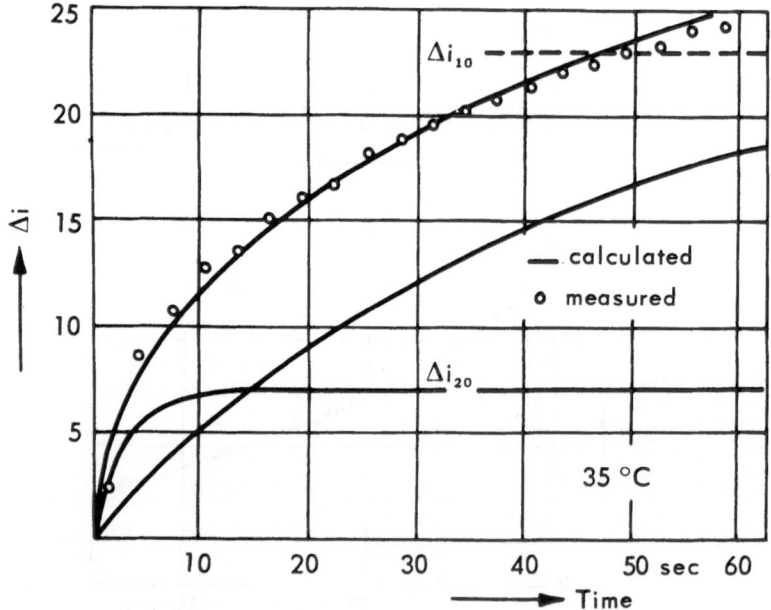

Fig. 17b. Comparison of measured and calculated rise curves.

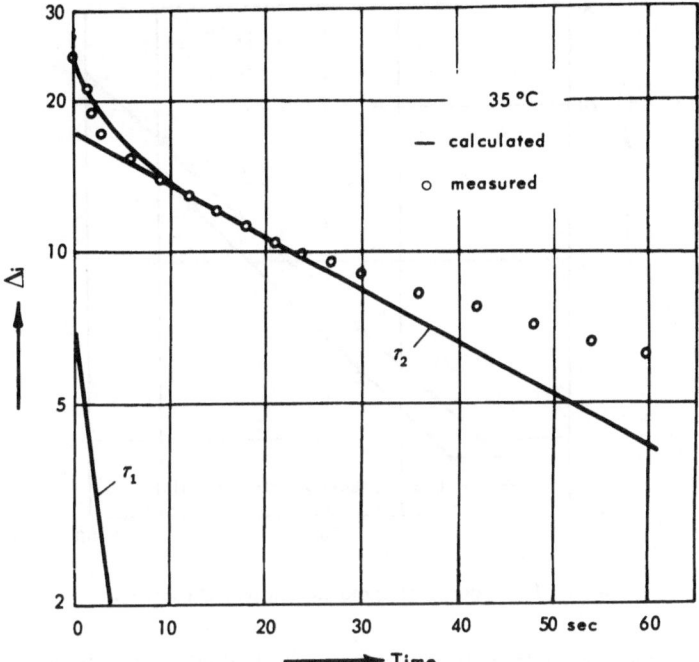

Fig. 18. Decay curve computed with the time constants $\tau_1$ and $\tau_2$ of the rise curve. (See Figs. 16 and 17a.)

from the slope of the new curve. Figure 17a gives the result for the curve of Fig. 16a. We find two characteristic time constants for the rise of photoconductivity, namely, $\tau_1 = 3.1$ sec and $\tau_2 = 42.2$ sec. Figure 17b compares the measured and the calculated curves. The numerical agreement is reasonably good.

If the traps are far from saturation and are not emptied by light, it is possible to calculate the decay curve with the experimental values of $\tau_1$, $\tau_2$, $\Delta i_{10}$, and $\Delta i_{20}$. It is shown in Fig. 18 that the initial decay of the photocurrent is adequately described by

$$\Delta i = \Delta i_{10} \, e^{-t/\tau_1} + \Delta i_{20} \, e^{-t/\tau_2}$$

The deviations beyond 30 sec may be caused, for instance, by traps with large time constants.

Figure 19 gives the photocurrent as a function of light intensity after 60 sec of illumination; $\Delta i$ is proportional to light intensity within the investigated intensity and temperature ranges. Since the rise and decay curves can be normalized to one single function by

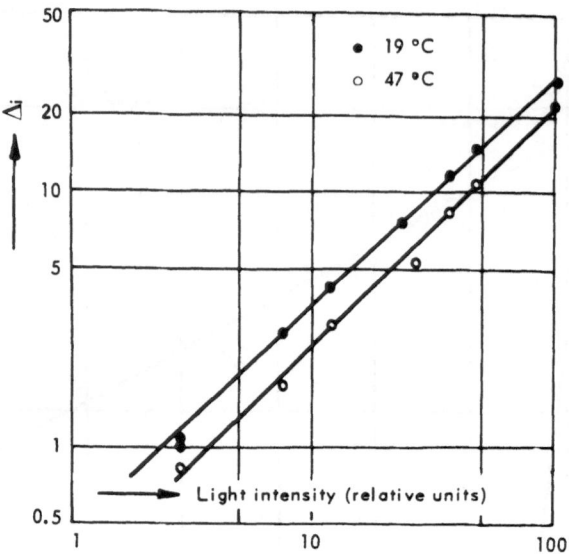

Fig. 19. Dependence of photocurrent on light intensity.

the corresponding filter factors, we conclude that the time constants are independent of light intensity. Hence, there is no multiple trapping process. The time constants are practically independent of temperature between 15 and 55°C (Fig. 20).

Using our experimental results, we are able to develop a trapping model, which fits the measured curves completely. Our traps act as recombination centers with relatively long trapping times. The

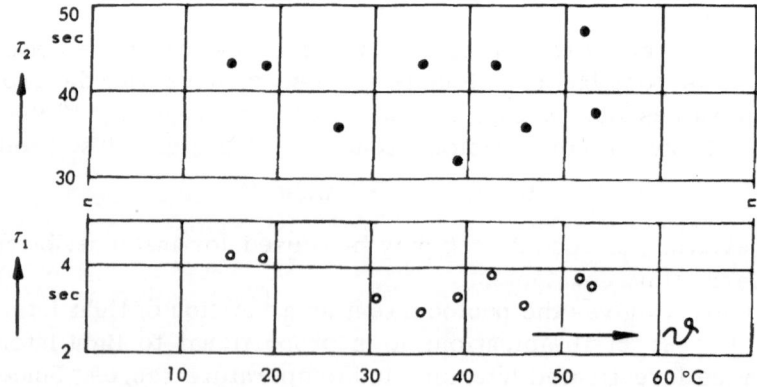

Fig. 20. Temperature dependence of the time constants $\tau_1$ and $\tau_2$.

recombination process is governed by the trapped carriers [14].

The temperature dependence of the photocurrent $\Delta i$ (after illumination for 60 sec) is given by Fig. 21a. $\Delta i$ can be split into the parts $\Delta i_t$ arising from trapping effects and $\Delta i_r$ caused by fast rise and decay processes (Fig. 21b). Since both quantities do not change very much within the measured temperature range, it follows from our model [14] that the mobility of the photocarriers is only weakly temperature-dependent.    This result is to be expected, if the transport of the carriers occurs in a band.    While our photo-conductivity results indicate band conductivity at high temperatures, this does not contradict our model of a hopping (dark-) conductivity at low temperatures (below 500°K).    As our Hall measurements indicate, the mobility must be very small in the high-temperature range also, but since a Hall measurement in the intrinsic temperature range detects only the difference between the two mobilities, mobilities at high temperatures could possibly somewhat exceed the limit given above in the second section.

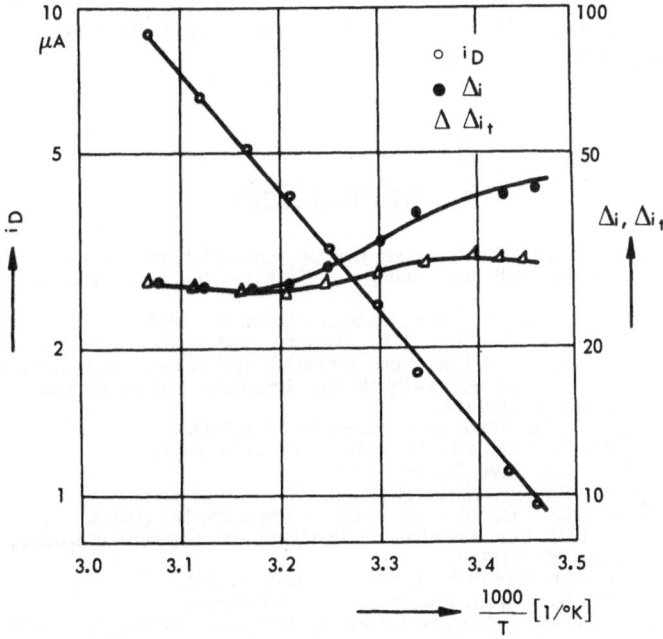

Fig. 21a.  Temperature dependence of the total photocurrent $\Delta i$, the partial photocurrent $\Delta i_t$ (due to trapping), and the dark current $i_D$. (See Fig. 21b.)

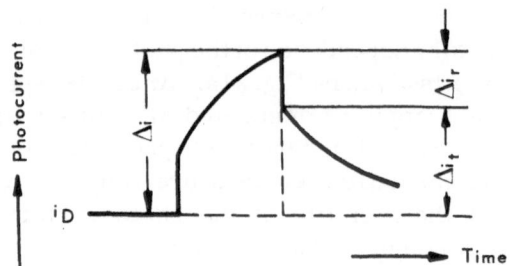

Fig. 21b. Separation of the photocurrent $\Delta i$ into parts related to fast recombination ($\Delta i_r$) and to trapping ($\Delta i_t$).

## ACKNOWLEDGMENT

We are very grateful to Prof. H. Haken for helpful discussions, especially on hopping conductivity. E. Grobe and F. Gschwend gave much valuable assistance in discussions and treatment of the data. We are also indebted to G. Schottky for critical remarks concerning the trapping model. We are indebted to Hermann C. Starck, Goslar and Wacker-Chemie, Munich, Germany, who made pure boron available to us. This work has been supported by the Deutsche Forschungsgemeinschaft.

## REFERENCES

1. Greiner, E.S., and J.A. Gutowski, J. Appl. Phys. 28:1364 (1957) and 30: 1842 (1959).
2. Geist, D., Vortrag auf der Tagung des Fachausschusses Halbleiter, Bad Pyrmont, 1963.
3. Hagenlocher, K.A., Dissertation, Stuttgart, Germany, 1958.
4. Gaulé, G.K., J.T. Breslin, J.R. Pastore, and R.A. Shuttleworth, "Optical and Electrical Properties of Boron and Potential Application," in J.A. Kohn, N.F. Nye, and G.K. Gaulé (eds.): Boron—Synthesis, Structure, and Properties, Plenum Press (New York), 1960, p. 159–174.
5. Pollak, M., Bull. Am. Phys. Soc., Series II: 6,5 (1961).
6. Pollak, M., and T.H. Geballe, Phys. Rev. 122: 1742 (1961).
7. Mott, W.F., Can. J. Phys. 34: 1356 (1956).
8. Conwell, E.M., Phys. Rev. 103:51 (1956).
9. Heikes, R.R., and W.D. Johnston, J. Chem. Phys. 26: 582 (1957).
10. Price, P.J., IBM Res. Develop. 2:123 (1958) and Appendix of Koenig, S.H., et al., J. Chem. Solids 2:268 (1957).
11. Dore, M., and C.H. Carmichael, Nature 191:486 (1961).
12. Pollak, M., and D.H. Watt, Phys Rev. 129: 1508 (1963).
13. Gaulé, G.K., and R.R. Patty, APS Meeting, Evanston, Illinois, June 1962.
14. Neft, W., Dissertation, Stuttgart, Germany, 1964.

## DISCUSSION

GAULÉ: You assume a hopping conductivity up to 500°K, with an activation energy of approximately 0.4 eV. Both figures seem to exceed those for other semiconductors.

NEFT: Below 500°K, we assume hopping conductivity for the dark current only. We believe that the photocurrent in this temperature range is a band current, possibly using the same band as the high-temperature intrinsic current. Neither we nor, to our knowledge, other workers have direct evidence for band conductivity in any case. The band model apparently loses its validity for mobilities of 1 $cm^2/V$-sec or less. This and our AC measurements make the conclusion of a hopping model in the specified ranges inevitable. Our main difficulty is in the relatively high activation energy. In view of analogous work on NiO, it is believed, however, that a theoretical justification for these high energies can be found in the future.

# Trap-Dominated Electrical and Optical Effects in Crystalline Boron

## G. K. Gaulé, J. T. Breslin, and R. R. Patty*

*Institute for Exploratory Research*
*U. S. Army Electronics Command*
*Fort Monmouth, New Jersey*

At temperatures below 200°K, high-purity crystalline beta-rhombohedral boron exhibits unusual persistent changes of conductivity and of optical absorption in the near infrared after illumination with visible light. Since boron is opaque for wavelengths shorter than 0.8 $\mu$ , the photoexcitation must occur in a thin surface layer of the crystal. Typical experiments are performed in three phases, namely, (1) photoexcitation, (2) observation of fast and slow decay processes, and (3) thermal stimulation of the excess conductivity and subsequent restoration to equilibrium by rapid heating. The results indicate that the excess optical absorption is caused by trapped electrons, while the excess conductivity is a consequence of trapped holes. In isothermal situations, the concentrations of both trapped carriers are proportional to one another, but, under rapid heating, the "evaporation" of the trapped holes precedes that of the trapped electrons. The shape of the thermal stimulation curve for the photocurrent in phase 3 of low-temperature experiments depends on the color of the visible light used for the excitation during phase 1. This is probably related to the increase of the absorption coefficient (which was remeasured) in the intrinsic (opaque) region from red to blue. The probable cause for the ineffectiveness of infrared quenching during any of the three phases of our experiments is discussed. Trapping occurs also at very low temperatures ($\approx 15$°K), but no persistent photoconductivity is observed unless the material is later brought to a higher temperature. Theoretical curves for several trap models were numerically computed on the basis of the differential equations for the essential transitions and compared with the experimental curves. This procedure appears promising. The virtual absence of rectification effects in boron and the insensitivity of boron against doping with most other elements are explained in terms of the same trap model.

## INTRODUCTION

There are many indications for a large number of imperfections on the atomic scale in single crystals of beta-rhombohedral boron. The most obvious clue for this comes from the chemical analysis; even material of the highest purity available today contains at least 10 ppm of carbon and at least 1 ppm (individually) of several other impurities. Because of its complex crystal structure, boron, when grown from the melt or otherwise formed at a high temperature,

*Present address: North Carolina State University, Raleigh, North Carolina.

will also have imperfections in the form of misplaced boron atoms, especially vacancies. The introduction of certain foreign atoms may cause the formation of additional boron vacancies, as discussed below. (For a comprehensive treatment of atomic imperfections, see Kröger [1].) The absence of rectification effects in boron–metal contacts seems to indicate a large number of recombination (and generation) centers which apparently prevent the formation of a depletion layer. The small carrier mobilities consistently observed in boron seem to indicate a high number of scattering centers, or, especially for low temperatures, a conductivity based on the hopping of electrons from center to center [2].

The small number of mobile carriers usually found in purified boron at low and intermediate temperatures can be reconciled with the large number of atomic imperfections only by assuming that most of the imperfections acting as acceptors (making the material $p$-type) are compensated by other imperfections acting as donors. Boron is remarkably insensitive to doping [3,4]. It appears that the introduction of certain donor elements, for example, carbon or silicon, is accompanied by the creation of so many additional acceptors as to keep the acceptor surplus nearly unaltered. One would expect that foreign atoms may not be simply substituted for boron atoms because of the small size and the peculiar binding character of the latter. Possibly some boron sites are made vacant upon the introduction of foreign atoms. The boron vacancies would probably act as threefold acceptors, thus compensating three simple donors. For example, three silicon atoms might replace four boron atoms, leaving the total number of valence electrons unchanged.

The most interesting, and challenging phenomena, however, arise from imperfections acting as traps for electrons or holes in nonequilibrium situations, for example, during or after illumination, or during rapid temperature changes. In this, boron has unique properties. It responds to illumination with visible (nonpenetrating) light not only with an increase in conductivity, but also an increase of the infrared absorption in the near infrared. At low temperatures, the two increases persist for a long while after the illumination has ended. The conductivity and the optical absorption both respond in a characteristic and mathematically predictable manner to rapid heating. The next section treats those equilibrium optical properties of boron needed for the discussion of the nonequilibrium experiments.

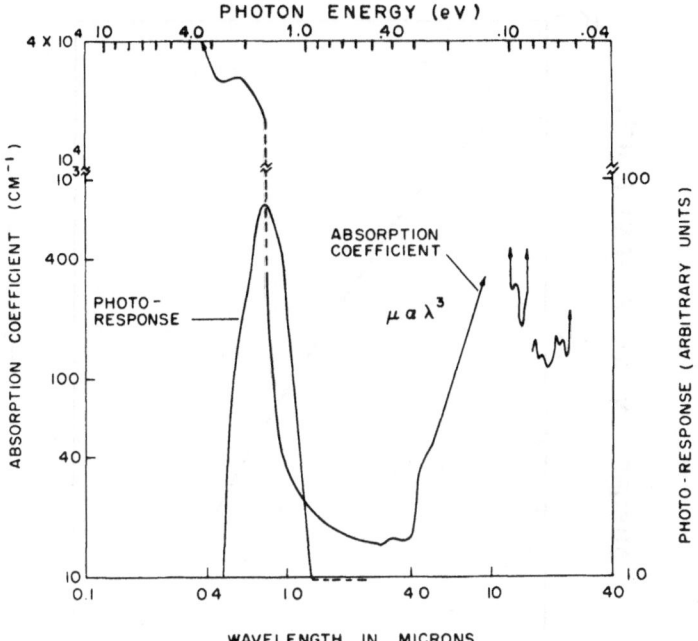

Fig. 1. Optical absorption (in double logarithmic scale) and photoresponse of conductivity of beta-rhombohedral boron crystal at room temperature. The steepest part of the absorption edge near ≈ 0.8 μ defines the optical band gap ( ≈ 1.55 eV). Light normal to (10Ī1) plane.

## THE EQUILIBRIUM OPTICAL ABSORPTION OF BETA-RHOMBOHEDRAL BORON

The following points were given special attention in our determination of the optical absorption of boron: (1) As shown below, the absorption coefficient is under certain conditions increased by illumination. A weak, and nearly monochromatic light beam was thus used whenever possible in our measurements. (2) The absorption depends on the orientation of monocrystalline regions with respect to the light beam [5,6]. In our experiments, the beam was usually in the [10Ī1] direction. (3) Several definitions for the optical band gap of a semiconductor are in use [9]. We take the band gap value from the steepest part of the absorption curve [3].

Figure 1 shows, in double logarithmic representation, the absorption coefficient of high-purity crystalline boron as function of wavelength λ and also of photon energy. The opaque region ( λ < 0.8 μ ) shows an increase of absorption towards the shorter wavelength. A peculiar step is also observed. A similar step, but

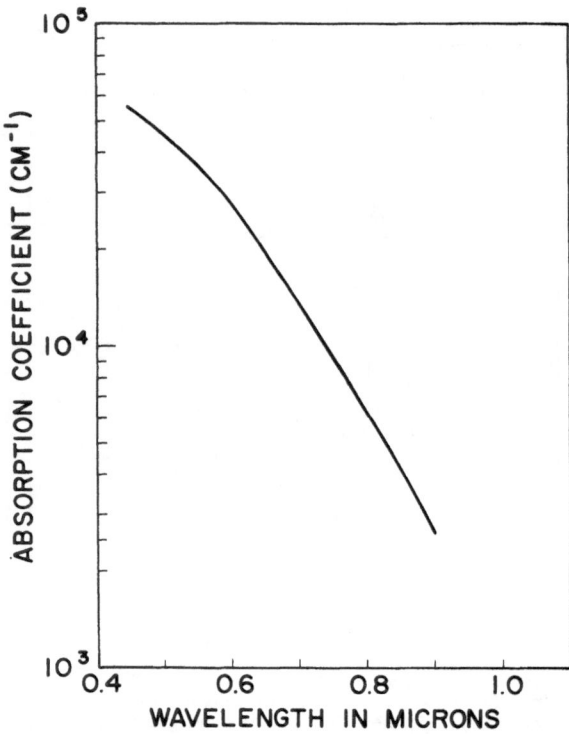

Fig. 2. Optical absorption of boron film (0.8 μ thick) deposited on quartz plate.

not as pronounced, has been observed with germanium and is discussed by Hannay [10]. Comparative measurements with a boron film (Fig. 2) did not reveal such a step. Our film was deposited on a quartz plate and had a thickness of $0.8\mu$. The film did not have a well-ordered structure. This may explain the absence of the step, which could be characteristic of well-crystallized material. The absorption edge for pure crystalline boron is very steep, in fact steeper than that for germanium [10]. As discussed by Gaulé et al. [3], impurities tend to shift the absorption edge towards longer wavelengths. Figure 3 demonstrates this on the basis of more recent measurements with boron of varying carbon content. It also shows inferior transmission for polycrystalline material, probably because of orientation and grain boundary effects. This is in disagreement with results by Brungs [5], who finds an increased transmittance in polycrystalline samples. The absorption edge (steepest part of curve) for pure boron seems to be at $0.8\mu$, corresponding to an optical band gap of 1.55 eV. Birnbaum [7] observed recombination radiation from apparently indirect transitions with a photon energy of 1.53 eV. Dietz and Herrmann [6] attribute the lower, less steep part of the absorption edge to indirect transitions, following a $\mu \propto \Delta \lambda^{-2}$ power law, whereas Gebhart and Jacobsmeyer [8] assume direct transitions obeying a $\frac{3}{2}$ power law in this range. In our double logarithmic plot (Fig. 1), we do not find a straight part below the steep part to suggest any particular $\mu \propto \Delta \lambda^{-n}$ law. However, such a straight part could be masked by absorption from impurity levels near the band edges. Further investigation of the lower, less steep part of the absorption edge with extremely pure and perfect material should clarify this point. The minimum absorption coefficient of presently available boron, which occurs near $3\mu$ (Figs. 1 and 3), is relatively large, namely, $\approx 10$ cm$^{-1}$, probably also due to impurities. The comparative value for germanium is smaller than one.

Beyond $4\mu$, all high-purity samples investigated showed a substantial increase of the absorption. This cannot be attributed to free carrier absorption for the following reasons: (1) The absorption drops again beyond $10\mu$ to exhibit two additional "windows;" (2) in the straight part, the absorption increases as $\lambda^3$, and not as $\lambda^2$, as in the free carrier case; and (3) the absorption does not vanish with vanishing conductivity at low temperatures.

Becher [11] finds an absorption band near $9.3\mu$, and another, weaker one near $13.3\mu$ for pulverized beta-boron. These bands are probably due to lattice vibrations. Our curve from mono-

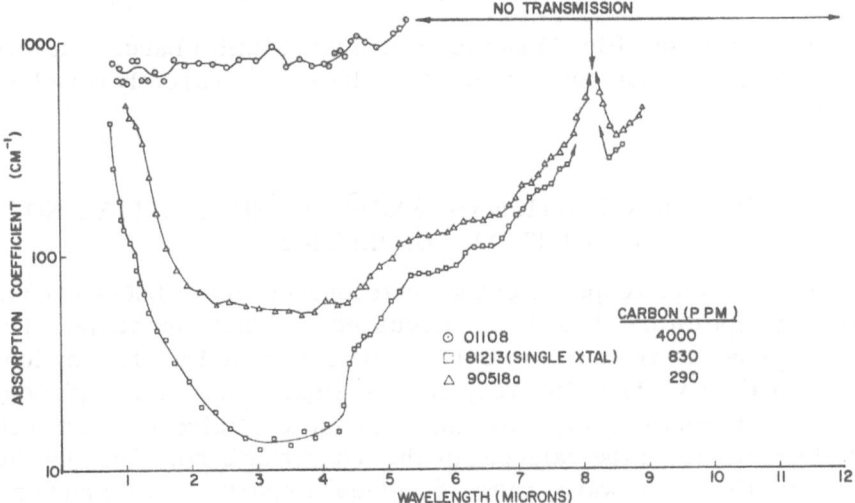

Fig. 3. Optical absorption of beta-rhombohedral boron as function of carbon concentration. Note that the transmission through the single crystal is superior in spite of only medium purity.

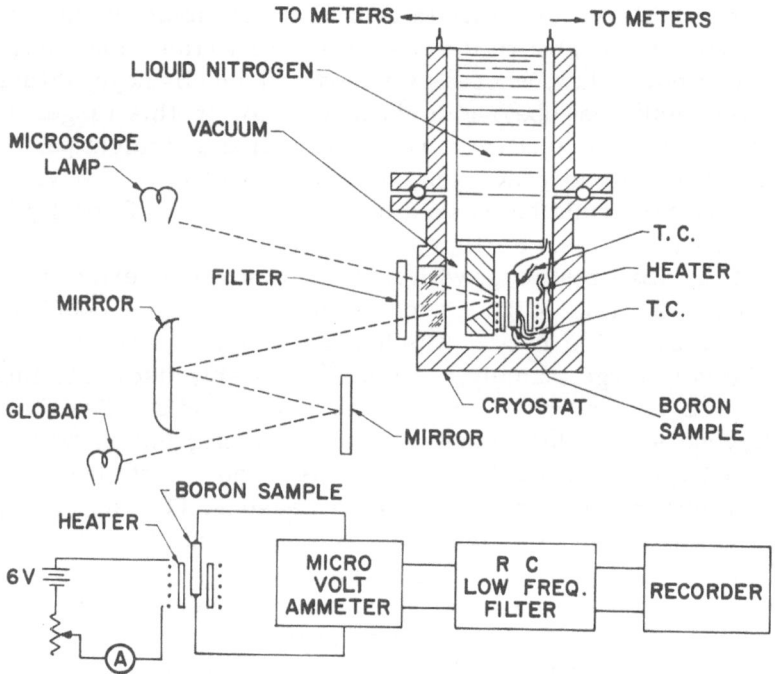

Fig. 4. Low-temperature apparatus for simultaneous optical and electrical measurements. The sample can be rotated and replaced by an equivalent aperture. The system as shown is set up for Peltier effect (T.C. is thermocouple). For the measurement of the optical absorption, a second window and a detector were installed to the right.

crystalline boron (Fig. 1) seems to indicate similar bands. Beyond $40\,\mu$, the absorption coefficient of beta-boron is apparently very high ($>10^3$ cm$^{-1}$).

## THE PHOTOEXCITATION OF EXCESS CONDUCTIVITY AND OPTICAL ABSORPTION

The spectral response of the photoconductivity in beta-boron at room temperature has been discussed by Gaulé et al. [3]. For convenience, a typical curve is again shown in Fig. 1. For long wavelengths ($\lambda > 1\,\mu$), the response is small because the photons are not energetic enough to cause transitions (direct or indirect) of electrons from the valence into the conduction band. The absorption in the long wave range involves impurity levels and only infrequently leads to the generation of a free carrier. On the short wave side of the absorption edge, the photoresponse also drops sharply. Here, most of the electron–hole pairs are generated within a surface layer of shallow depth (roughly the inverse of the absorption coefficient, $1/\mu \approx 2 \cdot 10^{-5}$ cm). Possibly most of these pairs recombine via surface recombination centers before making a contribution to the photocurrent [3]. Room temperature photoresponse curves similar to ours have been reported by Brungs [5].

Above, as usually, the term photoresponse refers to the saturation value reached by the photocurrent after a certain rise time for a given light intensity. This is quite satisfactory for beta-boron at room temperature, but at low temperatures the rise times become impractically long. This has thus far prevented the determination of the usual photoresponse for low temperatures. We did investigate instead the rise and the decay of the photocurrent at various temperatures and for various spectral distributions of the exciting light. We used the system shown in Fig. 4. As indicated, the system could also be used to heat one end of the sample to perform thermoelectric measurements. All the dark currents and photocurrents measured indicated $p$-type conduction. Our values for the Seebeck voltage roughly agree with those of Dietz and Herrmann [6].

On the basis of a simple model for $p$-type boron, we now discuss characteristic changes of the density of free holes, $p$, and thus of the conductivity, with time. Figure 5 describes the equilibrium, or dark situation for the model for three characteristic temperature ranges. As pointed out in the introduction, the presently available boron is strongly compensated, the concentration of the acceptors being only slightly larger than that of the donors. The $p$-type con-

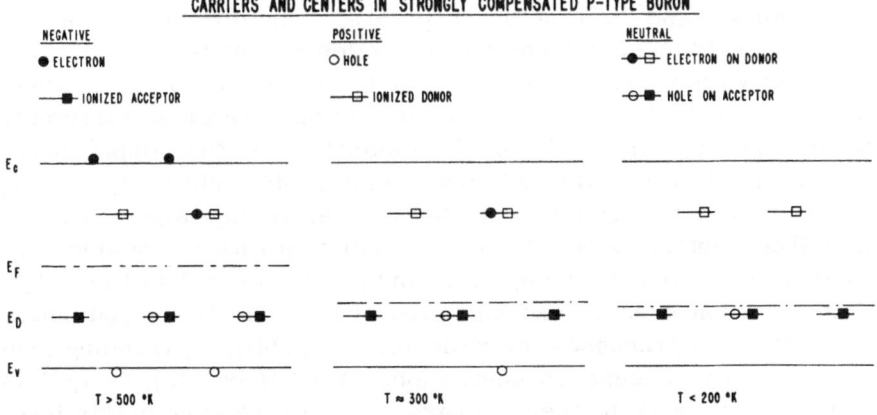

Fig. 5. Equilibrium populations in compensated $p$-type boron for three characteristic temperature ranges, assuming simple acceptor levels.

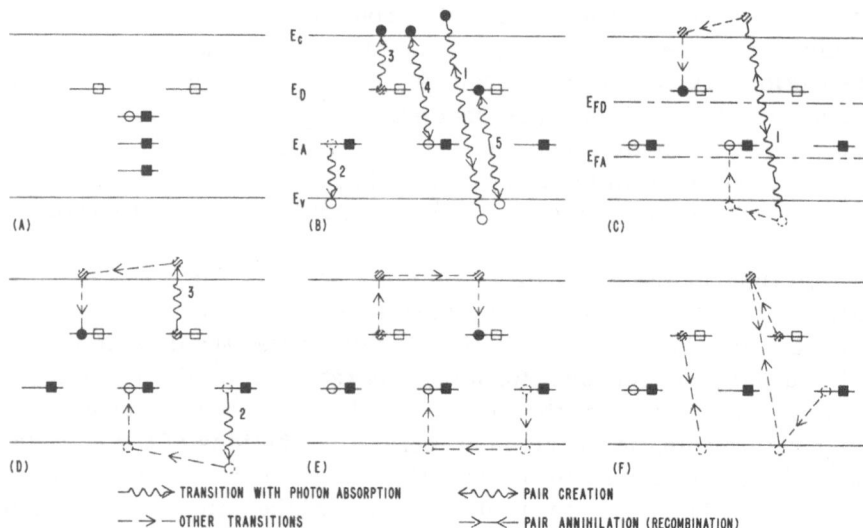

Fig. 6. (a) Model similar to Fig. 5, both with threefold acceptors (vacancies ?). (b) Possible transitions involving the absorption of one photon. (c) Creation of a free electron–hole pair with subsequent trapping. (d) Possible mechanisms for the excess infrared absorption through trapped carriers. (e) Thermal excitation causing excess conductivity when traps are filled. (f) Recombination processes which restore equilibrium.

ductivity in the intrinsic range (high temperatures) is possibly due to a hole mobility which is higher than that of the electrons. In the extrinsic range, the position of the Fermi level is as indicated because of the compensation [10]. This gives but a small concentration of free carriers, leading to the usual high resistivity of boron. At low temperatures in particular, virtually all the electrons from the donors come to rest on acceptors. The few uncompensated acceptors seldom send holes into the valence band because of their large activation energy (in the order of 0.3 eV, see below). Thus, the dark conductivity of boron at low temperatures is extremely small. The possible effects of illumination on the simple model are outlined in Fig. 6 (b). Although some photoconductivity arising from infrared absorption was observed at room temperature [3], we will concentrate on the effect of the radiation with a photon energy equal or greater than the optical band gap [process 1 in Fig. 6(b)]. This radiation creates excess pairs of free electrons and holes. Recombination processes, as indicated in Fig. 6(f), and trapping tend to reduce the excess concentration. With Bube [12], we call an energy level a trap when a charge carrier resting on this level will more likely return to its band, rather than make any other transition.

For simplicity, we assume that the concentration of trapped holes $p_t$ changes only by transitions from and to the valence band; thus,

$$\dot{p}_t = Cp - Pp_t \tag{1a}$$

$$p = (\dot{p}_t + Pp_t)/C \tag{1b}$$

The term $Cp$ describes the capture of holes from the excess population of the valence band $p$ into the traps. This occurs with a probability $C$. The term $Pp_t$ gives the inverse process, having a probability $P$. If the number of optically created pairs per unit time and unit volume is $A$, we have for the free excess holes

$$\dot{p} = A - \dot{p}_t - Rp \tag{2}$$

The second term describes the exchange with the traps; the third, the loss of holes through recombination, occurring with a probability $R$. $R$ usually depends in a complicated fashion on the electron concentration, the temperature, and other factors. For simplicity, we consider it a constant here and an empirical function of temperature further below.

Elimination of $p$ from equations (1) and (2) yields

$$\ddot{p}_t + (P + R + C)\dot{p}_t + RPp_t = AC \tag{3}$$

Observing that $p$ and $p_t$ must vanish for $t = 0$ (onset of illumination), we obtain the solution

$$p_t = [AC/PR \, (g - f)] \, [g(1 - e^{-ft}) - f(1 - e^{-gt})] \tag{4a}$$

where

$$\left.\begin{matrix} g \\ f \end{matrix}\right\} = \tfrac{1}{2} \{P + R + C \pm [(P + R + C)^2 - 4 \, PR]^{\frac{1}{2}}\} \tag{4b}$$

Because of equation (1), $p$, $\dot{p}$, etc. now also can be expressed with the two exponential functions and appropriate constants. Typical thus computed rise curves are shown in Fig. 7. To bring out the dominant physical processes, we discuss the three characteristic branches of the curve in an approximate fashion, noting that in our and also in Neft's and Seiler's [2] experiments, the two decay constants $g$ and $f$ have different orders of magnitude, which simplifies equation (4b) to

$$g \approx P + R + C$$
$$f \approx PR/(P + R + C) \tag{4c}$$

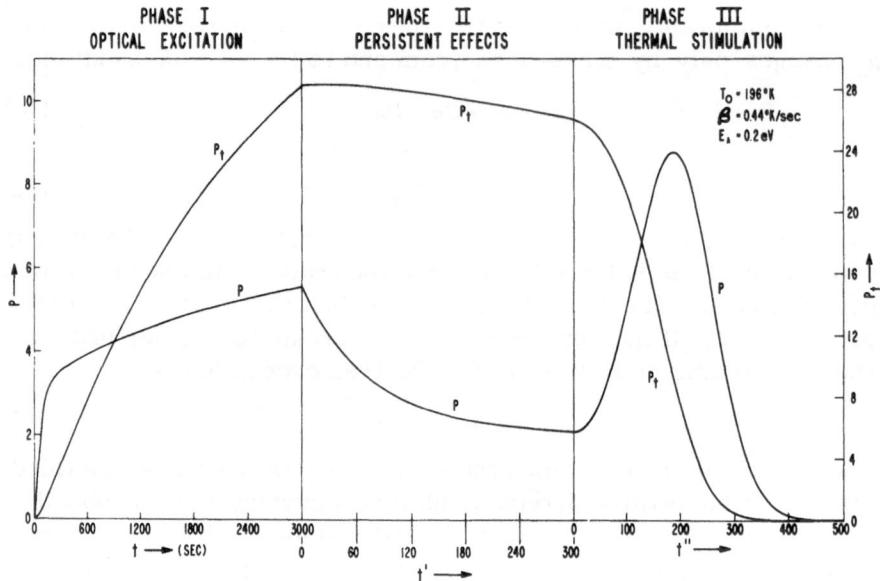

Fig. 7. Excess density of free holes $p$ and of trapped holes $p_t$ for the three characteristic phases of the photoconductivity in beta-rhombohedral boron, as computed from equations (3), (3'), and (1).

Immediately after the onset of the illumination, for

$$0 < t \ll 1/g \ll 1/f \tag{5a}$$

we find

$$p_t \approx ACt^2/2$$
$$p \approx At \tag{5b}$$

and

$$\dot{p}_t/\dot{p} \approx Ct \ll C/g < 1 \tag{5c}$$

which means that, in the beginning, most of the excess holes remain in the valence band. This is changed drastically (see Fig. 7) when

$$1/g \ll t \ll 1/f \tag{6a}$$

During this period, both populations increase nearly linearly.

$$p_t \approx ACt/g$$
$$p \approx A(1 + Pt)/g \tag{6b}$$

The growth rates compare as follows:

$$\dot{p}/\dot{p}_t \approx P/C \tag{6c}$$

The probability $P$ for a hole to become thermally excited into the valence band is (see, for example, Bube [12])

$$P = v_{th} N_v S_t \exp\left(- E_A/kT\right) \tag{7a}$$

where $N_v$ denotes the effective density of states in the valence band, $S_t$ is the capture cross section, $E_A$ is the activation energy of the traps, and $v_{th}$ is the thermal velocity of free holes. On the other hand, the capture probability $C$ is given by

$$C = v_{th} N_t S_t \tag{7b}$$

where $N_t$ denotes the trap density. Thus,

$$P/C = (N_v/N_t) \exp\left(- E_A/kT\right) \tag{7c}$$

where $N_v$ is usually [10] of the order of $10^{19}$ cm$^{-3}$. Since boron has approximately $1.3 \cdot 10^{23}$ atoms/cm$^3$, and assuming that the traps are related to one of the impurities in the 10-ppm range, we would expect $N_v/N_t$ to be of the order of 100 or less. $P/C$ is, however, kept small for low temperatures by the Boltzmann factor in equation (7c). $P/C \approx 10^{-1}$ appears to be typical for dry ice ($\approx 200°$K) experiments. From

$$P/C \ll 1 \tag{7d}$$

and from equation (6c) we recognize immediately that during the second range [equation (6a)] of the rise curve most of the optically created holes are fed into the traps rather than into the valence band. From equation (4c) it also follows that

$$1/f \gg 1/R \tag{7e}$$

meaning the rise time is many times the recombination time. Equation (6b) indicates that $p$ and $p_t$ approach thermal equilibrium with one another when

$$p/p_t \approx P/C \tag{8a}$$

This would be fully accomplished for

$$t \gg 1/f \tag{8b}$$

a case which we did not experimentally realize. An important consequence of equation (8b) is

$$p \approx A/R \tag{8c}$$

This is the conventional photoresponse of the free carrier density. It is independent of the traps; the traps only cause the long

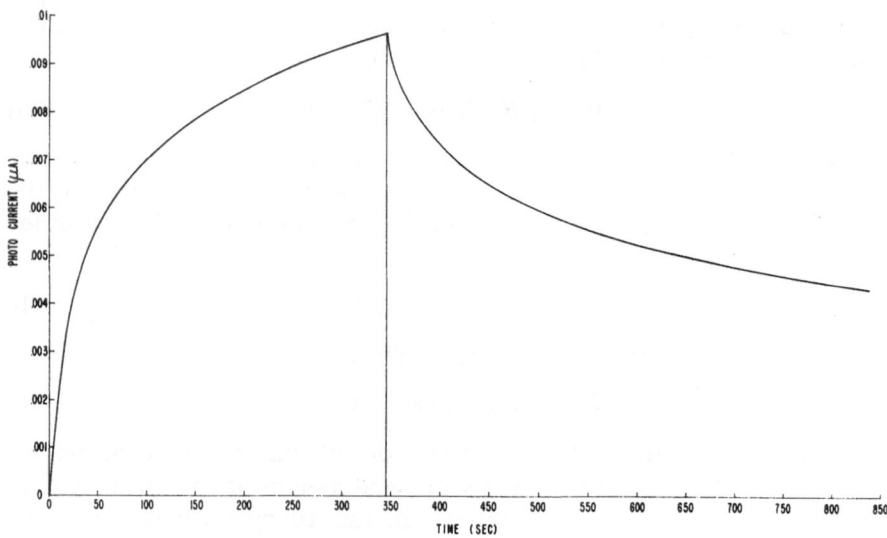

Fig. 8. Experimental rise and decay curve for the photocurrent through a beta-rhombohedral boron crystal, taken at 196°K. Both branches of the curve can be described by two exponential functions. (See text and Fig. 7.)

($\approx 1/f$ sec) delay in attaining equation (8c). It should be emphasized that the steady state characterized by equation (8b) and also by

$$\dot{p} = \dot{p}_t = 0 \tag{8d}$$

does not imply trap saturation; in fact, we believe that the assumption $p_t \ll N_t$ is always justified, unless light intensities much higher than those applied by us are used.

The extremely long rise times prevented us from observing the conventional spectral response curve for beta-boron at low temperatures. We can, however, report on the spectral response of the photosensitivity which we define for this work in a very special sense as the rate of increase of sample conductance for a given intensity $J$ of illumination (in our experiments, $J$ is roughly $10^{19}$ photons/cm²-sec).

The so-defined photosensitivity is at 77°K and at 196°K much larger (by a factor of approximately three) for blue than for red light; for the region of the maximal photoresponse at room temperature ($\lambda \approx 0.8\,\mu$; see Fig. 1), it practically vanishes. This behavior follows closely the absorption curve in the opaque region (Fig. 1). Because of the high absorption coefficient $\mu$, no light is absorbed in the interior of the sample, and the photoexcitation is limited to a surface layer of thickness $1/\mu \approx 10^{-5}\text{--}10^{-4}$ cm, with the smaller value, of course, applying to blue light. The observed change in conductance, however, should still be proportional to the total number of excess holes created in the sample, unless there is a superlinear increase of the conductivity with the number of photons absorbed in a given volume. The latter case would explain the higher sample sensitivity for blue light. No evidence for super-linearity effects is found when the photoconductivity is taken as a function of time with constant excitation. This may be deduced from the experimental curve of Fig. 8, which is in excellent agreement with equations (2) and (4a). An essentially linear re-sponse was also observed by Neft and Seiler [2] when they varied the light intensity in their room-temperature photoconductivity and trapping experiments. Since the absorption of the exciting light takes place in such a thin layer near the surface, it should depend critically on the structure of the surface, and possibly also on surface states. We could destroy the low-temperature photo-sensitivity by mechanically damaging the surface. Etching (with $KNO_3$ at 700°C) dulled the optical polish, but had little effect on the photosensitivity. Polycrystalline material with small crystallites showed little or no coil temperature photosensitivity. These observations indicate that the observed photocurrents were not surface currents through the first few atomic layers of the sample.

The peculiar shift of the photosensitivity of beta-boron to short wavelengths, as the temperature is decreased, motivated us to search for possible changes of the optical absorption with tempera-

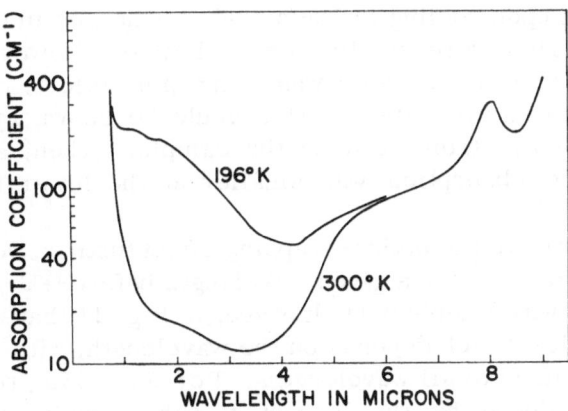

Fig. 9. Optical absorption of beta-rhombohedral boron crystal when sample is illuminated with white light. At the lower temperature, the white light causes a substantial excess absorption. Source, tungsten filament.

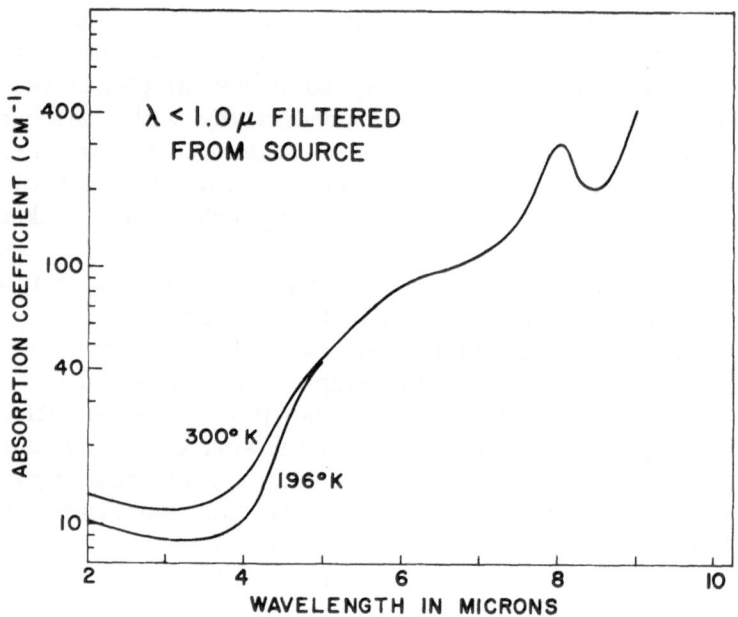

Fig. 10. Repetition of the experiment of Fig. 9 with the short wavelengths ($\lambda < 1.0$) kept off the sample. The transmittance for the lower temperature is greater, as expected.

ture. Usually, the causes for optical absorption by a semiconductor (free carriers, lattice vibrations) are diminished as the temperature decreases. In our first experiment (Fig. 9), however, with white light falling on the sample, boron displayed just the opposite behavior; the absorption in the window range $1-5\,\mu$ increased considerably upon cooling to 196 or 78°K. Changes in other ranges of the spectrum were smaller or negligible. However, even the change in the window range was a trapping effect, rather than a genuine temperature effect. This could be shown by preventing light with $\lambda < 1\,\mu$ from striking the sample. Then, as originally expected, the absorption was smaller at the lower temperature (Fig. 10).

To investigate the optical trapping effect further, we determined the transmittance $\tau$ for a given wavelength before ($\tau = \tau_0$) and during illumination with visible light. It is seen in Fig. 11 that $\tau$ approaches a steady value which depends on the wavelength, with a time delay which is common to all wavelengths. For $\lambda = 1.55\,\mu$, the transmittance $\tau$ is reduced by the largest factor, namely, five. Boron behaves as an optically controlled (through blue or white light) filter for the window range. At liquid-nitrogen temperature, the transmittance is depressed somewhat more than at dry ice temperature, but this requires much more time ($\approx 10^4$ sec); at room temperature, the reduction is almost instantaneous, but limited to $\approx 5\%$ (Fig. 12). If white light is incident during cooling to liquid-nitrogen temperature, a high absorption coefficient is obtained rapidly. In all the low-temperature experiments, the increased absorption and the increased conductivity persist; they decay only very slowly after the exciting light is turned off. This is discussed in detail below.

In our experiments we must, experimentally and conceptually, separate the exciting and the measuring (or sampling) light. The sampling light is weak, nearly monochromatic (usually $\lambda = 1.5\mu$), and chopped. It is produced by passing radiation from a globar infrared source through a chopper and then through a Perkin-Elmer 83 monochromator. From the exit slit of the monochromator, the sampling light penetrates the sample, which is located in a cryostat, and reaches the detector which responds to chopped light only. The exciting light is produced by a tungsten filament and is of high intensity. It is not chopped and, thus, does not influence the detector. To influence the spectral distribution of the exciting light, filters were used. (A monochromator would have reduced the intensity too much.)

Usually only the face of the sample farther from the exit slit of the monochromator was illuminated with exciting light. In one

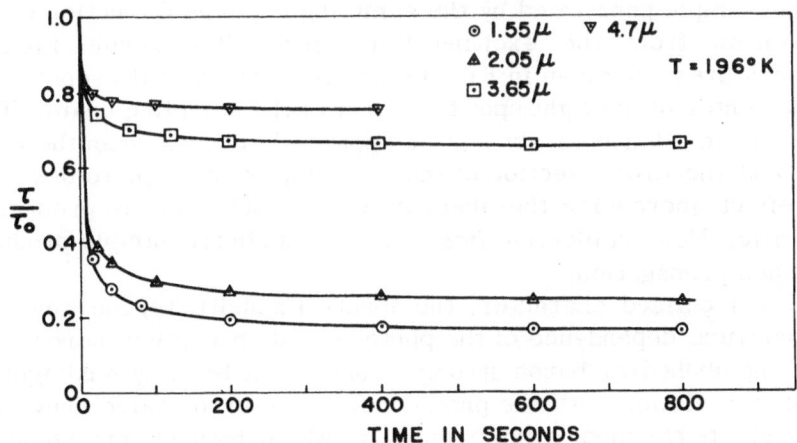

Fig. 11. Experiment similar to that described by the top curve in Fig. 9. The influence of white light on the transmittance is shown as function of time for several wavelengths for $T = 196°$K. Note that time constants are equal for all wavelengths.

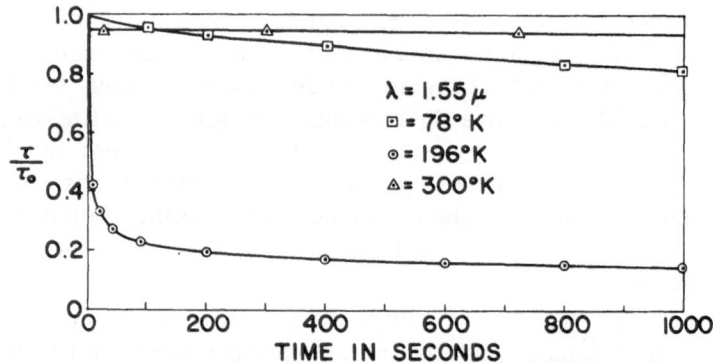

Fig. 12. Experiment similar to that given by Fig. 11, but with temperature as parameter. For $T = 78°K$, the curve will eventually (after $10^4$ sec) go below the 196°K curve.

case, however, we illuminated both faces with exciting light. This produced an additional increase in absorption, suggesting that the effect does not penetrate or diffuse deep into the sample, but is limited to a depth $(1/\mu$, the inverse absorption coefficient) below an illuminated surface.

In another experiment, we sent only a narrow beam of sampling light through the sample. Onto the face of the sample farther from the exit slit of the monochromator we focused an equally small image of a source of exciting (white) light. By scanning the sample face with this spot of white light while keeping the beam of sampling light stationary, we could observe the transmittance of that region of the sample penetrated by the sampling beam as a function of its separation from the exciting light spot. The results for two temperatures are given in Fig. 13, where $\Delta$ indicates the separation of the center of the light spot from that of the sampling beam. The onset of an additional absorption apparently occurs when the light spot and the cross section of the sampling beam begin to overlap; the effect increasing the absorption does not seem to propagate laterally. Nor did electric fields or currents in the sample produce any such propagation.

The localized character, the spectral sensitivity, the time and temperature dependence of the photoexcited absorption increase in beta-rhombohedral boron strongly suggest (at least by hindsight) a close correlation with the photoconductivity in the same material. We thus performed experiments in which both effects could be measured simultaneously. Samples for these experiments were prepared from material with a purity and structural perfection as described previously [3] by first grinding and then polishing (with diamond paste) to form two plane and parallel surfaces. The sample thickness was kept small ($\approx 150\,\mu$), so that at least a fraction of a percent of the thickness would be influenced by the exciting light. The sample faces normal to the sampling light beam were only a few square millimeters. Contacts of silver paste were placed on the ends of the sample. The sample was then mounted over a small aperture in a cold finger extending from the inner portion of a cryostat (compare with Fig. 4) such that either the sample or an identical aperture could be rotated into the sampling beam, which was focused on the sample. The sampling wavelength was $1.55\,\mu$ for the reasons given above and the sample temperature was 77, 196, or 300°K. The middle value, 196°K, was usually preferred because the trapping effects are quite pronounced at this temperature, while the rise times are not as unreasonably long as for still lower temperatures. The electric field applied through the contacts while measuring the photoconductivity was approximately 7 V/cm.

From basic optical considerations, it is reasonable to assume that the increment of the absorption coefficient $\Delta\,\mu$, which describes

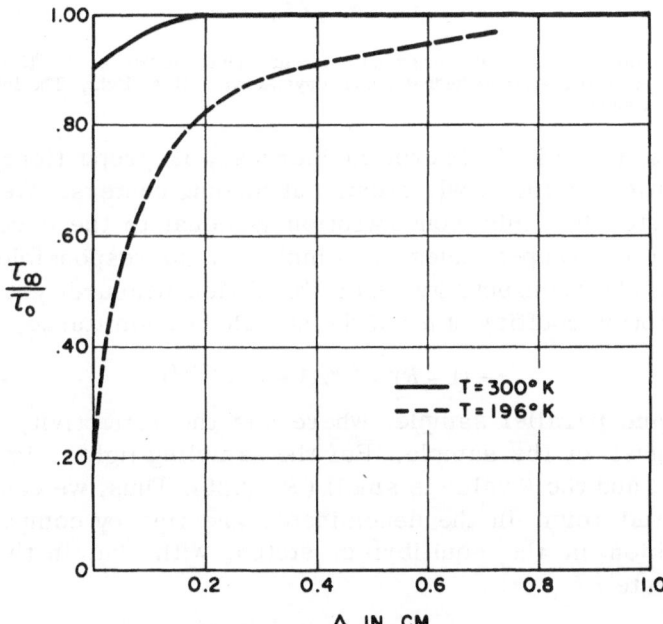

Fig. 13. Decrease of the transmittance $\tau$ as function of the lateral separation between the exciting white light and the sampling light ($\lambda = 1.5\,\mu$). The distance of the center of a spot of white light from the center of the sampling beam is given by $\Delta$. $\tau_0$ gives the transmittance before, and $\tau_\infty$ after very long excitation.

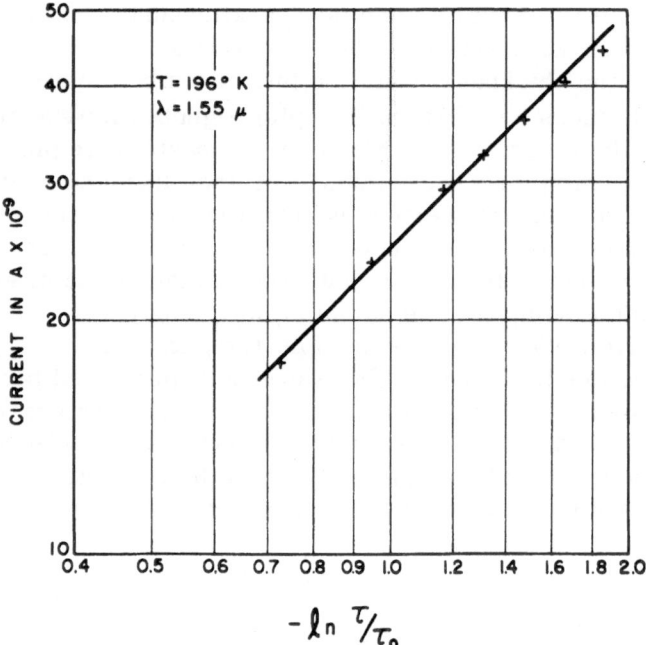

Fig. 14. Transmittance and photocurrent measurements taken at 196°K during the illumination of a beta-rhombohedral boron crystal with white light. The initial phase ($\tau \lesssim \tau_0$) is not shown.

the optically excited absorption increase, is proportional to the concentration of the newly created absorbing centers. We wish to demonstrate that this concentration is equal to the excess concentration of trapped holes $p_t$, which is also responsible for the persistent photoconductivity. Our absorption measurements give us the absorption coefficient $\mu$ via the sample transmittance, which is

$$\tau = (1 - R)^2 \, e^{-\mu \, d} / (1 - R^2 \, e^{-2\mu \, d}) \tag{9a}$$

for a plane parallel sample, where $R$ is the reflectivity and $d$ is the thickness of the sample. For the sampling light, $\mu$ is 15 cm$^{-1}$ or higher, and the $R$ value is small ($R \approx 0.25$). Thus, we can neglect the squared term in the denominator and find by comparing the transmission in the equilibrium state $\tau_0$ with that in the photo-excited state $\tau$

$$\ln \tau_0 / \tau = - \ln \tau / \tau_0 = (\mu - \mu_0)d = (\Delta\mu)d \tag{9b}$$

This is the abscissa of Fig. 14; this figure represents transmission and photocurrent values taken at 196°K after long periods of photoexcitation. From the 45° slope of the curve in Fig. 14 and because the photocurrent is proportional to $p$, the excess hole density, we deduce that $\Delta \mu \propto p$. On the other hand, we know from equation (8a) that $p$ becames nearly proportional to $p_t$ after long times of photo-excitation, so that indeed $\Delta \mu \propto p_t$. We can thus measure the occupation of the traps by purely optical means. This is done without producing any changes in the boron by the measurement, because the sampling light itself, being on the long wave side of the optical band gap and of low intensity, produces no photoexcitation. Infrared quenching of the trapping effects in boron by the sampling light should also be negligible, since it could not be induced by us when using much stronger intensities. The simplest physical model explaining our observation that $\Delta \mu \propto p_t$ is based on the assumption that either the trapped holes, or the trapped electrons with a concentration $n_t \propto p_t$, cause the excess absorption. This is discussed in detail in a later section, in connection with heating effects.

During our investigation of the photoconductivity of beta-boron at various temperatures, we noticed a bulk photovoltaic effect in one sample. A small voltage ($\approx 1$ mV) was observed when the bulk of the sample was illuminated with two microscope lamps at room and at liquid-nitrogen temperatures. Since the contacts were shielded from the light and since the voltage did not change when a light spot traversed the surface of the sample, influence from contact or internal barriers can be excluded. The spectral response from 0.75 to 1.2 $\mu$ corresponded to penetrating light. The rise and decay times were only a fraction of a second at either temperature. Thus, the bulk photovoltaic effect does not appear to be linked to trapping effects.

A concentration gradient of some dominant impurity [13] is probably responsible for the bulk photovoltaic effect. The voltage changed sign upon cooling to liquid-nitrogen temperature; no well-founded explanation for this is presently available.


## THE DECAY OF EXCESS CONDUCTIVITY AND OPTICAL ABSORPTION AT CONSTANT TEMPERATURE

The photoconductivity and the optically excited additional absorption in beta-boron are persistent; they decay only very slowly (within hours) at low temperatures. Experimentally, we find that the decay can be described by two exponential functions. This is also predicted by our theory. The two decay constants $f$ and $g$ are

## TABLE I

Estimates for Probabilities and Decay Constants From the Evaluation of Measured Rise and Decay Curves

| Quantity | Unit | $T = 300°K$ | $T = 196°K$ | $T = 78°K$ |
|---|---|---|---|---|
| $1/g$ | sec | $\approx 1\ (3.1)^{*}$ | 71.5 | 285 |
| $1/f$ | sec | $\approx 20\ (42.2)$ | 1820 | 5700 |
| $R$ | sec$^{-1}$ | $3 \cdot 10^{-1}$ | $0.73 \cdot 10^{-2}$ | $\approx 10^{-3}$ |
| $C$ | sec$^{-1}$ | $10^{-1}$ | $0.67 \cdot 10^{-2}$ | $\approx 10^{-3}$ |
| $P$ | sec$^{-1}$ | 0.05 | $0.8 \cdot 10^{-3}$ | $\approx 10^{-4}$ |

*Values in parentheses are taken from Neft and Seiler [2].

already known from equations (4b) and (4c). To find the two concentrations $p$ and $p_t$ during the decay period (Phase II in Fig. 7), we have to set $A = 0$ in equation (3), and establish the new initial conditions. We denote the time at which the illumination is turned off by $t = t_I$ and introduce, for convenience, a new time scale by

$$t' = t - t_I \tag{10a}$$

We abbreviate the new initial values as follows: For $t = t_I$ or $t' = 0$,

$$p = u_I,\ p_t = v_I,\ \dot{p}_t = \dot{v}_I,\ \text{etc.} \tag{10b}$$

For physical reasons, $p$ and $p_t$, and because of equation (1), $\dot{p}_t$ must be continuous at $t = t_I$, whereas $\dot{p}$, because of equation (2), makes a jump. It is easily verified that these conditions are met by

$$p_t = (v_I + v_s)\ e^{-ft'} - v_s\ e^{-gt'}$$
$$v_s = (fv_I + \dot{v}_I)/(g - f) \tag{10c}$$

and

$$p = (u_I - u_s)\ e^{-ft'} + u_s\ e^{-gt'}$$
$$u_s = (g - P)\ v_s/C \tag{10d}$$

All the quantities represented are positive. From experimental decay curves for $p$ (Fig. 8), estimates for $f$ and $g$ are readily obtained. Hereby, we always find that $g \gg f$. Further information is then gained by comparing the short-lived part of $p$, which is proportional to $e^{-gt'}$, with the persistent part, which is proportional to $e^{-ft'}$. With this, and by using equations (4c) and (6b), we can estimate the probabilities $C$, $R$, and $P$ in equation (3). Some results are given in Table I. The values in parentheses are, for comparison, taken from the work of Neft and Seiler [2]. It is very difficult to account for the changes of the parameters with temperature. For this, measurements at intermediate temperatures are necessary. In equation (7a), we describe the temperature dependence of $P$ through a Boltzmann factor. This could account for the changes in $f$, for which $P$, because of equation (4c), gives an upper limit. No value for the trap depth $E_A$, however, fits the $P$ or $f$ values for all three temperatures. Possibly, shallower traps dominate the persistent

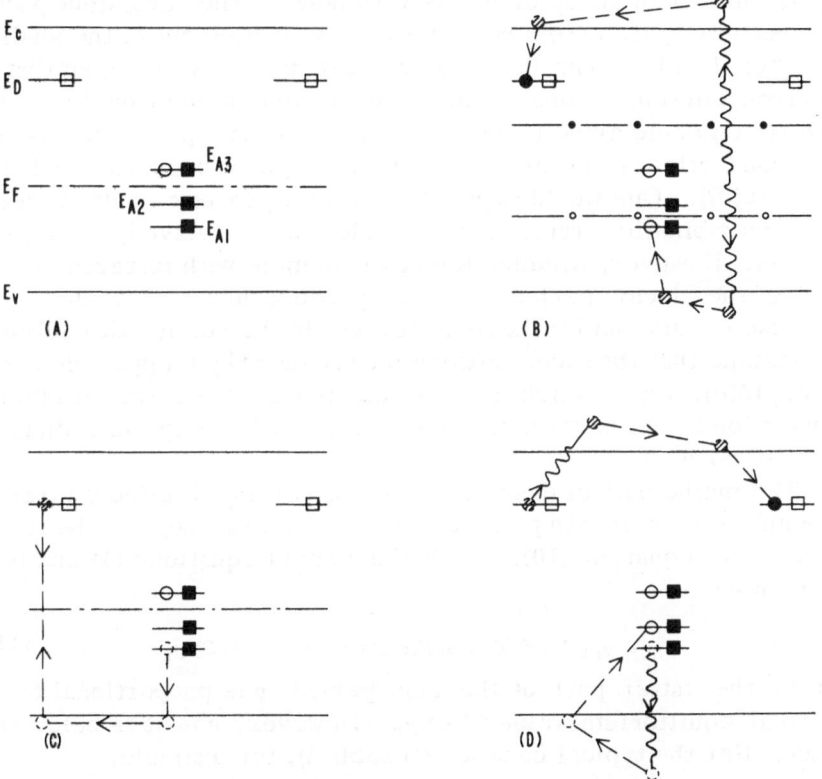

Fig. 15. Tentative model for beta-rhombohedral boron with threefold acceptor levels (vacancies). (a) Equilibrium at low temperature. (b) Trapping after photoexcitation; populations are described by pseudo-Fermi levels. (c) Recombination via donor. Recombination via acceptor is unlikely because of the net negative charge of the acceptor. (d) Processes possibly causing excess infrared absorption.

photoconductivity as one goes to lower temperatures. This is also suggested by experiments at even lower temperature discussed below. Similarly, our empirical (and very crude) values for C and R may represent different transitions as the temperature is lowered. In Fig. 15, we present a tentative model for compensated p-type boron with a threefold acceptor level. At equilibrium at low temperatures [Fig. 15(a)], all the lower levels and a large fraction of the top acceptor levels are filled by electrons from the compensating donor levels. The threefold acceptor has, therefore, as a rule, a net negative charge. This is so, even when one or two holes rest on it. The net negative charge repels electrons and, thus, retards recombination of electrons with holes via the acceptor site. Optical excitation will fill some of the donor levels with electrons and an equal number of acceptor levels with holes. The persistent photoconductivity arises from holes thermally excited out of the shallow acceptor levels. The donor levels are probably deep, so that no electron current is observed. The persistent additional infrared absorption could arise from the excitation of trapped electrons into the conduction band or from the analogous process for holes [Fig. 15(d)]. One would expect that such a process would increase the recombination rate, and, if holes are involved, the excess current. However, illumination of the sample with infrared ($\lambda > 1 \mu$) during the decay period failed to produce any noticeable effect. Because of this and for the reasons given in the section that follows, we assume that the excess absorption is caused by trapped electrons [Fig. 15(d), top], which become excited by the infrared into the conduction band and then return to a trap, apparently via radiationless transitions.

The mathematical description of the persistent effects in terms of $p$ and $p_t$ is very simple. For $t' \gg 1/g$, we can neglect the terms with $e^{-g t'}$ in equation (10). With the help of equations (1) and (4c), we then have

$$p \approx p_t (P - f)/C \approx p_t PR/C (R + C) \propto (P/C) p_t \qquad (11a)$$

As in the latter part of the rise period, $p$ is proportional to the thermal equilibrium value $(P/C) p_t$. However, now it is below this value. For the typical case $R \approx C$ (Table I), for example,

$$p \approx \tfrac{1}{2} (P/C) p_t \qquad (11b)$$

The simultaneous observation of the persistent absorption change and of the persistent photocurrent thus yields results similar to those of Fig. 14 except for the lower proportionality factor.

From basic considerations [10] and also from Table I, one would expect that $P$ would decrease more rapidly with temperature than $C$. At very low temperatures, $P/C$ thus would be very small and, because of equation (11a), no persistent photocurrent is observed even when many traps are charged. To test if we can charge traps at very low temperatures, we illuminated a sample mounted on a cooling finger cooled with liquid helium. The sample temperature was then approximately 10°K and the dark current only $\frac{1}{100}$ that at liquid-nitrogen temperature. Probably because of the very small dark current, we could now observe a small instantaneous photocurrent. No persistent photocurrent was observed, however, unless the sample was warmed to liquid-nitrogen temperature after the illumination was completed (compare with Figs. 19 and 20). This proves that the traps are capable of capturing carriers at near helium temperature. Absorption measurements at that temperature should confirm this.

## EXCESS CONDUCTIVITY AND OPTICAL ABSORPTION UNDER RAPID HEATING

When a boron sample in the state of persistent excess conductivity and optical absorption is rapidly heated, the excess conductivity vanishes after going through a steep peak, while the excess absorption vanishes monotonically, as shown in Fig. 16. It is noted that the experimental currents of Fig. 16 include the dark current, which becomes substantial at high temperatures. By subtracting the dark current, we could indeed show that the excess current vanishes. The peaked current curves are often called electrical glow curves, because of the analogy to the light emission from heated luminescent materials. We prefer the expression thermal stimulation peak [12]. The thermal stimulation is a consequence of increased evaporation of holes from the traps according to equation (1)

$$\dot{p}_t = Cp - Pp_t \tag{1}$$

where $P$, because of the Boltzmann factor in equation (7a), increases rapidly. An estimate of the excess hole density $p$ may be obtained from equation (2)

$$Rp + \dot{p} = -\dot{p}_t \tag{2}$$

which simplifies to

$$p \approx -\dot{p}_t/R \qquad \text{for} |\dot{p}/p| \ll R \tag{2a}$$

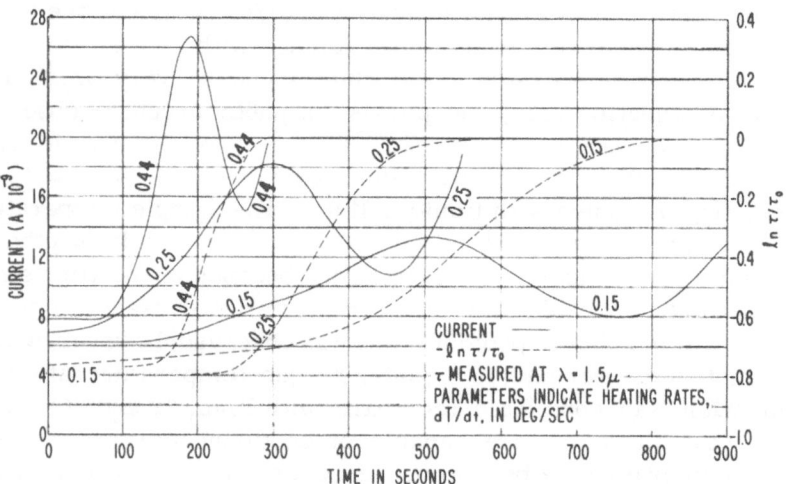

Fig. 16. Photocurrent and infrared (at $\lambda = 1.5\,\mu$) transmittance $\tau$ simultaneously measured during rapid heating. The initial values for the current and for $\tau$ were obtained by
3000-sec illumination with white light. (Compare with Fig. 7.) Note the delay in the
restoration of $\tau$ to its equilibrium value $\tau_0$. Note also that the peaks of the current curves
coincide with the inflection points of the $\ln(\tau/\tau_0)$ curves. The latter probably represent
$n_t$, the density of trapped electrons. The increase of the current beyond the valley is not
a trapping effect, but is due to the dark current which becomes dominant at high temperatures. (Compare also with Table II.)

The condition on the right, requiring that the relative change of $p$
corresponding to a time interval $1/R$ be small, is not always fulfilled
in the cases of most rapid heating (compare Fig. 16 and Table I),
but equation (2a) is still useful in discussing physical trends.

For convenience, we define for the onset of the heating period

$$t = t_{II}, \quad p = u_{II}, \quad \dot{p} = \dot{u}_{II}, \quad p_t = v_{II}, \text{ etc.} \tag{12a}$$

and choose a new time scale according to

$$t'' = t - t_{II} \tag{12b}$$

By integrating equation (2a), we then have

$$\int_0^\infty p \, dt'' = v_{II}/\bar{R} \tag{13}$$

We took advantage of the fact that obviously $p_t \to 0$ as $t'' \to \infty$. In view
of the symmetry of the experimental conductivity peaks of Fig. 16,
the contribution of $\dot{p}/R$ to the integral would be small, justifying the
omission of $\dot{p}/R$. The conductivity peaks of Fig. 16 correspond to
$T = T_{max} \approx 270°K$ for all three heating rates (Table II). According to
Table I, we must assume that $R$ increases during the trap discharge.
We, therefore, introduced an average value $\bar{R}$ into equation (13).

Since the excess current is proportional to $p$, equation (13) predicts that the area under the curve for the excess current should be nearly independent of the heating rate. We find the excess current by substracting the dark current from the curves of Fig. 16. The empirical areas $Q$ are given in Table II. The smaller areas for the faster heating rates could indicate an increased effective $R$ during fast heating, probably due to liberation of many trapped electrons at or near the peak time. Within this time range, the excess current or $p$ is roughly proportional to the slope of ln $(\tau/\tau_0)$, but this is not true for other parts of the empirical curves of Fig. 16, in particular, not for the beginning of the heating, where the optical absorption reacts with a considerable delay. In other words, equation (2a) is not well obeyed when we assume that the excess absorption arises from the trapped holes, because this would require $p_t \propto -\ln (\tau/\tau_0)$ for all times, as explained previously. In isothermal experiments, the densities $n_t$ and $p_t$ of trapped electrons and holes are usually nearly equal, or at least proportional to one another. This is no longer so in the case of rapid heating if, as we assume, the traps for the electrons are deeper than those for the holes. In that case, the emptying of the electron traps starts after (at a higher temperature) that of the hole traps, in agreement with Fig. 16 if we assume that

$$n_t \propto - \ln (\tau/\tau_0) \tag{14}$$

We also note in Fig. 16 that the inflection points (steepest slopes) of the transmission curves coincide with the peaks for the conductivity. This could indicate that the recombination of electrons is enhanced when many free holes are present. Our basic differential equation (3), modified for the case of rapid heating, is

$$\ddot{p}_t + (P + R + C)\, \dot{p}_t + (PR + \dot{P})p_t = 0 \tag{3'}$$

where now $P$ and $R$ become functions of time via their dependence on

## TABLE II

### Evaluation of the Thermal Stimulation Curves of Fig. 16 (Dark Current Subtracted)

| Quantity | Unit | Heating rate, $\beta$ (deg/sec) | | |
|---|---|---|---|---|
| | | 0.15 | 0.25 | 0.44 |
| Temperature, $T_{max}$ | deg Kelvin | 270 | 270 | 280 |
| Empirical areas, $Q$ | arbitrary | 0.26 | 0.23 | 0.2 |

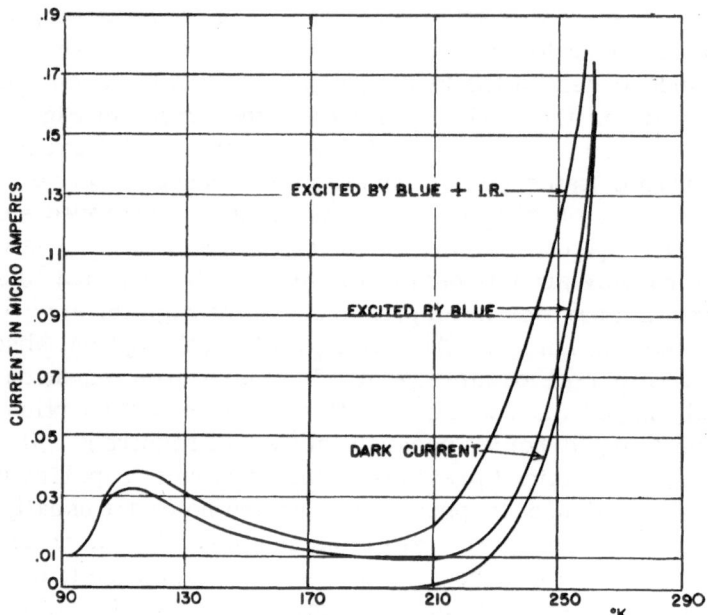

Fig. 17. Thermal stimulation curves obtained by illuminating for 5400 sec with blue (4000–5000 A) light, with and without infrared radiation present. The initial current is 40 times the dark current. The latter is not shown to scale. Note that abscissa is now temperature, and not time, as in Fig. 16. Heating rate is approximately 0.5°K/sec.

temperature ($R$ probably also via $n$ or $n_t$, but this is neglected). The temperature is given by

$$T = T_0 + \beta t''  \qquad (15)$$

where $T_0$ is the starting temperature and $\beta$ is the heating rate. Physical considerations give the initial conditions which require that $p, p_t$, and their first derivatives be continuously carried over from the preceding persistent phase (Fig. 7). Approximate solutions for equation (3') are usually obtained by a method which is roughly equivalent to neglecting $\ddot{p}_t$ [12], but even then, machine computations are usually necessary. Furthermore, our transmission curves of Fig. 16 together with equation (14) indicate that $\ddot{n}_t$ is not always negligible, and $\ddot{p}_t$ should be of comparable magnitude. We thus obtained numerical solutions of $p_t$, and through equation (1), of $p$, directly from equation (3'), using the temperature dependence of $P$ given by equation (7a), and making plausible assumptions on the temperature dependence of $R$. A typical result is shown in Fig. 7. It is seen that the drop in $p_t$ sets in immediately, and that $p_t$ has decreased to less than 50% of its initial value ($v_{II}$) at the peak for $p$.

This is in contrast to the empirical $\ln(r/r_0)$ curve of Fig. 16, which shows an initial lag and only an approximately 25% drop at the conductivity peak. This confirms our opinion that the latter curves represent $n_t$, and not $p_t$. Stimulation curves for $T_0 = 90°$K are shown in Figs. 17 and 18. The heating rate was always approximately 0.5 deg/sec. It is noted that the abscissa is now temperature, rather than time, as in Fig. 16. The sample was usually optically excited until the persistent photocurrent was forty times the dark current. (The low-temperature dark current was only $2.5 \cdot 10^{-10}$ and is not shown to scale in Figs. 17 and 18.) This persistent photocurrent was obtained after 5400 sec for blue light (4000–5000 A), and after 14,400 sec for red light (6000–7000 A), using comparable light intensities. Infrared light alone never produced more than a 2.5-fold increase, but it has a significant effect when used together with visible light, as seen in Fig. 17. The stimulation peak is higher if infrared was present during the illumination, but the valley following the peak is less pronounced. A deep valley is possibly the result of a peak in the density of free electrons which occurs after that for the holes (see Fig. 17) and causes the rapid recombina-

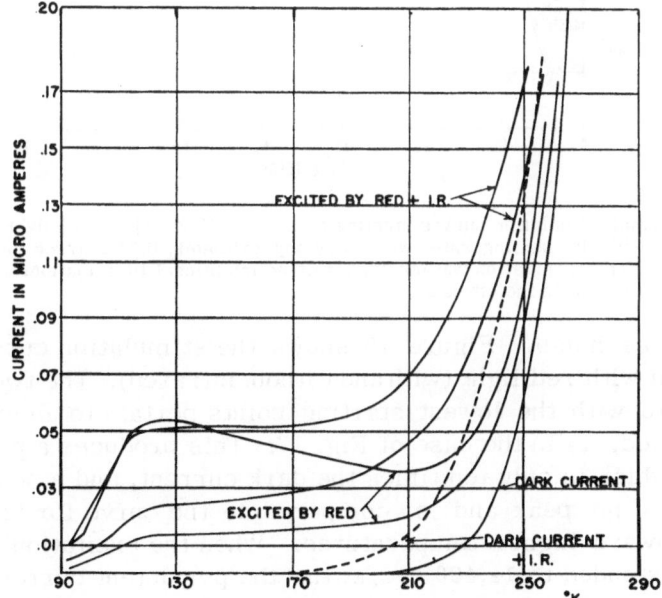

Fig. 18. Thermal stimulation curves obtained under conditions similar to those for Fig. 17, but using various combinations of red (6000–7000 A) with infrared radiation, and various degrees of initial excitation.

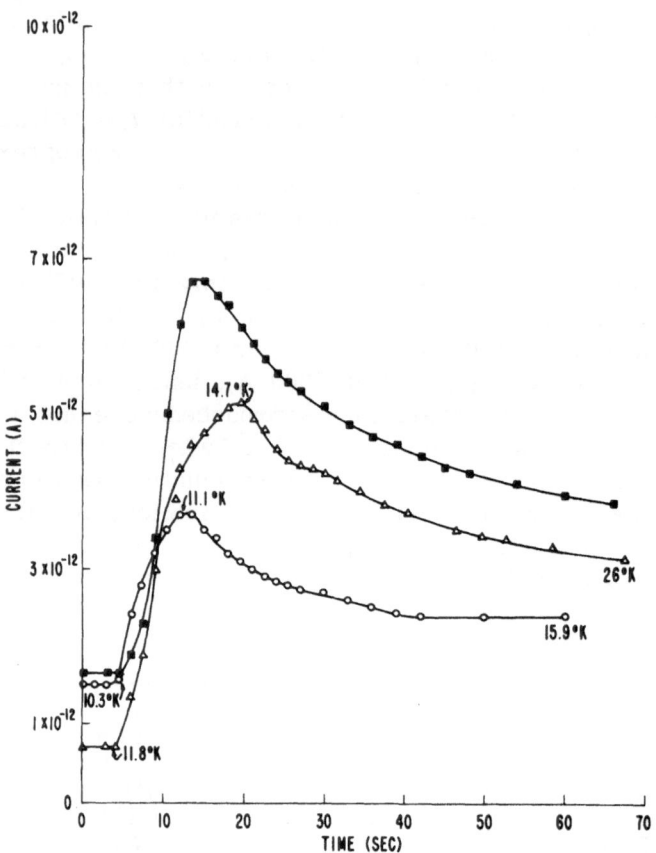

Fig. 19. Thermal stimulation curves starting from $T_0 \approx 10°K$. The peak height increases, as expected, with the heating rate, which was approximately 0.25 deg/sec for the curve in the middle. The stimulation was obtained after no intentional illumination, but infrared radiation was present in the cryostat.

tion of free holes.   Figure 18 shows the stimulation curves after excitation with red light (with and without infrared). The two curves in Fig. 18 with the lowest starting points pertain to an excitation of 5400 sec, as in the case of Fig. 17.   This produces a persistent current which is only ten times the dark current, and a stimulation which has no peak and is, compared to the curve for blue light, moved toward higher temperatures.   When the excitation with red light is extended to 14,400 sec, so that the persistent current equals that in the experiment with blue light, a high stimulation peak is observed, followed by a shallow valley.   Adding infrared in this experiment makes the valley disappear.

The color memory of the boron crystal is apparently based on the distribution of electrons and holes over the traps, which is strongly dependent on the spectral range of the optical excitation, whereby red, and to some degree also infrared light, seems to favor the deeper levels. To obtain a model explaining the stimulation curves, it will be necessary to simultaneously observe the infrared absorption. It is already known from purely optical experiments (Fig. 7) that an excess optical absorption with the expected time dependence is present at 78°K. Furthermore, it should be tested if

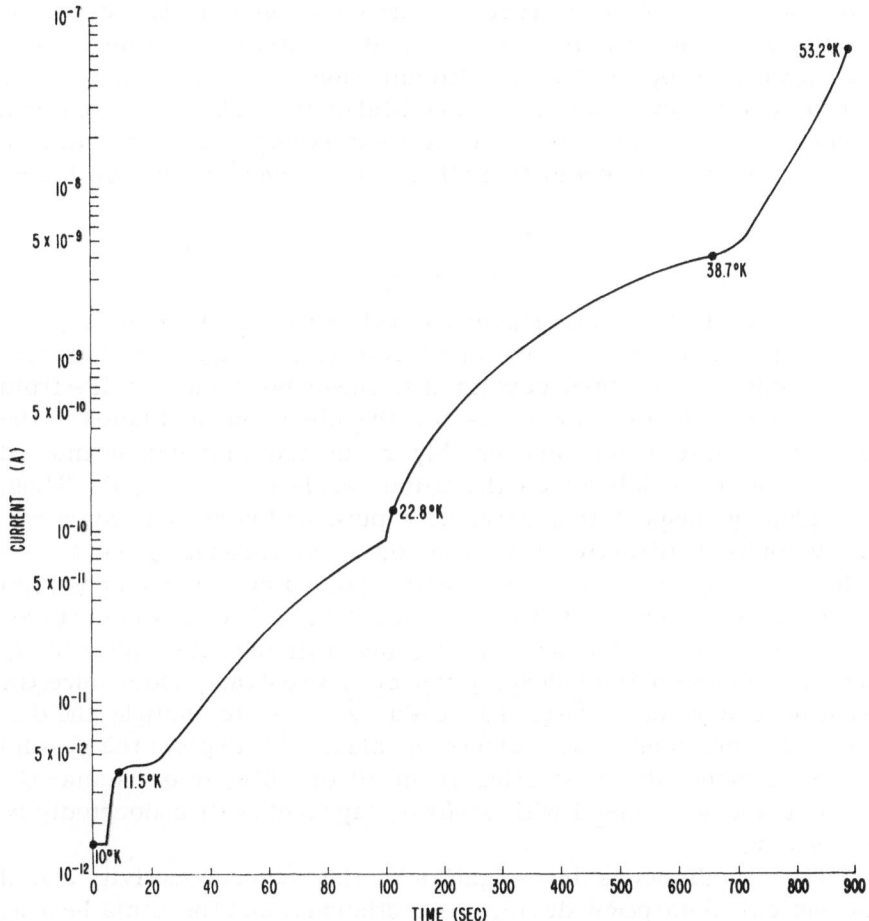

Fig. 20. Thermal stimulation experiment, starting from 10°K, after illumination with white light. Comparison with Fig. 19 shows that the white light did not produce any persistent current at 10°K. Note change of scale for $t = 100$ sec.

the peak at 270°K, found in the experiments starting from 200°K, can be reproduced when starting from much lower temperature. The stimulation curves starting from 90°K usually follow the area rule discussed above [equation (13)].  As in all the other experiments, infrared quenching was ineffective, but thermal cycling in the dark through a high enough temperature ($\approx 300°K$) removed all non-equilibrium effects. Some rapid heating experiments were started at $T_0 \approx 10°K$.  Figure 19 shows stimulation peaks near 12°K. The peak height increases as expected with increasing heating rate. No intentional illumination was necessary to produce these peaks; there was, however, some infrared radiation present in the cryostat.

Figure 20 shows the result of a stimulation experiment which was preceded by intentional illumination with white light. The initial peak is now replaced by an initial step which is followed by a monotonical increase and one further step.  The experiments starting from 10°K seem to indicate a sequence of shallow traps.

## SUMMARY

The crystalline beta-rhombohedral boron used in our experiments displayed very unusual and fascinating optical and electrical phenomena.  It appears certain that these phenomena arise from local energy levels, or traps for the electrons and holes in the crystal.  These levels presumably result from foreign atoms and from structural defects on the atomic scale in the crystal.  Thus, we strongly suspect that extremely pure and perfect boron would behave quite differently. Unfortunately, such boron is not yet available, and trapping effects are still reported by other groups [2,6] using the best material presently available. The experiments at, or starting from, dry ice temperature will probably get a fairly accurate mathematical description by a two-trap model once the treatment developed here has been expanded to include the differential equations of the electrons also.  To explain the results of experiments at or starting from 10 or 90°K, in particular the color memory, a model with several trap levels will undoubtedly be necessary.

The simultaneous investigation of the photoconductivity and of the optical absorption during the isothermal and the rapid heating phases of our experiments has proven a very fruitful research tool. Further experimentation in that direction is probably more important than to establish detailed trap models on the basis of the presently available data. In particular, experiments such as ours should be extended over a wider temperature range, and properties, such as magnetoresistance, and possibly, luminescence, also should be observed simultaneously with the conductivity.

## ACKNOWLEDGMENT

The authors are indebted to many of their colleagues for stimulating or clarifying comments—in particular to Professor H. Kallmann, New York University, Professor K. W. Boer, University of Delaware, and Professor F. A. Kröger, University of Southern California. The help of Mr. H. Robinson, USAECOM, in numerically solving the differential equations with the Burroughs 5500 computer was invaluable.

## REFERENCES

1. Kröger, F. A., The Chemistry of Imperfect Crystals, North-Holland Publishing Co. (Amsterdam), 1964.
2. Neft, W., and K. Seiler, "Semiconductor Properties of Boron," this volume; see also W. Neft, dissertation, Technische Hochschule, Stuttgart, Germany, (1964).
3. Gaulé, G. K., J. T. Breslin, J. R. Pastore, and R. A. Shuttleworths, Boron—Synthesis, Structure, and Properties, Plenum Press (New York), 1960, p. 159.
4. Geist, D., "Electron Paramagnetic Resonance, Electrical Conductivity, and Impurity Diffusion in Doped Boron," this volume, p. 203.
5. Brungs, R., "Thermal and Optical Band Gaps of Monocrystalline Beta-Rhombohedral Boron," this volume, p. 119.
6. Dietz, W., and H. Herrmann, "Conductivity, Hall Effect, Optical Absorption, and Band Gap of Very Pure Boron," this volume, p. 107.
7. Birnbaum, M., and T. L. Stocker, "Recombination Radiation of Boron," this volume, p. 225.
8. Gebhart, F. L., and V. P. Jacobsmeyer, "The Optical and Electrical Constants of Beta-Rhombohedral Boron," this volume, p. 133.
9. Moss, T. S., Optical Properties of Semiconductors, Butterworths (London), 1959.
10. Hannay, N. B., Semiconductors, Reinhold Publ. Corp. (New York), 1959.
11. Becher, H., "Tetragonal Boron and Borides with Similar Structures," this volume, p. 89.
12. Bube, R., Photoconductivity of Solids, J. Wiley and Sons (New York), 1960.
13. Tauc, J., Photo and Thermoelectric Effects in Semiconductors, Pergamon Press (New York), 1962, p. 98.

## DISCUSSION

BILLIG: In your experiment I recognize two measurements. Is the transmission measured at the same time as the photoconductivity?

GAULÉ: Yes.

BILLIG: Do you use the same sort of light, or different wavelengths?

GAULÉ: We use two different kinds of light—(1) the exciting light, with a photon energy greater than the band energy ($\lambda < 0.8\,\mu$) and of high intensity, and (2) the measuring light with a wavelength of $1.5\,\mu$. The intensity of the measuring light is small, just sufficient to permit the measurement. Our transmission measurements are made after the exciting light is turned off. It is believed that the occupation of the traps is determined by the previous excitation with visible light, by the temperature, and by the time elapsed after the excitation and not by the measuring light (because of its small photon energy and intensity). As seen in Fig. 11, the transmission at $1.5\,\mu$ is most sensitive to the status of the traps.

BILLIG: For some reason the transition (causing the increased absorption near $1.5\,\mu$) must be quite high into the conduction band.

GAULÉ: This is an experimental result, as seen in Fig. 9, where we recognize a difference between the two transmittance curves, one corresponding to no excitation, and one corresponding to excitation. The effect is greatest for $1.5\,\mu$.

SEILER: Why do you have such a small effect? If the traps are so shallow, a wavelength of, for example, $4.7\,\mu$, should be as effective as $1.5\,\mu$.

GAULÉ: It appears possible that the absorbing process is determined by the density of the receiving states. There could be two overlapping bands favoring a process with the wavelength $\approx 1.5\,\mu$.

SEILER: Did you calculate the carrier density in the surface layer which becomes photoconducting? If I understand your Fig. 16 correctly, you have currents in the nanoampere range.

GAULÉ: Yes, indeed. The persistent photocurrent after typically 1-hr excitation is about 40 times the dark current. The latter is about $\frac{1}{100}$th the current at room temperature under similar conditions. The carrier concentration near the surface is $\approx 10^{11}$ cm$^{-3}$ before and $\approx 10^{15}$ cm$^{-3}$ after excitation, assuming $\frac{1}{100}$ of the total sample thickness is influenced.

SEILER: Do you need a strong light source

GAULÉ: Yes, but even so, we never approach saturation. I think this has to do with the optical situation. We use visible light for the excitation, for which boron is highly absorbing. The exciting photons are absorbed in a $10^{-4}$ cm thick layer near the surface. The new carriers which are generated by the photons presumably have a high probability to recombine at the surface and only a few will survive. Thus, the quantum efficiency is small and saturation not attainable even within a few hours.

HERRMANN: I want to make three points. We have also taken electrical glow curves, and we get a maximum near 130°K.

GAULÉ: The thermal stimulation curves which you saw in Fig. 16 start from 196°K. When we start from a lower temperature, for example, from nitrogen or helium temperature, we get maxima at approximately 110°K and even at lower temperatures.

HERRMANN: Could you not explain your excitation with polarons?

GAULÉ: I do not know if the characteristic energies in our experiments are compatible with a polaron model.

HERRMANN: In your mathematical model, you do not seem to account for the population of the recombination centers.

GAULÉ: We assume that the recombination centers are so numerous that they do not become saturated during the process. Conversely, we assume that the traps are not saturated by the initial excitation. We thus assume that the average retrapping time and the average recombination time do not change much during an isothermal process. This appears to be in agreement with the experiment.

HERRMANN: Did you apply varying intensities in your photoconductivity experiments?

GAULÉ: Yes, we did, but since for the reasons given above we never approached saturation, the results were qualitatively the same, except for very weak, or very short, illumination. In that case, the change in transmittance and the change in resistance vanish in the background, that is, in the dark properties.

# Electron Paramagnetic Resonance, Electrical Conductivity, and Impurity Diffusion in Doped Boron

## D. Geist *

*Physikalisches Institut*
*University of Cologne*
*Cologne, Germany*

Numerous impurities when present in large concentrations raise the electrical conductivity of pure boron at temperatures below the intrinsic range. Doped samples may be prepared by doping the melt. Iron and manganese are outstanding because of their high diffusivity, which enables them to enter solid boron below the melting point at a high rate. On the other hand, the electrically active element carbon does not diffuse faster than other elements (silicon, copper, gold, tungsten, and magnesium). Carbon and silicon, in contrast to germanium, cause electron paramagnetic resonance with a $g$-value about 2.003. A similar resonance is found in $B_{12}C_3$; in both cases, it is probably related to unsaturated bonds of a radical type induced by the foreign atoms.

## ELECTRICAL RESISTIVITY OF BORON

Boron has been known as an elemental semiconductor since early measurements at high temperatures by Weintraub [1] who prepared boron from the melt in an arc furnace between copper electrodes in the absence of carbon. By measurements of the low-temperature conductivity, differences in purity are easily detected (Fig. 1). The intrinsic conductivity begins above 200°C, as shown in Fig. 2. The activation energy $E_G$ is $1.4 \pm 0.1$ eV. The indicated error is the absolute error with due regard to possible systematic errors. For low-temperature measurements, safe contacts are indispensable. Alloyed gold contacts were found satisfactory with respect to mechanical and electrical properties.

## SURFACE BEHAVIOR AND HEAT TREATMENT

High-resistance semiconductors may always have surface layers with a conductivity higher than that of the bulk material.

*Present address: Institut für Angewandte Physik, Technische Hochschule Clausthal, Clausthal-Zellerfeld, Germany.

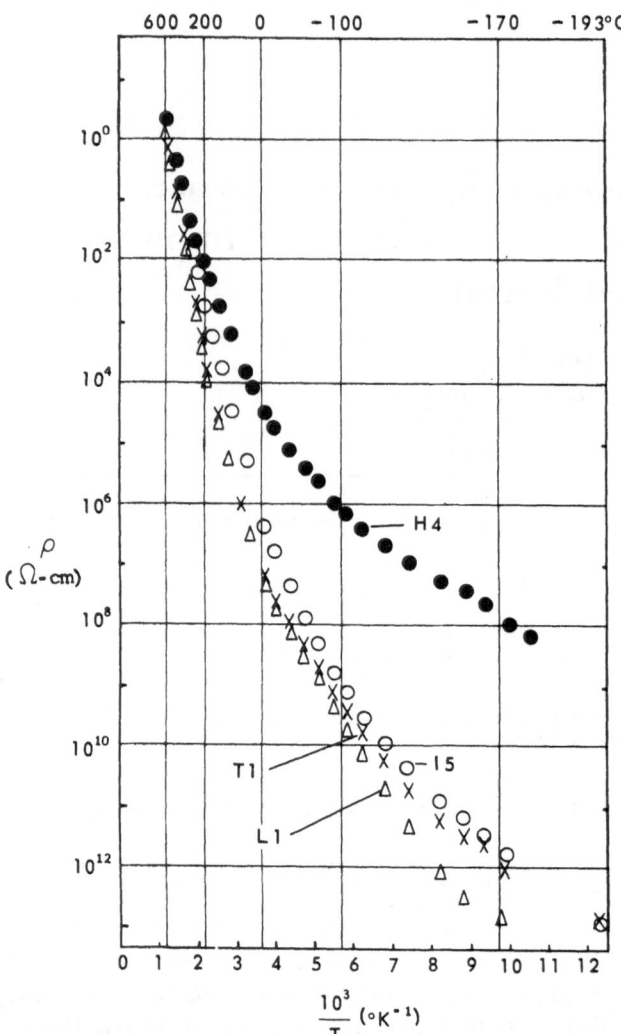

Fig. 1. Electrical resistivity of boron versus temperature. T 1, zone-refined (Wacker Co., Munich, Germany); L 1 (Light Co., Coelnbrock, Great Britain); I 5 and H 4 (Starck Co., Goslar, Germany).

Usually the properties of such layers depend on the ambient. Figure 3 shows that between 20–600°C the resistance of boron is independent of the ambient. Atmospheres of air, hydrogen, and nitrogen, or a vacuum of $10^{-3}$ mm Hg give the same values. A longer treatment (10 hr in air at 650°C) gives a decrease in resistance (Fig. 4). Etching with nitric acid brings back the starting value, which means that only a thin surface layer has been influenced. Hot air reacts with boron, as is seen from the etching effect in Fig. 5, which reveals the structure of the material.

## DIFFUSION OF FOREIGN ATOMS

Insofar as foreign atoms are effective as doping agents, their diffusion may be followed by electrical measurements [2]. Figure 6 shows measurements on the diffusion of manganese in boron. The experimental arrangement was so chosen that the diffusion problem

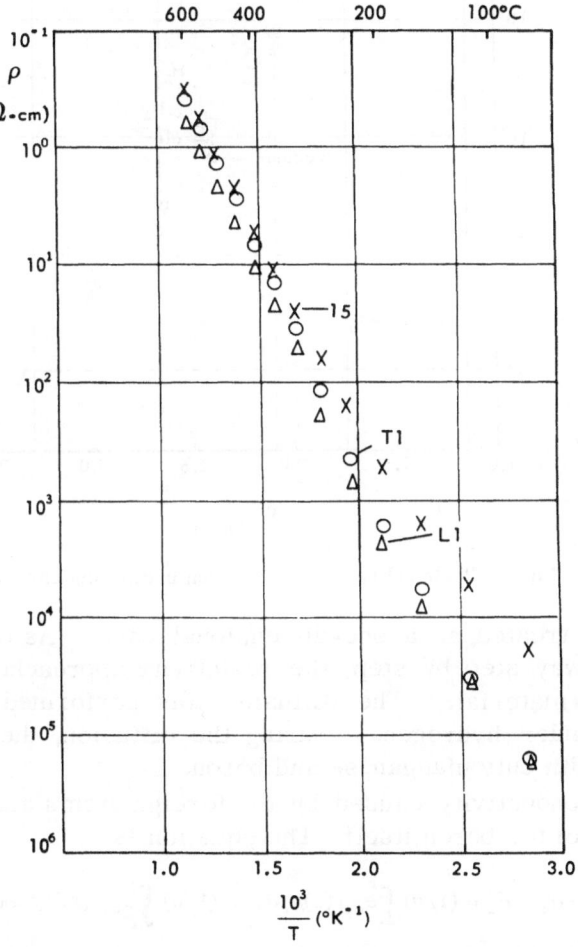

Fig. 2. The beginning of the intrinsic range of boron. (See Fig. 1 for notation.)

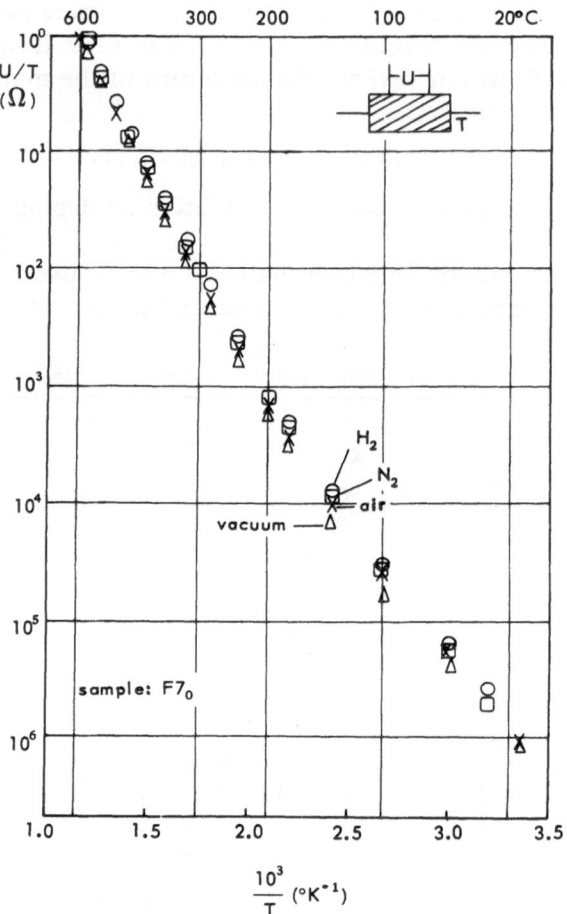

Fig. 3. Electrical resistivity of boron in different ambients.

could be treated as a one-dimensional case. As the material is ground away step by step, the resistivity approaches the value of the bulk material. The diffusion was performed for 15 min at 2035°C under hydrogen. During the diffusion, the boron was in contact with only manganese and boron.

The conductivity caused by the foreign atoms adds to the conductivity of the boron itself. The equation is

$$\sigma = \sigma_o + \sigma_z = (1/d) \int_x^b e_o \mu(x) n_o dx + (1/d) \int_x^b e_o \mu(x) N(x) a(x) dx \tag{1}$$

where the additional conductivity $\sigma_z$ is proportional to the total number of impurity atoms, $d = b - x$ is the sample thickness, $e_o$ is the elementary charge, $\mu$ is the mobility, $a$ is an empirical ionization factor, and $N$ is the concentration of the foreign (here, manganese) atoms. It was assumed for the calculations that $\mu$ and $a$ are constants and that the concentration of the foreign atoms follows a Gaussian distribution

$$N(x) = N_o \, (\pi Dt)^{-\frac{1}{2}} \, \exp \, (- \, x^2/4Dt) \tag{2}$$

Deviations from the computed conductance versus thickness de-

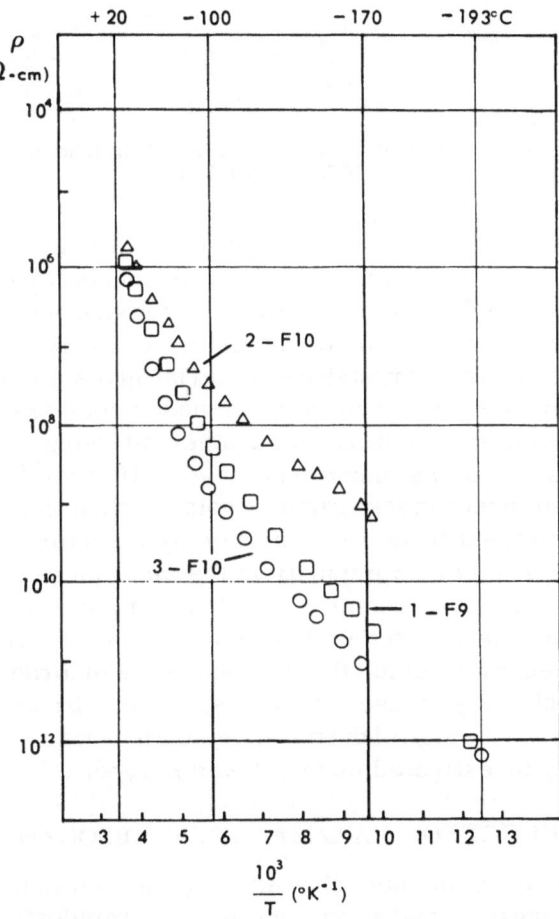

Fig. 4. Electrical resistivity of heat-treated boron (10 hr at 650°C in air). Curve 1, starting material before the heat treatment; curve 2, sample after the heat treatment, rinsed with deionized water; and curve 3, sample washed with boiling nitric acid.

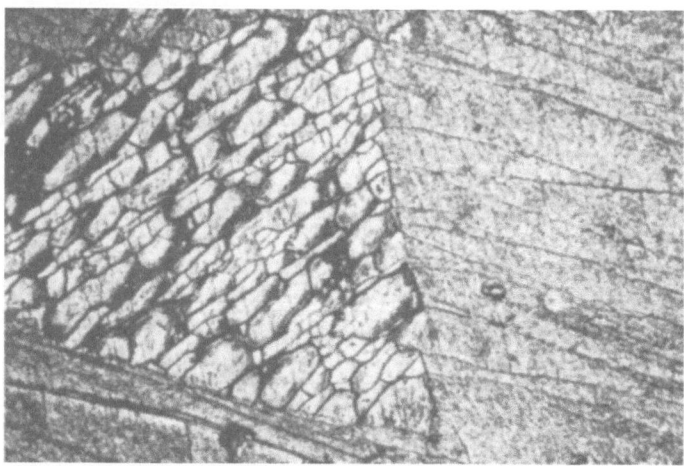

Fig. 5. Etched boron surface with structure visible (polished, hot-air etched, 10 min at 800°C, material T 1).

pendence were found only for the high-concentration range. For the evaluation of the diffusivity $D$, the low-concentration range was used. $D$ is $4 \cdot 10^{-6}$ cm$^2$/sec at 2035°C for manganese.

For fine crystalline material of the same type, this value is reproducible; coarse crystalline material gives a somewhat lower value. Figure 7 gives measurements under comparable conditions with fine crystalline material and other diffusing elements. Only iron diffuses as fast as manganese ($D = 5 \cdot 10^{-6}$ cm$^2$/sec at 2035°C). The other elements investigated — gold, carbon, silicon, tungsten, and magnesium—diffuse more slowly by a factor of at least ten. One may ask whether an impurity in the manganese or iron may be responsible for these results. Analysis of the results shows this not to be the case. Whether iron and manganese act as acceptors by themselves or whether they influence the disorder in the boron lattice, which may cause acceptors, cannot be answered. It is concluded from the sign of the thermoelectric power at low temperatures that all investigated material was $p$-type.

## ELECTRON PARAMAGNETIC RESONANCE

Alloying experiments showed carbon, silicon, germanium, beryllium, tungsten, and aluminum dissolve rapidly in molten boron. The strong conductivity increase caused by carbon is seen in Fig. 8 [3]. This material also was found to be $p$- type.

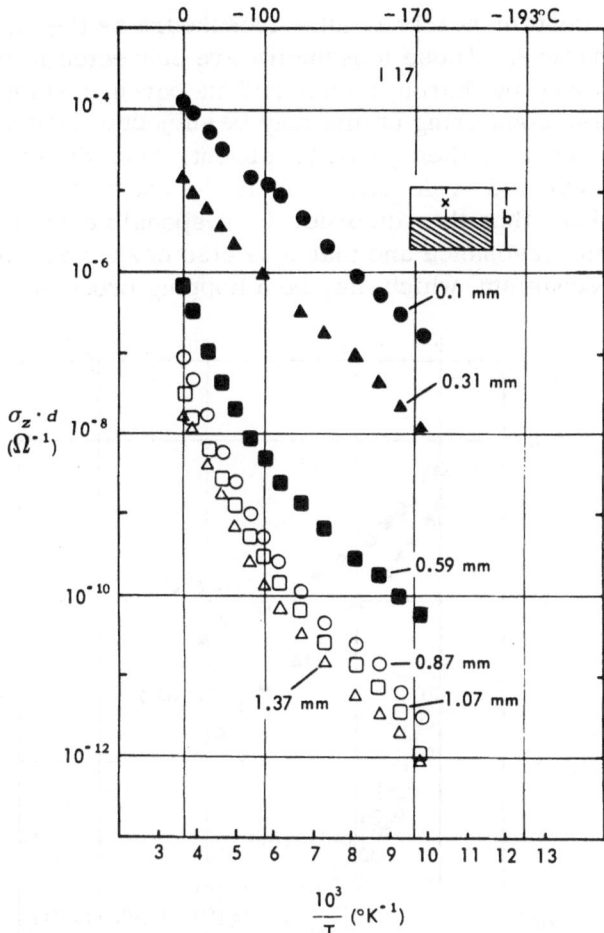

Fig. 6. Additional conductance $\sigma_z d$ of manganese-diffused boron, stepwise ground from the doped side. Diffusion 15 min at 2035°C under hydrogen. The thickness of the removed material, $x = b - d$, is indicated on the curves.

Electron paramagnetic resonance is often found in commercially available boron [4–6]. Figure 9 shows the resonance curves for 300, 90, and 4.2°K. The $g$-value is 2.003, which is nearly the free-spin value. Deliberate additions of carbon or silicon produce this resonance, as is seen for carbon and silicon in Fig. 10. There may be a close correspondence to the electron spin resonance in boron carbide. A single line with almost exactly the same $g$-value ($g = 2.0030 \pm 0.0002$ at 1.7°K) is found in this material [7]. Both

substances contain nearly regular icosahedra as the subunit in the
crystal structure. These icosahedra are connected in beta-rhombohedral boron by boron atoms and in boron carbide by carbon
atoms. These connecting atoms may be subject to disorder, and, in
particular, some of them may be absent. Such disorder in boron
carbide is revealed by deviations from the exact stoichiometry. It
is most likely that this disorder is responsible for the electron
paramagnetic resonance and that it is also connected with the conductivity mechanism, which may be a hopping process.

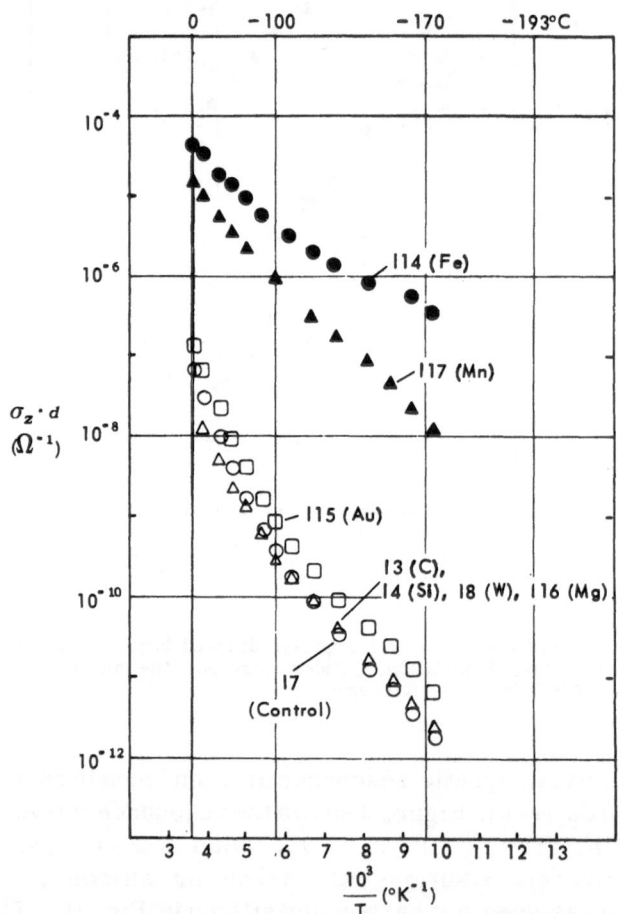

Fig. 7. Additional conductance $\sigma_z d$ of boron with diffused foreign atoms. Conditions are
the same as those for Fig. 6, except $b - d = 0.3$ mm.

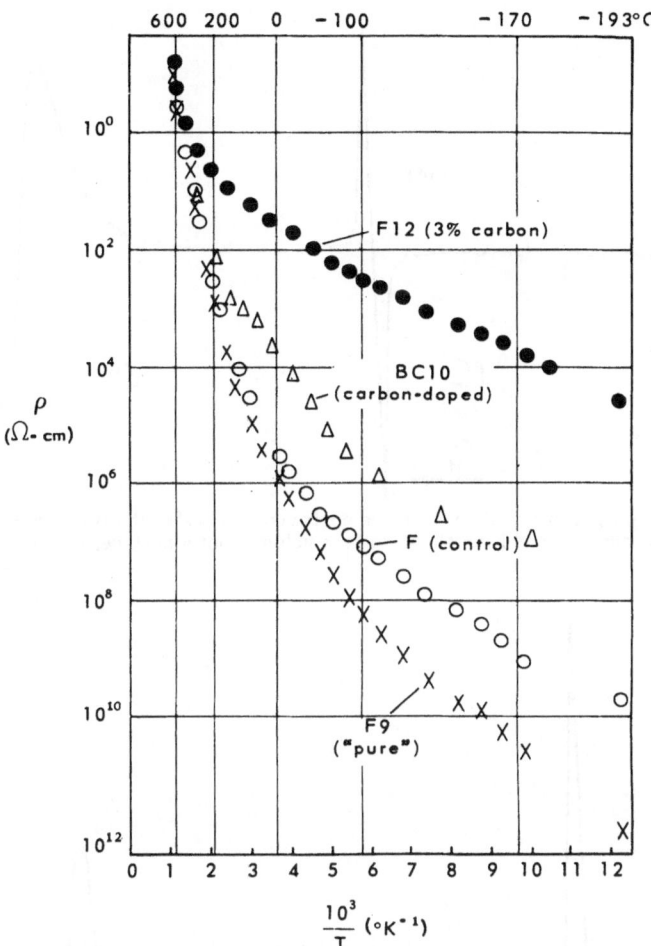

Fig. 8. Electrical resistivity of pure boron and of boron doped with carbon in the molten state. Compare these with the curve for boron molten without carbon.

## ACKNOWLEDGMENTS

The author wishes to thank Professor Jaumann of the University of Cologne, Cologne, Germany, who made this work possible. Grateful appreciation is expressed to the Consortium für Elektrochemische Industrie, which supplied some of the boron. Mr. Gläser helped with the resistance measurements and the diffusion experiments.

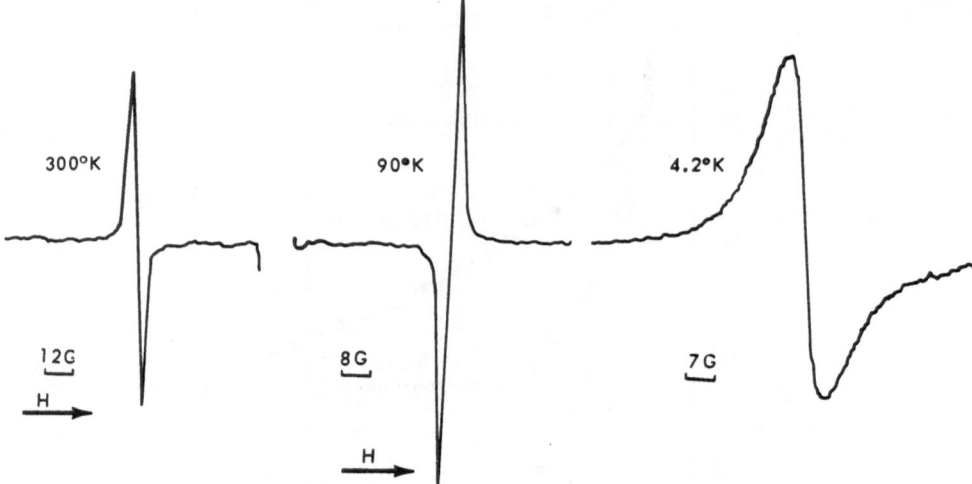

Fig. 9. Electron paramagnetic resonance in boron. At 4.2°K, the line width is greatest. The resonance had been shown to be a volume, not a surface, effect [4–6].

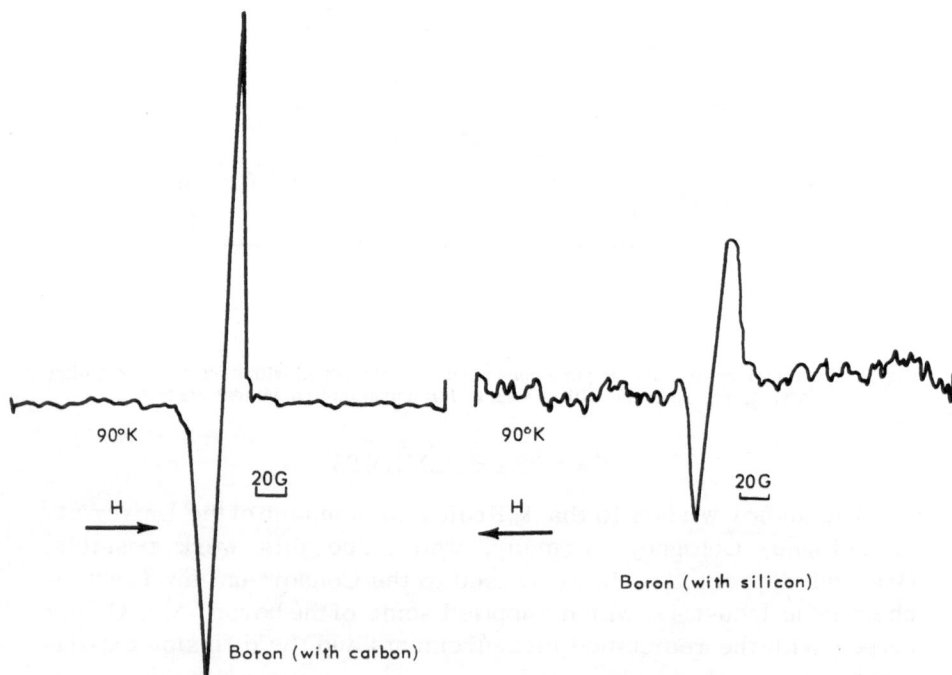

Fig. 10. Electron paramagnetic resonance in boron with carbon (left) or silicon (right) added.

## REFERENCES

1. Weintraub, E., J. Ind. Eng. Chem. 5:106,299 (1911).
2. Geist, D., and H. J. Gläser, Tagung, Bad Pyrmont, Germany, April 1963.
3. Hagenlocher, A., Ph. D. Dissertation, Technische Hochschule, Stuttgart, Germany, 1958; see also Hagenlocher, A., "Semiconductor Properties of Boron," in J. A. Kohn, W. F. Nye, and G. K. Gaulé (eds.): Boron—Synthesis, Structure, and Properties, Plenum Press (New York), 1960, pp. 128–134.
4. Geist, D., Phys. Soc. (London) Intern. Conf. Phys. Semicond. (Exeter), 1962, p. 633.
5. Geist, D., Festkörperprobleme II, Vieweg (Braunschweig), 1963, p. 93.
6. Geist, D., Phys. stat. solidi 5:217 (1964).
7. Geist, D., Conference on the physics of semiconductors, Dunod (Paris), July 1964.

## DISCUSSION

SEILER: In your formula, you used the concentration of the atoms in a linear term. That means that you suggest an excess conductivity and not a reserve conductivity. If you have a reserve conductivity, you have a dependence of the square root of the concentration of the impurities.

GEIST: This is true for otherwise pure material. For partly compensated material, which seems a realistic assumption here, a linear term mostly is adequate.

BILLIG: You have used purely additive conductivities and also a mobility independent of impurity concentration.

GEIST: This has been done to simplify the calculations, which give only the order of magnitude for the diffusion coefficient.

SEILER and BILLIG: It would be desirable to use other methods, for example, radioactive elements, to find out the diffusion depth.

GAULÉ: We noticed that the properties of boron at helium temperatures can be changed by light.* Was the sample kept in the dark in your spin resonance experiments?

GEIST: The sample was at the end of a closed waveguide about 80 cm long, immersed with its lower end in the liquid helium. The sample was, therefore, clearly in the dark. Only room-temperature radiation from the upper end of the waveguide was not strictly excluded.

SEILER: Have you an idea about the amount of impurity in your samples?

GEIST: Much lower than 1%.

*See Gaulé, Breslin, and Patty, "Trap-Dominated Electrical and Optical Effects in Crystalline Boron," this volume, p. 169.

BILLIG: Is the diffusion front even and not too rough to make your experiments ? Did you have statistical variations in impurity concentration ?

GEIST: The diffusion proceeds quite regularly and gives reliable and reproducible results.

SEILER: How large is the ionization factor, if you assume a constant mobility ?

GEIST: The value may be between 0.01 and 1.

# Magnetoresistance in Elemental Boron*

## W. P. Lonc, S. J.,† and V. P. Jacobsmeyer, S. J.

*St. Louis University*
*St. Louis, Missouri*

The magnetoresistance of crystalline, beta-rhombohedral boron and amorphous boron film was studied as a function of the magnetic field, the temperature, the specimen purity, and the crystal orientation. The monocrystals exhibited anomalous negative magnetoresistance for fields up to 5 kG, and a positive magnetoresistance for higher fields. The polycrystals showed positive magnetoresistance, with complete saturation for fields of 2 kG. The resistance did not exceed $3 \cdot 10^{-3}$ at room temperature, and depends upon the type of boron, the temperature, and the magnetic field. In the range from 200 to 400°K, $\Delta\rho/\rho$ is proportional to $T^{-5.6}$ for the polycrystalline samples, and to $T^{-3}$ for the monocrystalline. The magnetoresistance decreased with increasing specimen purity for the polycrystalline samples. The negative magnetoresistance in the monocrystals, the positive effect in the polycrystals, and the negative effect in the film are isotropic with respect to the angle between the directions of the electric current and the magnetic field. The positive effect in the single crystals, however, is anisotropic and may be described by the empirical formula $\Delta\rho/\rho = a + b \cos^2 \theta'$ where the coefficients depend on the specimen temperature, and $\theta'$ is the angle between the magnetic field and the [111] rhombohedral axis.

## INTRODUCTION

Reports of magnetoresistance measurements on high-resistivity elemental boron have suggested noticeable anomalous behavior, even at room temperatures. Using boron with a mosaic or poly-crystalline rhombohedral structure, Gaulé and co-workers [1] observed negative magnetoresistance, as well as an hysteresis effect. Carmichael and Dore [2], working with polycrystalline material of various impurity concentrations, also reported negative magnetoresistance, with saturation at approximately 3 kG.

In view of the interesting results reported by the above authors and in view of the occurrence of negative magnetoresistance in other materials over the past few years [3], an experiment was performed to measure magnetoresistance in single crystals, in

---

*A preliminary report of this work was presented at the Philadelphia Meeting of the American Physical Society, March 1964; Bull. Am. Phys. Soc. 9: 279 (1964).

†Present address: Regis College, Willowdale, Ontario, Canada.

polycrystals of various impurities, and in film. The magneto-resistance was measured as a function of magnetic field strength, ambient temperature, and orientation of the specimens with respect to the direction of the magnetic field. Results are given for dependence of magnetoresistance on fields up to 8 kG and for dependence on temperature from 100 to 450°K. Anisotropy in the magnetoresistance effect in the single crystals is shown to be a function of the angle between the [111] rhombohedral axis and the direction of the magnetic field.

## EXPERIMENTAL PROCEDURE

A two-probe method was used, with the boron specimen forming one arm of a DC bridge circuit. The resistivity of the boron is of the order of thousands of megohm-cms at lower temperatures, thus requiring a null detector with a very high input impedance. The two-electrode method was used due to the extreme difficulty of establishing low-noise contacts along the body of a specimen.

The electrodes consisted of small-diameter copper wire cemented to the end surfaces of the specimen by means of an epoxy-based conductive cement, Eccobond 58C. Following the recommended curing, the contact was found to be negligibly rectifying, sufficiently noise-free, and mechanically durable. The behavior of the contact was further tested by fitting a specimen of germanium with the same type of probes and measuring the magnetoresistance up to 8 kG. The results for germanium agreed with published values, allowing us to conclude that the cemented contacts were not introducing any spurious or anomalous effects. As a final check on the behavior of the cemented contacts, the resistance of the contacts was measured by a voltage divider technique which showed that the contact resistance for each contact was insignificant compared with the bulk resistance.

X-ray transmission patterns of the monocrystalline specimens revealed the orientation of the rhombohedral [111] axis with respect to the external geometry of the specimens. These specimens were consequently mounted so that for some runs the [111] axis would always be in the plane containing J and B, for all angles between J and B; for other runs, the [111] axis would always be perpendicular to the plane described above.

The resistance of boron depends quite strongly on temperature, and the magnetoresistance is small, thus requiring optimum temperature stability. In our experiments, it was not feasible to control temperature automatically; instead, it was more practical to surround the specimen with as much material as possible, resulting in a temperature drift slow enough to allow reliable magnetoresistance measurements.

The possibility of inhomogeneities in the magnetic field producing spurious or anomalous magnetoresistance effects, as discussed by Koppe and Bryan [4], was ruled out in our case by making a test run using a Varian magnet with a measured homogeneity of 1 part in 40,000. No measureable deviation from results obtained on a magnet with a measured homogeneity of 1 part in 500 was observed. A further possible source of spurious magnetoresistance, as discussed by Bate and Beer [5], arises from a combination of impurity-concentration gradient and point-probe placement. Since such a spurious effect should arise only in the case of point probes, we assume that our area contacts would mask any contribution from concentration gradients.

## SPECIMENS

The spectrographic data on the boron specimens are only approximate. Moreover, inability to measure the Hall effect prevented a calculation of the carrier concentration, which would have helped in the interpretation of the magnetoresistance data. However, the specifications supplied by the manufacturers of the various specimens serve as an indication of specimen quality.

The single crystals were cut from material [6] kindly supplied by G. K. Gaulé. They were prepared by a vacuum-fusion process, and are 99.9% pure, with major impurities reported to be silicon, carbon, iron, and manganese. An amorphous film, approximately 25,000 Å thick, was prepared by electron bombardment of some of the above material.

Four different types of polycrystalline boron were investigated in this experiment. Specimen No. 1, obtained from the Norton Co., Niagara Falls, Canada, was prepared [7] by reduction of boron trichloride in hydrogen atmosphere, the boron being deposited on hot graphite rods. The final product was found to be 97% boron and 2% carbon, with small amounts of iron and silicon. Specimen No. 2 was prepared in this laboratory by Friedrich [8] from pyrolysis of diborane gas. The purity of this boron sample is believed to be not higher that 99.99%. Specimen No.3, kindly supplied by J. Musgrave of the Eagle-Picher Co., Miami, Oklahoma, was prepared by vapor-phase method using $BBr_3$, with deposition on tantalum

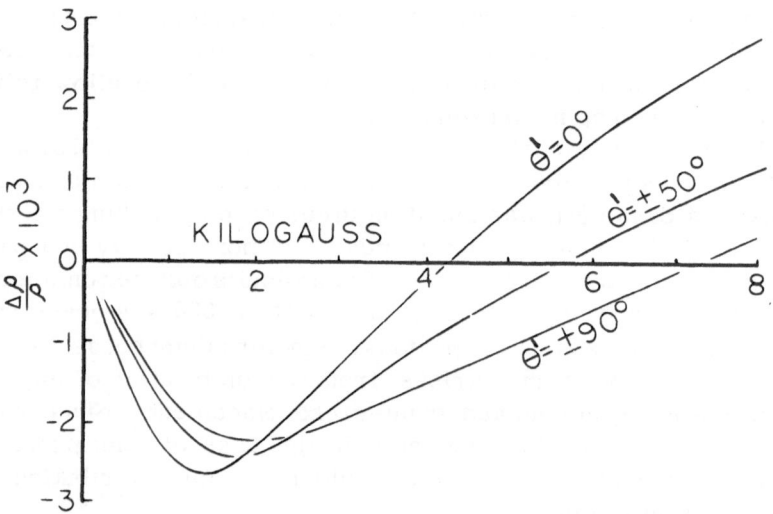

Fig. 1. Magnetoresistance as a function of magnetic field in monocrystalline boron. The [111] rhombohedral axis is in the plane of J and B; $\theta'$ is the angle between the [111] axis and B. Data taken at 300°K.

filaments [9]. The tantalum was removed, after which the boron rods were float-zoned. Spectrographic analysis indicated that the principal impurities of this specimen were 300 ppm carbon and 2 ppm silicon. Specimen No. 4 was obtained through the kind interest of R. Brungs from Hinz and Wirth [10] of Consortium für Elektrochemische Industrie, Munich, Germany. The material was prepared by zone-melting the product derived from pyrolysis of $B_2H_6$. Spectrographic analysis showed a carbon content of no more than 30 ppm, all other impurities being less than 1 ppm.

## EXPERIMENTAL RESULTS

Measurements were made of $(\rho - \rho_0)/\rho_0$ as a function of magnetic-field strength, of angle between magnetic field and current, and, finally, as a function of temperature. The quantity $\rho$ is the resistivity with applied magnetic field.

Figure 1 shows the magnetic-field dependence of the magneto-resistance effect in a boron single crystal. The behavior consists of a negative effect which is almost isotropic for orientation of the [111] rhombohedral axis with respect to the magnetic field, and a positive effect which is quite anisotropic. The isotropy of the negative effect and the anisotropy of the positive effect are shown in more detail in Fig. 2. When the [111] axis is perpendicular to the plane containing J and B, there is no anisotropy either in the negative or in the positive effects. However, when the [111] axis is in the plane containing J and B, then as the angle between J and B is varied, only a slight change in the magnitude of the negative effect is observed, but the positive effect is maximum at the point where the [111] axis is parallel to B. The two lower curves in Fig. 2 are the maximum negative excursions, occurring at approximately 2 kG. The two upper curves are the positive excursions at 8 kG.

The magnetic-field dependence of the polycrystals is shown in Fig. 3. In general, the polycrystalline specimens exhibit a rapid, positive increase in magnetoresistance for fields up to 2 kG. In specimen No. 3, the magnetoresistance was just observable above the noise, but, in No. 4, magnetoresistance was not greater than the background-noise level ( $\Delta\rho/\rho = 1 \cdot 10^{-4}$ ). It is noted that the magnetoresistance saturates for magnetic fields higher than 2 kG. In all

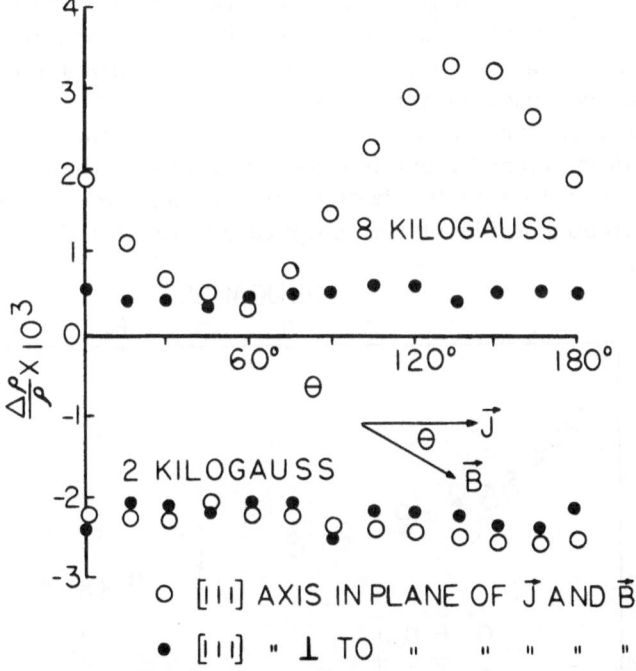

Fig. 2. Anisotropy of magnetoresistance as function of angle $\theta$ between current vector and magnetic-induction vector for boron monocrystals. Curves are typical for all monocrystal specimens. Data taken at 300°K. The anisotropy for fields above 5 kG can be approximated by the empirical formula $\Delta \rho/\rho = a + b \cos^2 \theta'$, where $a = 0.3 \cdot 10^{-3}$, $b = 2.7 \cdot 10^{-3}$, and $\theta'$ is the angle between the [111] axis and B.

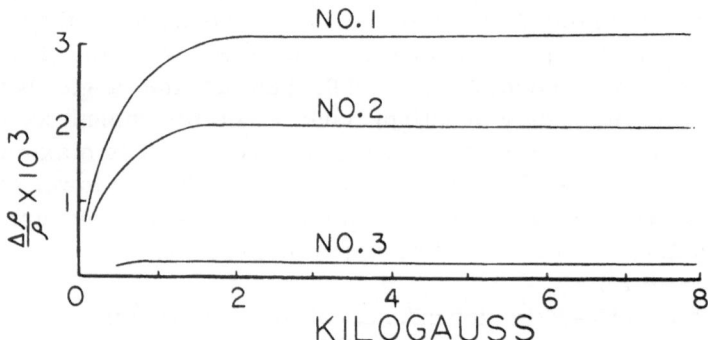

Fig. 3. Change in magnetoresistance as a function of magnetic induction for several boron polycrystals of varying purity. Curves are practically insensitive to change of angle between current vector and magnetic-induction vector. Data taken at 300°K.

specimens, the magnetoresistance was found to be independent of the angle between the current and magnetic-field directions.

Figure 4 shows the magnetic-field dependence of the magneto-resistance in the boron film, both for transverse and longitudinal orientations. For magnetic fields up to 8 kG, the magnetoresistance was entirely negative, and there was no significant difference between the two different orientations.

Temperature dependence of the magnetoresistance in the single crystals was measured from 500 down to 100°K; at these extremes, the signal was lost in the background noise. No measurements were attempted outside of the temperature range indicated above.

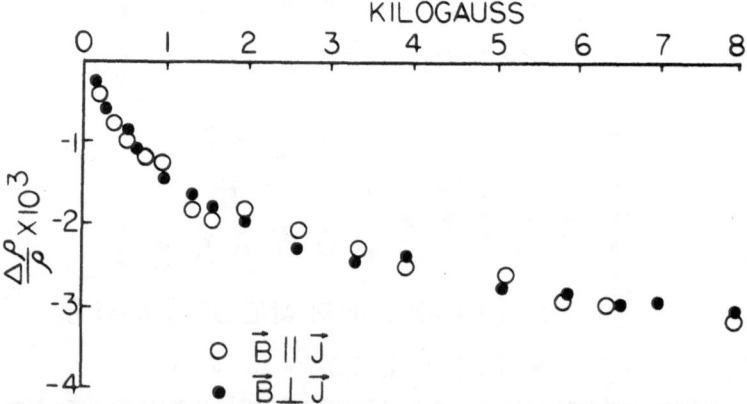

Fig. 4. Change in magnetoresistance as a function of magnetic induction for typical, thick boron film. Curve is insensitive to change in angle between current vector and magnetic-induction vector. Data taken at 300°K. The open circles represent the longi-tudinal magnetoresistance, and the darkened circles represent the transverse effect.

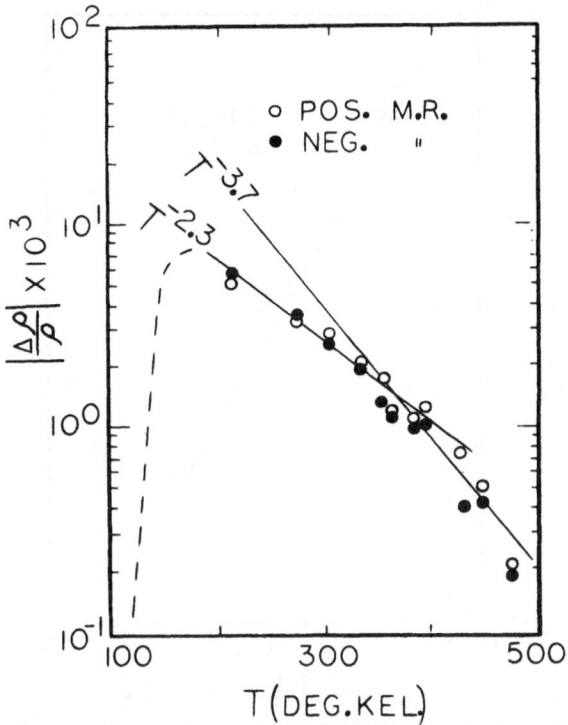

Fig. 5. Magnetoresistance as a function of temperature for boron monocrystals. The curves are interpolated between 100 and 200°K. The open circles represent the positive magnetoresistance at 8 kG, whereas the darkened circles represent the maximum negative magnetoresistance, always occurring at approximately 2 kG. The [111] axis is parallel to the magnetic field.

The results are plotted in Fig. 5, and indicate a power-law behavior of approximately $T^{-3}$ for temperatures between 200 and 450°K. There is a striking similarity in the temperature dependence between the negative and positive magnetoresistance effects. In the polycrystals, temperature dependence of the magnetoresistance is found to be stronger than in the crystals. Figure 6 indicates a power-law dependence of $T^{-5.6}$ in the temperature range 200 – 400°K. Excessive background noise prevented temperature-dependence measurements for polycrystals Nos. 3 and 4.

## ACKNOWLEDGMENTS

The authors are pleased to acknowledge the encouragement of Dr. W. A. Barker. Helpful assistance was kindly given by Dr. H. Roth of Raytheon and Dr. J. Mulhern, Jr., of the University of New

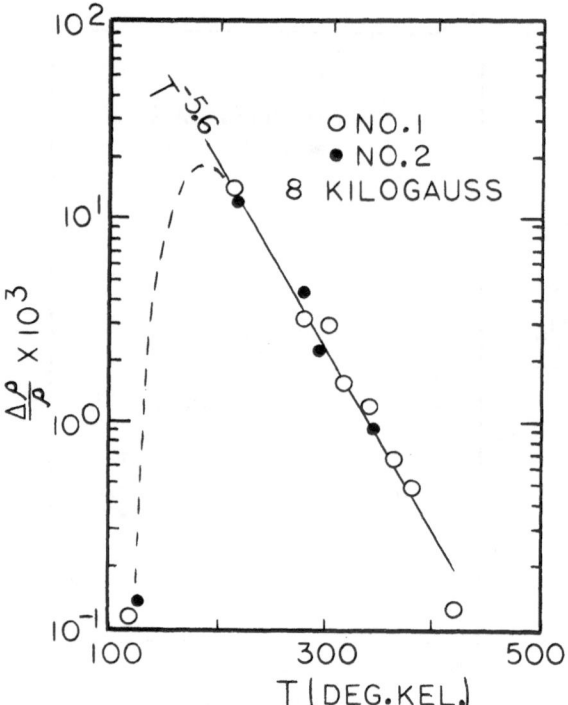

Fig. 6. Magnetoresistance as a function of temperature for the boron polycrystals Nos. 1 and 2. Curves are interpolated between 100 and 200°K.

Hampshire, Durham, New Hampshire. Specimens were graciously supplied by Dr. J. Levine of Raytheon, Mr. G. K. Gaulé of USAEL at Ft. Monmouth, New Jersey, Dr. Musgrave of the Eagle-Picher Co., Miami, Oklahoma, and Dr. R. Brungs of Woodstock College, Woodstock, Maryland. Assistance in X-ray work was generously given by Dr. G. Carron and Mr. Koenig at McDonnell Aircraft. The authors are also indebted to G. Ernst and M. Ashmont for expert shopwork.

## REFERENCES

1. Gaulé, G. K., J. T. Breslin, J. R. Pastore, and R. A. Shuttleworth, in J. A. Kohn, W. F. Nye, and G. K. Gaulé (eds): Boron—Synthesis, Structure, and Properties, Plenum Press (New York), 1960, pp. 159–174.
2. Carmichael, C. H., and M. Dore, Nature 191:485 (1961).
3. Woods, J. F., and C. Y. Chen, Phys. Rev. 135A:1422 (1964).
4. Koppe, H., and J. M. Bryan, Can. J. Phys. 29:274 (1951).
5. Bate, R. T., and A. C. Beer, Proc. Intern. Conf. Semicond. Phys. (Prague 1960), Czechoslovak. Akad. of Sciences (Prague), 1961, pp. 177–181.

6. Yannacakis, J., and N. P. Nies, "Preparation of crystalline boron," in J. A. Kohn, W. F. Nye, and G. K. Gaulé (eds.): Boron—Synthesis, Structure, and Properties, Plenum Press (New York), 1960, pp. 38–41.
7. Fetterly, G. H., "The manufacture of boron," in J. A. Kohn, W. F. Nye, and G. K. Gaulé (eds.): Boron—Synthesis, Structure, and Properties, Plenum Press (New York), 1960, pp. 15–26.
8. Friedrich, L. W., unpublished Ph. D. thesis, St. Louis University, 1953.
9. Bean, K. E., and W. E. Medcalf, "Utilization of boron filaments in vapor-phase deposition of boron," in J. A. Kohn, W. F. Nye, and G. K. Gaulé (eds.): Boron—Synthesis, Structure, and Properties, Plenum Press (New York), 1960, pp. 48–58.
10. Hinz, I., and H. Wirth, reported at the International Colloquium on Hyperpure Boron, Paris, July 1964. See also "Purity of Boron Produced by the Decomposition of Diborane and Subsequent Zone-Melting," this volume, p. 9.

# Recombination Radiation of Boron

## Milton Birnbaum and Tom L. Stocker

*Aerospace Corp.*
*El Segundo, California*

The band gap fluorescence of boron has been observed for the first time using Czochralski-grown, high-purity samples. A 30-mW, cw, helium—neon laser was used to excite the fluorescence and a high-speed, quartz-prism spectrograph was used to record the emitted radiation. The peak intensity of the luminescence (recombination radiation) at 80°K was found to occur at 1.55 ± 0.03 eV.

## INTRODUCTION

The optical and electrical properties of boron have been the subject of a number of recent papers. These data have provided estimates of the band gap of boron. At room temperature, values ranging from 1.27 to 1.51 eV have been reported [1–7].

Observation of band gap recombination radiation from boron would provide an accurate measurement of the band gap energy and at the same time provide a determination of the nature of the transition, namely, direct or indirect. In this paper, the first observations of the recombination radiation of boron are described.

## EXPERIMENTAL METHOD

A diagram of the experimental apparatus used to observe the recombination radiation is shown in Fig. 1. The major experimental problem encountered was that of detecting the very weak signals. In order to detect the recombination radiation, use was made of a prism spectrograph* of high speed ($f = 4.5$) together with photographic recording. A strong source was used to excite the fluorescence, namely, a 30-mW, cw, helium—neon laser† operating at 6328 Å. A narrow band filter‡ centered at 650 m$\mu$ was used to

*Model L-234-150 quartz monochromator equipped with an L-234-150A camera attachment, manufactured by the Gaertner Scientific Corp., Chicago, Illinois.
†Model 116 He-Ne laser manufactured by Spectra-Physics, Inc., Palo Alto, California.
‡Manufactured by Optics Technology, Belmont, California. Measurements showed this filter possessed an optical density greater than 2 in the spectral range 7000–9000 Å.

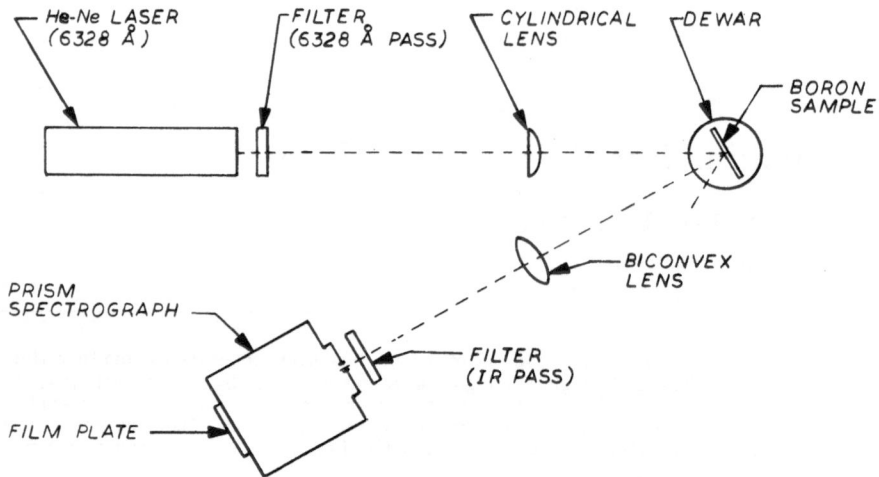

Fig. 1. Apparatus used for determination of recombination radiation of boron.

absorb the near infrared light emitted by the helium–neon discharge. Two filters* that absorb in the visible region but transmit infrared were used to exclude all visible light from the spectrograph.

In order to make the most efficient use of the laser light, it was brought to a fine focus by means of a planar cylindrical lens of 20-cm focal length. The recombination radiation is focused upon the entrance slit of the spectrograph by means of a double-convex lens of 9-cm focal length and *f*-number of approximately one. The alignment and focus of the optical system were easily checked with the aid of the small amount of 6328-Å light scattered by the damaged area of the semiconductor, which was also brought to a focus on the slit. When the scattered visible light was brought to a fine focus on the slit, the width of the image was approximately $300\,\mu$ and the height about 3 mm. After the sample was attached to the cold finger and properly positioned in the glass dewar, the image at the spectrometer entrance slit was approximately 1 mm in width. The diffuseness of the image when the sample was positioned in the dewar indicated the degradation in the optical performance introduced by passage of the light through the double-walled glass cylinder of the dewar. The specimens were cooled by attachment to a cold finger in contact with liquid nitrogen. Thus, liquid coolants were not in the optical paths (Fig. 1).

*Corning glass filters, type 7-69, manufactured by Corning Glass Corp., Corning, New York.

Because of the high dielectric constant of most semiconductors, a major fraction of the recombination radiation will be trapped by total internal reflection at the semiconductor surface. In the case of a ground or roughened surface, much more of the recombination radiation can be emitted from the sample. The semiconductor surfaces were initially polished to optical flatness, and two methods were used to roughen a small area of the semiconductor surfaces—scratching a fine line with a scribe and melting and spallation. The surface damage in the latter case is produced by focusing the output of a 5 J ruby laser on the semiconductor surface using the cylindrical lens shown in Fig. 1. This produced a damaged area in the form of a fine line. The helium–neon laser light is focused onto the scratch by means of the cylindrical lens (Fig. 1).

Kodak spectrographic plates, type 1-N, were used to detect the recombination radiation because of their high sensitivity in the near infrared. The plates were developed in Kodak D-19 following a standard procedure.

In order to calibrate the spectrograph, a series of calibration plates were prepared with the following lines (in Å): helium–neon laser, 6328; ruby laser, 6943; GaAs $p-n$ junction diode at 300°K, 9040; and GaAs $p-n$ junction diode at 77°K, 8440.* Spectrograph slit widths of 50–200 $\mu$ were used. The spectrograph has a low dispersion in the near infrared, namely, about 900 Å/mm at 8000 Å, thus necessitating use of slit widths not larger than approximately 100 $\mu$ in order to resolve the lines from the GaAs diodes. The wavelengths emitted from the GaAs junction diodes were determined with a recording infrared spectrophotometer[†] to a precision of about ± 10 Å. The lines from the GaAs diodes were used to determine the dispersion of the spectrograph in the spectral range very close to the boron recombination radiation.

Measurement of the separation of the lines was accomplished by scanning the plates with a recording microdensitometer.[‡] A densitometer slit of 700-$\mu$ length and 50-$\mu$ width was used. The reproducibility in the determination of the separation of the lines was about 10 Å in the near infrared.

Only a single reference line (6328) was put on the plates after

*Calibrated GaAs $p-n$ junction diodes were loaned to us by L. W. Aukerman and M. F. Millea, Electronics Lab., Aerospace Corp., El Segundo, California.

†Infrared monochromator, Mod. 99, manufactured by Perkin-Elmer Corp., Norwalk, Connecticut.

‡Spectroline microdensitometer, Model 22000, manufactured by the Applied Physics Corp., Monrovia, California.

exposure for the recombination radiation. The distance between the recombination radiation line and the 6328 line was compared with the distance between the 6328 and 8440 lines on the calibration plates.

In the case of the boron sample at 80°K, a slit width of 200 $\mu$ and a 14-hr exposure resulted in a well-defined line. Under similar conditions, a 6-hr exposure resulted in a barely visible line. The output of the helium—neon laser after passage through the band pass filter was approximately 20 mW. Under identical conditions, with the exception that the boron sample was at 300°K (liquid nitrogen eliminated from the dewar), no blackening was observed on the plates after a 16-hr exposure. These experiments were repeated a number of times to verify the consistency of the results.

Additional tests were performed to eliminate the possibility that a source other than boron recombination radiation was responsible for the plate blackening. A silicon sample was substituted for the boron sample in the dewar and no plate blackening was observed after a 24-hr exposure. With a 48-hr exposure under similar conditions, using a boron sample at 300°K (300-$\mu$ slit width), a barely visible line in the spectral region corresponding to boron recombination radiation was observed. A 62-hr exposure under similar conditions using silicon produced no blackening on the plate. Thus, the signal observed using boron at 300°K is attributed to recombination radiation.

In order to verify the entire measurement procedure, the wavelength of the peak in the recombination radiation of an $n$-type GaAs sample was determined at 80 and 300°K. At 300°K, an exposure time of 2 hr (200-$\mu$ slit width) resulted in an adequate plate density of the GaAs recombination radiation. This is to be contrasted with the 1-min exposure (50-$\mu$ slit width) required for an adequate negative density at 80°K.

## EXPERIMENTAL RESULTS

The boron crystal (99.9995% purity) was prepared by the Czochralski technique in the form of a rod approximately 2.5 cm long by 0.5 cm in diameter and was polycrystalline.* The ingot was sliced at an angle to the rod axis to provide slabs approximately 1 mm thick with a larger surface area (long dimension approximately twice the short dimension). A ruby laser was used to pro-

*Supplied by the Eagle-Picher Co., Miami, Oklahoma.

vide a roughened area in the shape of a fine line. At 80°K, the peak intensity in the recombination radiation was found to occur at $1.55 \pm 0.03$ eV. A very crude estimate of the half-width indicated a value of about 0.1 eV.

The recombination radiation emitted by an $n$-type slab of GaAs with a donor concentration (tellurium) of $3 \cdot 10^{18}$ cm$^{-3}$ was determined at 80 and 300°K. The sample was polished to a mirror finish and a roughened line area was obtained by scratching the surface with a scribe. At 300°K, the peak was found to occur at $1.44 \pm 0.03$ eV. At 80°K, the peak was found to occur at $1.51 \pm 0.03$ eV. These results for GaAs are in exact agreement with the values reported by Hill [8] for a similarly doped $n$-type sample of GaAs.

It is to be noted that the peak in the luminescence of boron at 1.55 eV occurs in the region of strong absorption, near the point of maximum slope in the absorption curve. Thus, the emission peak at 1.55 eV may correspond to a direct transition. More detailed studies of the luminescence of boron are in progress. The results of this study should be considered as a progress report until the more detailed measurements are completed.

## ACKNOWLEDGMENTS

We are indebted to L. W. Aukerman and M. F. Millea for helpful discussions and C. L. Fincher for valuable assistance in performing the experiments.

## REFERENCES

1. Uno, R., T. Trie, S. Yoshida, and K. Shinohara, J. Sci. Research Inst. (Tokyo) 47:216 (1953).
2. Lagrenaudie, J., J. Phys. Radium 14: 14 (1953).
3. Grenier, E. S., and J. A. Gutowski, J. Appl. Phys. 28: 1364 (1957).
4. Morita, N., J. Sci. Research Inst. (Tokyo) 48: 8 (1954).
5. Brungs, R. A., and V. P. Jacobsmeyer, J. Phys. Chem. Solids 25: 701 (1964).
6. Gebhart, F. L., and V. P. Jacobsmeyer, "The Optical and Electrical Constants of Beta-Rhombohedral Boron," this volume, p. 133.
7. Dietz, W. H., and H. Herrmann, "Conductivity, Hall Effect, Optical Absorption, and Band Gap of Very Pure Boron," this volume, p. 107.
8. Hill, D. E., Phys. Rev. 133: A866 (1964).

# On the Birefringence of Simple Rhombohedral Boron

F. H. Horn, E. A. Taft, and D. W. Oliver

*General Electric Research Laboratory*
*Schenectady, New York*

The results of an investigation of the birefringence of small rhombohedral crystals of $\alpha$-boron are reported. The method of preparing $\alpha$-boron crystals from a platinum melt is also presented.

## INTRODUCTION

The probable electronic and optical properties of simple $\alpha$-rhombohedral boron pose some interesting problems. The work of Kasper and Decker [1] indicates that, for the first time, we have an element whose crystal structure results from icosahedra rather than from atoms of boron at lattice sites. In $\alpha$-rhombohedral boron, as distinguished from the high-temperature $\beta$-rhombohedral boron, the structure is based on icosahedra only, which are in a closely packed arrangement. A model of the unit cell is shown on the left in Fig. 1. The rod has been placed in the c-direction of the rhombohedral unit cell. Bonds between the atoms of adjacent icosahedra in the c-direction may be considered essentially electron-pair or covalent bonds. In the model on the right, a section of a plane of icosahedra perpendicular to the c-axis is shown. In such planes, the conjunction of icosahedra results in a triangle of boron atoms. This bonding of icosahedra in planes perpendicular to the c-direction has been called delta or three-centered bonding, and has been described as formed from two electrons resonating between three positions. The $\alpha$-rhombohedral boron is thus not only anisotropic, but also the type of bonding differs in directions parallel or perpendicular to the rhombohedral c-axis. The electronic properties of $\alpha$-boron should reflect these structural differences.

231

Fig. 1. Model of α-boron structure. Metal rod has been placed in rhombohedral $c$-axis.

## PREPARATION OF α-BORON CRYSTALS

Our investigation is limited to optical studies, but before describing these, a word needs to be said about the α-boron crystals employed.

We have continued our previously reported studies on the crystallization of α-boron from high-purity platinum melts [2] in which we used our limited supply of zone-refined boron. When this boron was consumed, we found we were no longer able to form α-boron by recrystallization from platinum. We used commercial high-purity boron, some of which was prepared by floating zone. A nearly black coating on the outside was common for ingots recrystallized with commercial boron. Because we suspected that this coating was carbon which would freeze first since in boron it is an impurity for which the segregation coefficient is greater than 1 [3], we removed the coating by grinding about $\frac{1}{32}$ in. from the ingot surface. The cleaned ingot was then remelted and allowed to cool slowly from 1200°C to about 800°C, thence rapidly. Red crystals were then readily observed on the top surface of the ingot. Upon fracturing the ingot, they also could be seen, with reflected polarized light, as crystals embedded particularly in the upper portion of the platinum matrix. The crystals were recovered by electrolyzing away the platinum in concentrated hydrochloric acid. The crystals were recovered from the bottom of the beaker after repeated washings and decanting.

It appears that for the crystallization of α-boron from platinum, a pure source of boron is necessary. This also applies to the formation of α-boron by pyrolysis. We had previously estimated that our zone-refined boron contained on the order of $10^{18}$ carriers per cm$^3$. It may be that specific impurities, such as carbon, are particularly active in altering the course of $a$-boron formation.

## EXPERIMENTAL RESULTS

Crystals recovered from platinum melts are very small; typically, the largest-sized crystals have a maximum dimension of about 0.5 mm. The habit of the crystals recovered from the surface of a melt is well-formed rhombuses characteristic of the unit cell. These crystals, typical 100 $\mu$ long and 30$\mu$ thick, could be easily oriented optically and were, therefore, used in the present studies.

In crossed polarized light, the crystals cause extinction every 90° when rotated about the rhombohedral c-axis. The crystals are, thus, optically anisotropic, as expected. Crystals observed with transmitted white light appear yellow when the plane of polarization of the light is perpendicular to the c-direction of the crystal; an orange-red appearance is observed with the plane of polarization parallel to the c-direction. Thus, the crystal is not only anisotropic, but also dichroic.

The optical properties have been investigated more quantitatively by measuring the optical transmission using a quartz-prism monochromator with a polarizer at the input slit. These data have been converted into units of optical absorption (cm$^{-1}$) using the measured thicknesses of the crystals.

The optical absorption as a function of energy is shown in Fig. 2 for light with the electric vector parallel to, and perpendicular to, the c-axis. We see that for an absorption constant of approx. 500 cm$^{-1}$ the two curves differ by 0.1 eV or more, and this explains the observed dichroism. The slope of the absorption curve at lower photon energies is less with light polarized perpendicular to the c-axis than with light parallel to the axis. The gradual slope of the absorption curve suggests that the electronic-band structure of $a$-boron in the direction perpendicular to the c-axis is complex. This is in the direction of the three-centered bond, which has not been previously encountered in semiconductors. In the direction parallel to the c-axis, the absorption curve is sharper and actually steepens at lower wavenumbers. Both curves level off at about 1000 cm$^{-1}$, considerably short of the absorption constant (which is two to three decades larger) usually characteristic of intrinsic semiconductors. We can only speculate that the absorption may be limited by impurities or that an absorption band at a higher energy

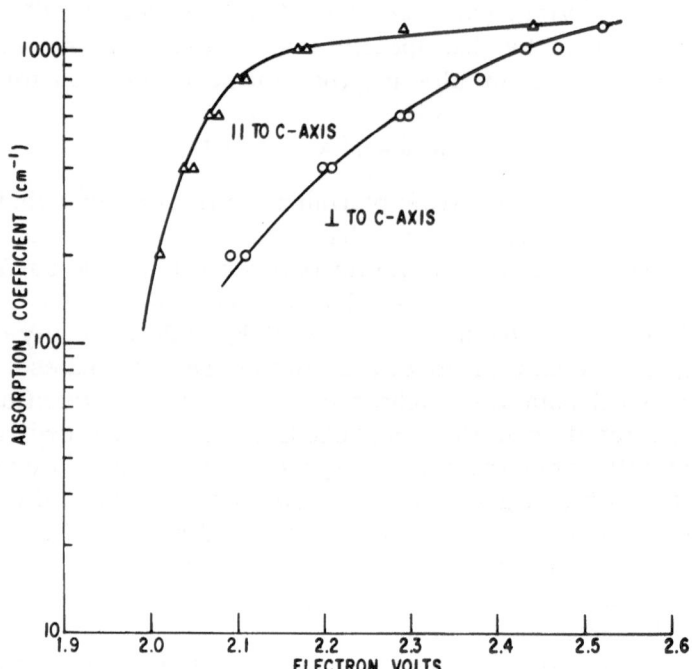

Fig. 2. Optical absorption coefficient for a-boron with light polarized parallel to and perpendicular to the rhombohedral c-axis.

will be revealed when very thin specimens are studied. More data are also necessary for the lower energies before one may hazard a guess on plausible mechanisms for optical absorption.

## CONCLUSION

Although early reports treated a-boron as isotropic, later investigations of the optical properties of a-boron qualitatively reflect its true structure; i.e., it is optically anisotropic, it is dichroic, and the optical absorption indicates different possible mechanisms for light absorption due to the different bonding in different directions. It is hoped that somewhat larger crystals of a-boron will make it possible to further the optical studies and to determine the electron-transport properties of this scientifically interesting semiconductor.

## REFERENCES

1. Decker, B., and J. Kasper, Acta Cryst. 12: 503–506 (1959).
2. Horn, F. H., J. Electrochem. Soc. 106: 905 (1959).
3. Horn, F. H., J. Appl. Phys. 30: 1612–1613 (1959).

# Vacuum-Deposited Amorphous Boron Films

Charles Feldman, Fred Ordway, and
William Zimmerman III

*Physical Electronics Laboratory*
*Melpar, Inc.*
*Falls Church, Virginia*

and Kishin Moorjani

*Catholic University of America*
*Washington, D. C.*

The properties of vacuum-deposited thin boron films are described. The layers, formed on fused silica substrates by electron bombardment, remain amorphous in structure up to approximately 900°C. Reaction with electrodes may occur, however, before this temperature is attained. Optical measurements indicate a minimum band gap of about 1.32 eV and a possible direct transition at 2.08 eV. Electrical behavior of the layers is a function of field, temperature, and thickness. Nonlinear conductivity is observed under high field ($>10^4$ V/cm) conditions. This nonlinear behavior is believed to be due to impact ionization of acceptor states situated about 0.1 eV above the valence band. An empirical expression describing the electrical properties is formulated.

## INTRODUCTION

This report describes the results of a limited study, extending over a period of several years, on the properties of amorphous boron films. The effort is part of a larger program dealing with materials and effects which may be useful in thin-film monotronic* circuits. The significance of amorphous layers arises from their ease of fabrication and their unique electrical properties. Amorphous films of a wide variety of materials may be formed by vacuum-deposition techniques. Boron, because of its semiconducting properties, high melting point, low density, and ability to form a hard, durable film, is of particular concern in this respect [1,2]. Theoretically, interest

---

*Monotronics—a word coined to represent electronics completely fabricated by a single technique (vacuum deposition).

in amorphous boron arises from the lack of long-range order and its effect on the properties of the films [3].

The two prime techniques for forming boron films are vapor-phase decomposition [4,5] and electron-bombardment vacuum deposition [6]. Vapor-phase techniques appear most useful for forming crystalline layers, since the substrates must be held at elevated temperatures (> 800°C), which encourages crystallization, to effect the necessary decomposition or reaction. Vacuum deposition, on the other hand, may be carried out at low substrate temperatures resulting in continuous amorphous deposits. The low substrate temperature also impedes any possible reaction with the substrate. The requirement for pure bulk boron as the source material in the electron-bombardment deposition technique has until recently, however, hindered studies of pure amorphous layers.

The research described here is by no means complete. There exist many unanswered questions in every phase of the program. The information presented on the structural, optical, and electrical properties of the films is thus primarily a guide for future work.

## SAMPLE PREPARATION

The films were deposited in an 18-in. diameter stainless steel belljar, evacuated by an oil diffusion pumping system. The system was equipped with appropriate mechanical and diffusion pump cold traps. Pressure during deposition was approximately $2 \cdot 10^{-5}$ torr. No material other than boron was deposited in the unit during the study. Fused silica substrates were used throughout the examination of electrical properties. (Other types of substrates were used as required for structural studies.) The fused silica substrates were outgassed at 500°C and allowed to cool to room temperature before the deposition was carried out. As stated in the introduction, deposition took place at low substrate temperatures to ensure obtaining amorphous films. This temperature is not critical, since amorphous boron films appear to be formed at substrate temperatures as high as 800°C. By way of comparison, germanium films in a similar vacuum system begin to crystallize at about 350°C, while silicon films begin to crystallize at approximately 700°C. During the deposition, the substrate temperature always rose to over 100°C, the exact temperature depending on the duration of the deposition. The source-to-substrate distance was approximately 14 cm.

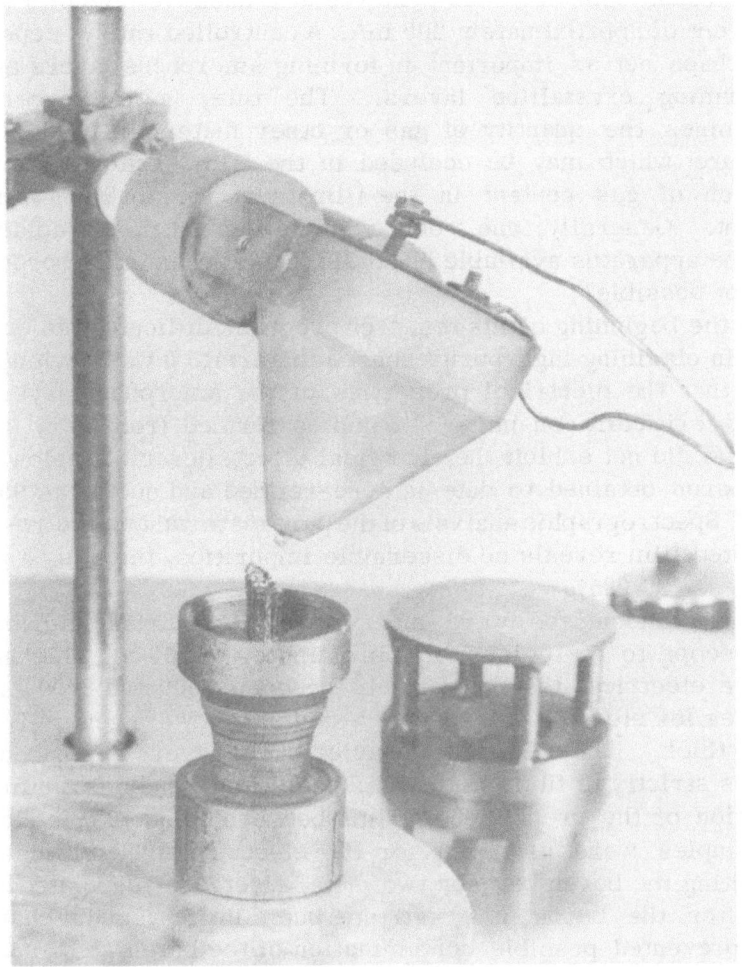

Fig. 1. Boron deposition source, illustrating carbon pedestal, boron rod, and electron gun.

The deposition source consisted of a zone-refined boron rod approximately 5 mm in diameter which rested on a high-purity carbon pedestal. The boron was heated directly by a small, focused electron beam. Care was taken not to bombard the carbon and to heat only a portion of the boron. Figure 1 shows a view of the pedestal, electron gun, and boron rod. Effort was made to maintain a constant deposition rate of 100–200 A/min throughout all the runs. This rate was achieved at a beam voltage of 6 kV and a current of approximately 200 mA. A controlled rate of deposition is perhaps not as important in forming amorphous layers as it is in forming crystalline layers. The rate, however, certainly determines the quantity of gas or other materials of high vapor pressure which may be occluded in the film. Unfortunately, the question of gas content in the film must remain unanswered at present. Generally, one would wish to deposit more rapidly, but with the apparatus available and the high melting point of boron, this was not possible.

In the beginning of this research program, difficulty was encountered in obtaining high-purity source material. It was obvious very early that the electrical properties of the amorphous layers depended critically on purity. Samples formed from very impure material did not exhibit the electrical effects described below. The best boron obtained to date is zone-refined and quoted as 99.999% pure.* Spectrographic analysis of the pure material and the resulting deposited film reveals no discernible impurities, implying a purity of at least 99.995%.

Film thicknesses were measured with a Zeiss interference microscope to a precision of approximately ± 150 A. Thicknesses for the electrical tests were held between 1000 and 8000 A. The samples for electron-microscope examination were generally about 300 A thick. The electrical behavior of the layers described here applies strictly to films in the thickness range studied. Future exploration of the properties of films between 1 and 10 $\mu$ is planned.

Samples were prepared for the electrical measurements by depositing the boron between two metal electrodes deposited before and after the boron in a separate but similar vacuum chamber. This prevented possible contamination of the boron. The samples were thus exposed to the atmosphere following both the first electrode deposition and the boron deposition. Outgassing was carried out on the uncoated silica substrates at 500°C and on the substrates bearing the first electrodes at 400°C. No outgassing was carried out between the boron deposition and the second, or top, electrode deposition. Generally, aluminum electrodes were used; however, other materials including chromium and gold gave similar results. The depositions were made through molybdenum masks

---

*This boron is supplied by United Mineral and Chemical Corporation, New York, New York and Koch-Light Laboratories, Ltd., Colnbrook, Bucks, England; however, the actual producer of the material is not known.

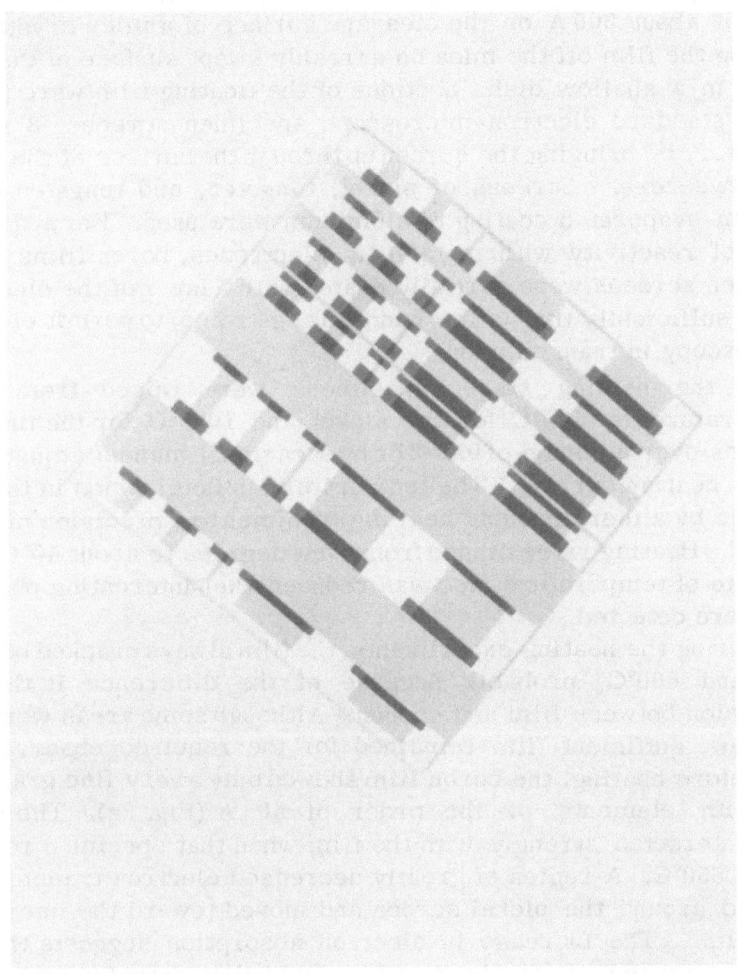

Fig. 2. Boron sample with aluminum electrodes deposited on fused silica.

designed to produce electrode areas from $1 \cdot 10^{-1}$ to $7 \cdot 10^{-4}$ cm$^2$. A photograph of a typical sample is shown in Fig. 2.

## STRUCTURE

High-temperature electron microscopy was used to investigate the structure of boron films and the interactions between boron and electrode materials. The observations were made with a JEM-6A electron microscope equipped with heating stage.

The specimens were prepared by evaporating boron to a thickness of about 300 A on the cleavage surface of a mica crystal and floating the film off the mica on a freshly swept surface of distilled water in a shallow dish. Sections of the floating film were picked up on standard electron-microscope specimen screens, 3 mm in diameter, by bringing the screen up through the surface of the water with tweezers. Screens of nickel, tungsten, and tungsten with a vacuum-evaporated coating of aluminum were used. For additional tests of reactivity with metal-film electrodes, boron films on the tungsten screens were partially coated with a layer of the electrode metal sufficiently thin (a few hundred angstroms) to permit electron microscopy in transmission.

In the heating stage, specimens were raised from room temperature to 590°C for the nickel and 1000°C for the tungsten screens over a period of 0.5–2 hr by occasional manual adjustments of the heating current. The temperature is determined in this apparatus by a thermocouple near the specimen to a precision of about ±10°C. Heating rates ranged from a few degrees to about 40°C/min; the rate of temperature rise was reduced when interesting phenomena were detected.

During the heating experiments, the film always cracked between 400 and 600°C, probably because of the difference in thermal expansion between film and screen. Although some areas were lost to view, sufficient film remained for the required observations.

Before heating, the boron film showed only a very fine granularity with elements of the order of 50 A (Fig. 3a). The nickel grid interacted strongly with the film when that specimen reached 530 to 550°C. A region of greatly decreased electron transmission formed around the nickel screen and moved toward the interior of the film. The increase in electron absorption suggests that the nickel migrated into the boron film. Behind the interface, presumably in a region of somewhat higher nickel concentration, crystallization apparently occurred to produce an irregular lacy structure. Figure 3b shows the unchanged boron film at the lower right, the darker region migrating inward, the crystallized region, and a portion of the nickel grid at the upper left.

In a specimen partially coated with aluminum, there was marked movement and coarsening of the metal film beginning at about 380°C. The appearance of the uncoated portion of the boron did not alter until at least 650–700°C had been reached. The specimen on the aluminum-coated tungsten screen formed lacy crystalline regions near the screen, as shown in Fig. 3c. These observations suggested that aluminum may nucleate crystallization of the boron above approximately 700°C. Neither the tungsten grid nor a partial coating of platinum, on the other hand, appeared to affect the structure of the boron film. In each case, the boron began to recrystallize only at about 900°C, and appeared to behave the same near the screen or metal coating as it did far away. Recrystallization of the boron took place within a few minutes at 920-980°C, yielding a coarser-grained structure (Fig. 3d).

The changes on heating were observed by electron diffraction as well as by electron microscopy. The diffuse rings of the original amorphous boron film (Fig. 4a) were found to sharpen into well-defined lines at the same time as the visible changes in morphology occurred.

The intensity distribution curve for the electron scattering by the amorphous boron film was approximated by exposing plates for approximately 0.5, 10, and 60 sec, recording microdensitometer curves for all three plates, and fitting the usable portions of the three curves together with appropriate scale factors. The form of the curve is indicated by Fig. 5. There are maxima in the molecular scattering at $(\sin \theta)/\lambda = 0.35$, $0.54-0.60$, and $0.7$, in addition to those at 0.12 and 0.20 previously reported from X-ray diffraction experiments [1]. Interpretation of the scattering curve and calculation of the corresponding radial distribution function are being carried out.

The radial distribution function for the amorphous material, in which only the near-neighbor distances are important, represents a Fourier transform of the scattering intensity curve. If the local atomic ordering is similar in the crystals, then the powder pattern might be expected to show a modulation resembling the shape of the amorphous scattering intensity distribution. Such a relationship appears marked in the pattern (Fig. 4b) obtained after crystallization of the boron film on the aluminum-coated tungsten screen or the film coated with aluminum. A more complex pattern (Fig. 4c) was obtained from the pure boron films on plain tungsten screens, with a change in the general envelope of the diffracted intensity suggesting a significant difference in the local ordering. Comparison of the electron diffraction patterns with X-ray patterns for the known polymorphs of boron [10,11] and the possible compounds, such as oxides, nitrides, boric acids, and metal borides, has revealed no similarities pointing to an identification of the products

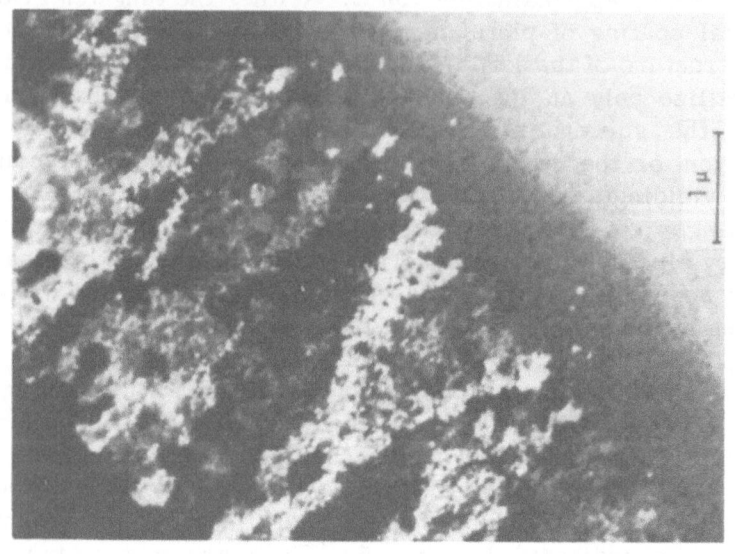

Fig. 3b. Electron micrograph of boron film ($\simeq 300$ A thick). Migration and crystallization of electron-absorbing region from nickel screen (upper left) into film (lower right) after heating to 590°C.

Fig. 3a. Electron micrograph of boron film ($\simeq 300$ A thick). Structure of original film.

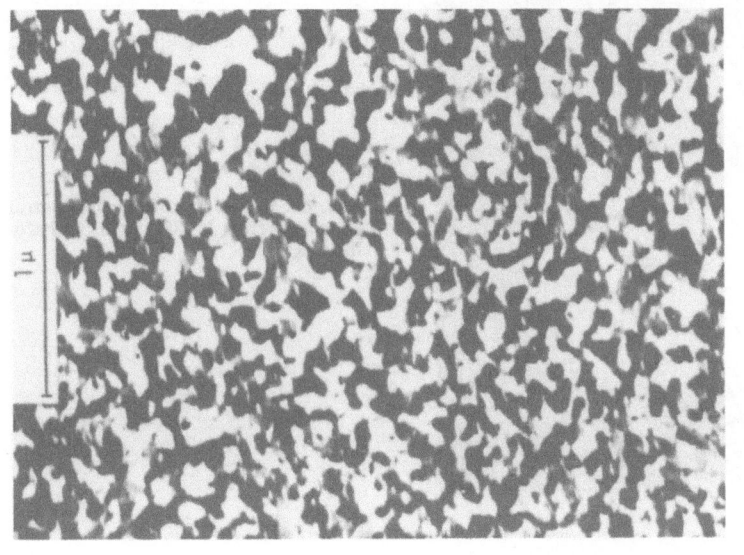

Fig. 3d. Electron micrograph of boron film ($\simeq$ 300 A thick). Film on tungsten screen, after heating to 920°C.

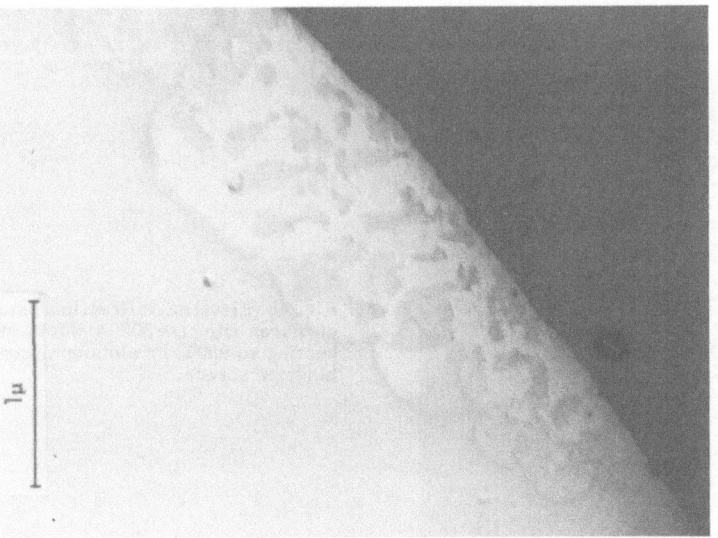

Fig. 3c. Electron micrograph of boron film ($\simeq$ 300 A thick). Crystallization of film near aluminum-coated tungsten screen, after heating to 710°C.

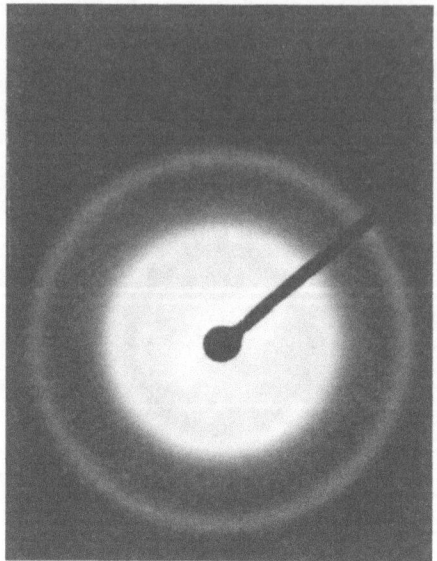

Fig. 4a. Electron diffraction pattern of boron film ($\simeq 300$ A thick) before heating.

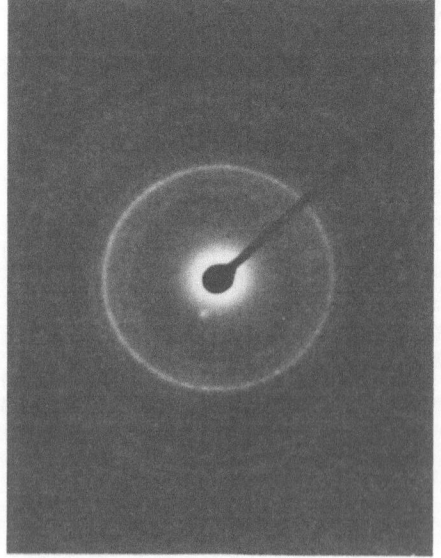

Fig. 4b. Electron diffraction pattern of boron film ($\simeq 300$ A thick) after heating to 980°C on aluminum–coated tungsten screen.

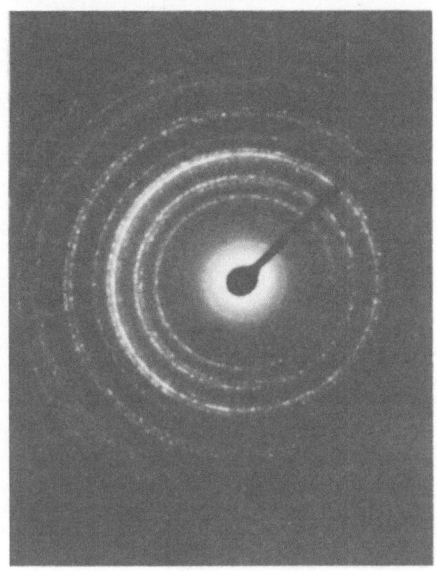

Fig. 4c. Electron diffraction pattern of boron film ($\simeq$ 300 A thick) after heating to 920°C on tungsten screen.

of crystallization. The difficulties inherent in such comparisons are well-known, but the case of pure boron appears sufficiently favorable to permit the conclusion that none of the known polymorphs was produced. Hoard and Newkirk [10] have suggested that boron may tend to form a variety of thermodynamically unstable crystalline structures as overgrowths on stable boride structures. Elucidation of the structural changes will require not only the study of the present results, but also the examination of other specimens with known structures and known contaminants.

## OPTICAL MEASUREMENTS

Optical measurements were carried out on a Cary 14 spectrophotometer in the range of 0.3–2.5 $\mu$. Films were deposited over the entire (1 by 1 in.) fused silica substrate and were made sufficiently thin to prevent the formation of optical interference fringes. This allowed easier observations of the band edge. The absorption and reflectance spectra of a film 0.37 $\mu$ thick are shown in Figs. 6 and 7, respectively. The absorption spectrum (Fig. 6) shows a considerable rise in absorption starting at approximately 1.32 eV, exhibits a shoulder in the region 1.85 eV, and finally

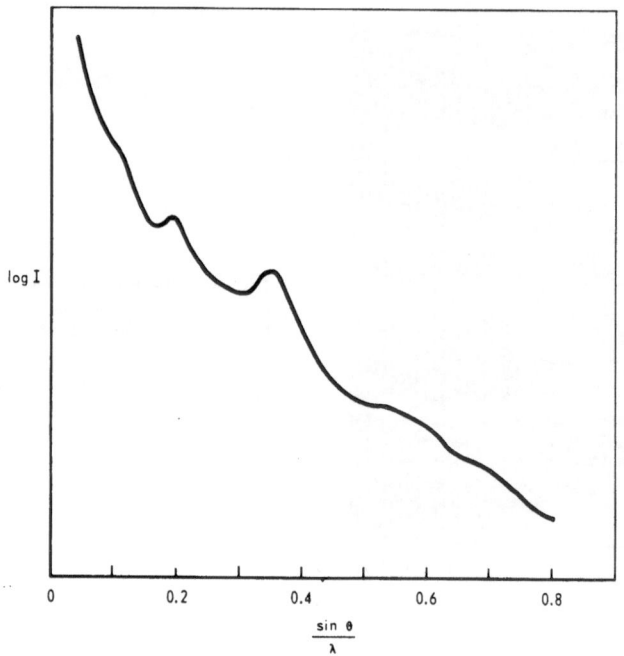

Fig. 5. Electron diffraction scattering intensity for amorphous boron film.

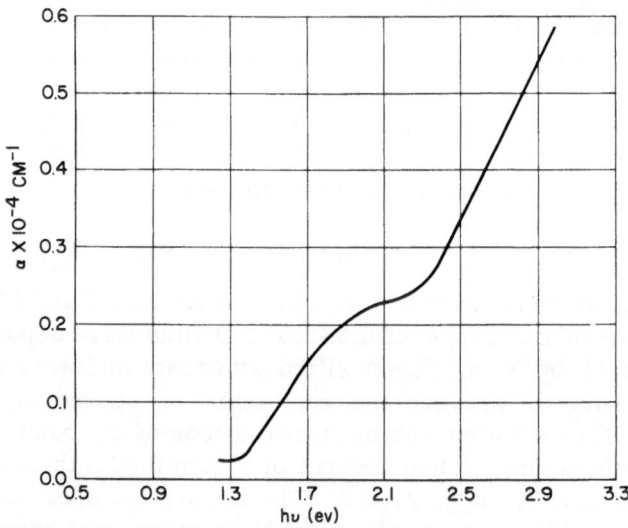

Fig. 6. Absorption spectrum of a boron film, $0.37\,\mu$ thick, deposited on a fused silica substrate.

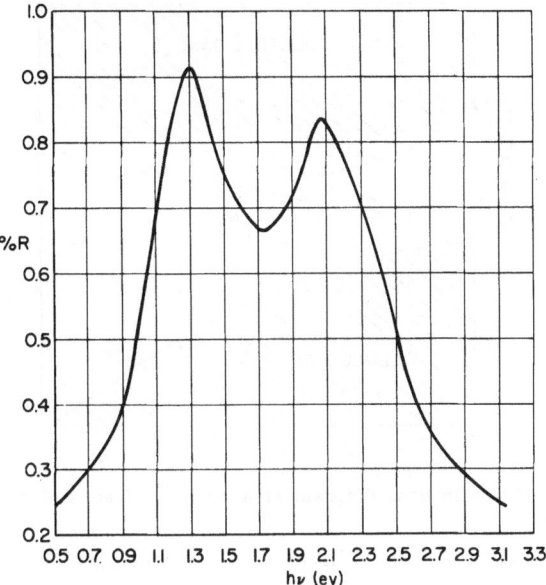

Fig. 7. Reflectance spectrum of a boron film, 0.37 $\mu$ thick, deposited on fused silica substrate.

rises sharply beyond 2.08 eV.  The curve thus differs from previously reported curves in the literature [12,13] and structure is observed in the band edge.  The difference is possibly due to the improved purity of the samples formed in this laboratory.  The spectrum shows general features which resemble those of group-IV elements [14], though the absorption in boron is less than in germanium or silicon.  One may interpret the results in terms of direct and indirect transitions [12].  The lower portion of the curve (1.34– 1.85 eV) may be due to indirect transitions between the valence and the conduction bands.  This allows the assignment of 1.32 eV to the minimum energy band gap, in general agreement with the values reported in the literature [12,15].  Direct absorption then begins at about 2.08 eV, and this value may correspond to the vertical gap at $k = 0$.   The shoulder between 1.85–2.08 eV probably corresponds to the region of superposition of direct and indirect transitions.

The interpretation of the absorption spectrum is in agreement with the peaks observed in the reflection spectrum of Fig. 7. Tauc [17] and others [18] have argued that, in the region where the reflectivity is controlled by the real part of the dielectric constant,

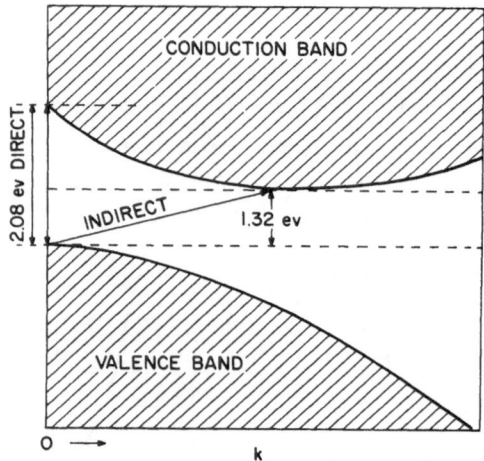

Fig. 8. Tentative schematic band diagram showing both direct and indirect transitions.

the positions of the peaks in the reflection spectra correspond to the values of energy gaps. These peaks are observed at 1.3 and 2.1 eV, values which agree quite well with the values of the gaps quoted above. A possible schematic band picture derived from these optical results is presented in Fig. 8.

## ELECTRICAL PROPERTIES

The current in the boron layers is observed to be a nonlinear function of voltage, temperature, and thickness. These functions first will be described separately and then combined to form a complete expression describing the observed phenomena.

The most striking electrical property of the boron films is their nonlinear current response to voltage. A typical oscilloscope trace illustrating the effect is shown in Fig. 9. The effect is symmetrical about the origin, indicating a symmetrical sample structure. (The intentional addition of an insulating or semiconducting layer to one side of the boron causes a nonsymmetrical current–voltage curve to appear.) The following empirical expression describes the observed behavior:

$$j = KV^n \quad \text{for } V_0 < V < V_b \tag{1}$$

where $j$ is the current density, $V$ is the voltage applied to the sample, $V_0$ is the threshold voltage, $V_b$ is the breakdown voltage, $T$ is the temperature, $K = K(T)$, and $n = n(T)$.

Note that this expression is valid only in a particular voltage range. In the region of small voltages and currents, below a critical voltage $V_0$, ohmic behavior seems to prevail. At high voltages ($V \geq V_b$), disruptive breakdown takes place. Neither the low nor the high voltage behavior has been explored in detail at this time. Disruptive electrical breakdown in the sample occurs at definite small spots on the surface and results in a direct electrical short. This short can usually be opened, and the sample restored to its original condition, by the application of a suitable high-voltage pulse. Presumably, the current is somewhat smaller because of the absence of a very small portion of the film electrode. When $V_b$ is approached slowly, heating of the sample occurs, and a type of thermal breakdown follows. The existence of $V_0$ will become apparent from the model presented below. The exponent $n$ in equation (1) is a function of the thickness, temperature, and sample purity. Generally, it has been observed that the more nearly pure the sample, the higher the value of $n$.

The behavior described by Fig. 9 and equation (1) also has been observed in this laboratory for amorphous films of silicon and germanium, and has been observed in other laboratories for films of selenium [19] and $As_2S_3$ [20] and for bulk SiC [21]. Silicon carbide or thyrite varistors operate by virtue of equation (1). Attempts at an explanation of this phenomenon are usually based on the existence of an amorphous high-resistance layer. In the case of SiC, the amorphous layer exists between the silicon carbide grains. Studies performed here on germanium films indicate that the effect disappears when the structure changes from amorphous

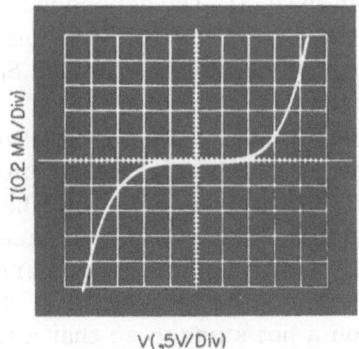

Fig. 9. Oscilloscope display of current vs. voltage in a thin boron sample, 0.64 $\mu$ thick and $10^{-2}$ cm$^2$ in area.

to polycrystalline. Note, however, that this change in structure is also accompanied by a large reduction in resistivity.

Mention should also be made of experiments carried out on field emission from thin films. In an early series of experiments, Malter [22] observed that the secondary emission from an electron-bombarded aluminum oxide layer followed a relation similar to equation (1). He pointed out the similarity between his results and those found in SiC, and postulated that both effects represented field emission due to the high field existing across the layers. Jacobs [23] studied the same effect in MgO layers and concluded that this enhancement in emission was due to an electron avalanche process created in the layer by the high induced electric field. The present study tends to support this view.

A clue to an understanding of the phenomena lies in the thickness dependency of the current. As the sample thickness increases, the current at constant field also increases. A plot of log $j$, at constant field, as a function of thickness $d$ produces a straight line. A family of curves at different fields is shown in Fig. 10. The phenomena illustrated by Fig. 10 may be described by

$$j = j_0 \exp{(ad)} \qquad (2)$$

where $j_0 = j_0(T)$, $a = a(E, T)$, $E$ is the electric field, and $d$ is the thickness. Equation (2) is similar to that found in avalanche multiplication and, therefore, $a(E)$ could be interpreted as an ionization coefficient similar to Townsend's first coefficient in a gaseous discharge. A curve of $a$ vs. $E$ is given in Fig. 11. As will be shown later, $a$ is proportional to log $E$. This is, however, a considerably slower rising function of $E$ than that observed in other solid-state multiplication phenomena [23]. The dependency on thickness tends to eliminate, as an explanation of the phenomenon, space-charge limited currents as described by Lanyon and Spear [25] and Hartke [26] for amorphous selenium.

Due to the high resistance of the boron films, it was difficult to determine the sign of the charge carriers from the Hall effect. The usual thermoelectric measurement, in which hot and cold probes are positioned close together on the same surface of the sample, also proved inconclusive. A slightly different thermoelectric technique was tried, however, with excellent results. In this technique, the sample was placed on a hot surface so that a temperature gradient was created perpendicular to the plane of the film. A gold wire probe was held on the bottom conducting electrode close to the

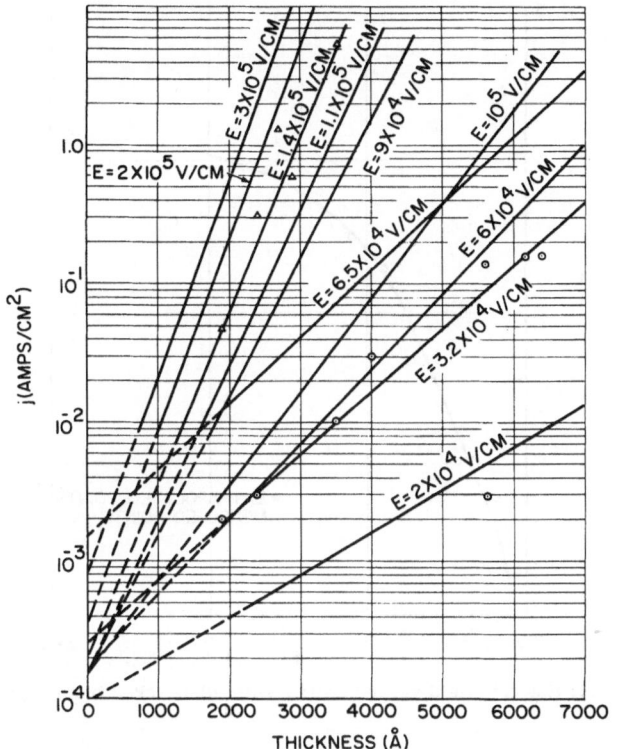

Fig. 10. Current density vs. film thickness for several fields. Experimental points to illustrate the scatter are plotted for only two fields to avoid confusion.

boron sample under study, and another electrode was placed directly on the top of the boron sample. When the temperature of the hot plate was less than approximately 200°C, a galvanometer connected to the probes gave a deflection indicating p-type conductivity. When the temperature rose above approximately 200°C, the galvanometer deflected in the opposite direction indicating a change to n-type. The effect was reversible; that is, cooling the sample to below 200°C again resulted in p-type conductivity. Shorted samples always exhibited n-type conductivity.

In order to determine the temperature dependency of the conductivity, the samples were placed in small stainless steel cans and evacuated to approximately $10^{-7}$ torr. Measurements were made while the cans were on the pumping system at temperature intervals up to 500°C. Measurements at higher temperatures were

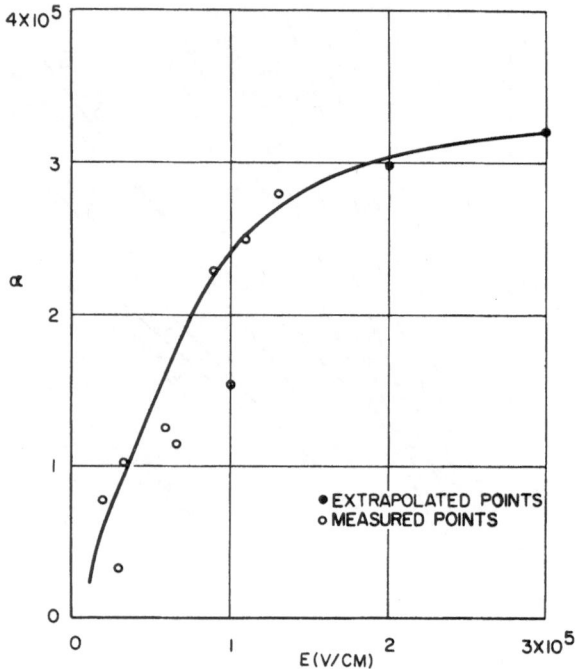

Fig. 11. Electric field vs. the coefficient $\alpha$ (data taken from Fig. 10).

difficult due to electrode insulation leakage. The results of the measurements are presented in Fig. 12 for $n$ vs. $1/T$ and Figs. 13 and 14 for log $j$ vs. $1/T$. Data from two different samples are used in the figures. Sample B86 is one of the films included in Fig. 11 and Fig. 15. Sample B163 was deposited at a later date from a different lot of the same purity. The samples illustrate the spread obtained in electrical characteristics. This divergence may be caused by variations in either deposition conditions or source material. The curves in Figs. 13 and 14 show a break in the neighborhood of 250°C which probably corresponds to the change in carrier type as revealed by the thermoelectric tests. If this transition temperature is $T_t$, one may write a formal expression for $n$ from Fig. 12.

$$n = \frac{\chi}{kT_t} + \epsilon_l \left( \frac{1}{kT} - \frac{1}{kT_t} \right) \text{ for } T \leq T_t \tag{3a}$$

$$n = \frac{\chi}{kT_t} + \epsilon_h \left( \frac{1}{kT} - \frac{1}{kT_t} \right) \text{ for } T \geq T_t \tag{3b}$$

where $X/kT_t$ is the value of $n$ at the transition temperature and is written in this form for convenience; $X$ is a constant and $\epsilon_l$ and $\epsilon_h$ are constants representing slope values at "low" and "high" temperatures, respectively.

The family of log $j$ vs. $1/T$ curves for different fields in Figs. 13 and 14 reveals two definite activation energies, both of which vary with applied field. The current at constant field, however, follows the usual semiconductor relationship with respect to temperature:

$$j = C_l \exp\left[-\theta_l (E)/kT\right] \text{ for } T < T_t \tag{4a}$$

$$j = C_h \exp\left[-\theta_h (E)/kT\right] \text{ for } T > T_t \tag{4b}$$

where $\theta_l (E)$ and $\theta_h (E)$ are activation energies and $C_l$ and $C_h$ constants for $T < T_t$ and $T > T_t$, respectively. Note that these equations involve $\theta/kT$ and not $\theta/2kT$; this allows the interpretation of $\theta$ to be left until later.

Equations (1)–(4) may be combined to form a single expression. Since equations (1) and (2) describe the same current in the same

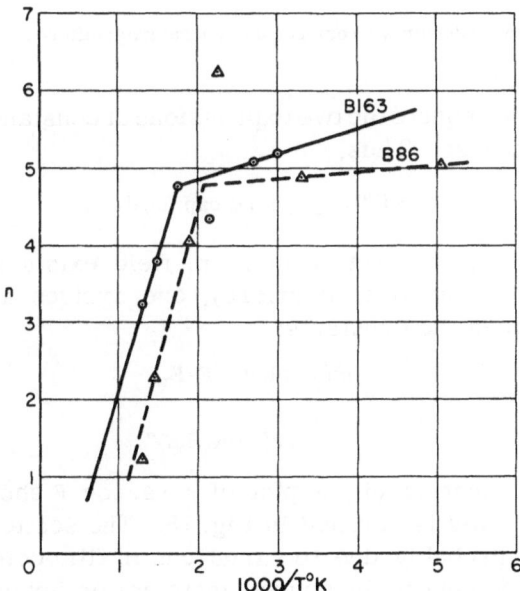

Fig. 12. Temperature dependence of the parameter $n$ for boron samples B86 ($\simeq 5600$ A) and B163 ($\simeq 5300$ A).

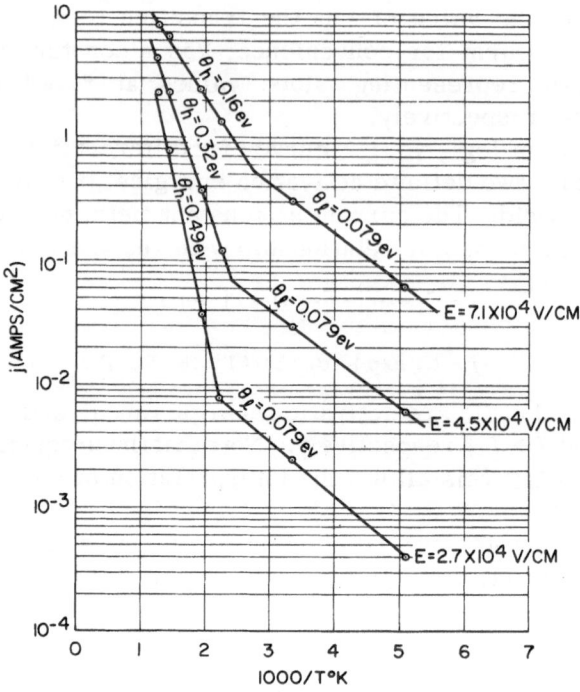

Fig. 13. Plot of current density vs. reciprocal temperature at different applied fields (B86).

sample, one may equate the two expressions at constant temperature, field, and thickness. Thus,

$$(KV^n)_{E,d,T} = [j_0 \exp (ad)]_{E,d,T} \tag{5}$$

With the assumption that a uniform field exists in the samples (i.e., no space charge or barriers), one arrives at the following relations between the constants:

$$ad = n \log_e (E/E_0) \tag{6}$$

$$j_0 = KV_0^n = K(E_0 d)^n \tag{7}$$

According to equation (6), a plot of $a$ vs. $\log_e E$ should result in a straight line. This is verified in Fig. 15. The scatter of the points in Fig. 15 is probably due to variations in film composition from run to run. A change in source material or impurity content or both changes $n$ and, of course, $a$. From equation (6), $a = 0$ at $E = E_0$; therefore, equation (7) defines threshold values below which ohmic behavior is obtained. This low-field behavior has been observed but not studied in detail. Combination of equations (2), (3), and (6) yields

$$j = j_0 \exp [n \log_e (E/E_0)] \tag{8}$$

or

$$j = j_0 \exp \left[ \left( \frac{\chi - \epsilon_{l,h}}{kT_t} + \frac{\epsilon_{l,h}}{kT} \right) \log_e (E/E_0) \right] \tag{9}$$

Use of equation (4) and the boundary conditions of ohmic behavior at $E = E_0$ and rearrangement of terms yield

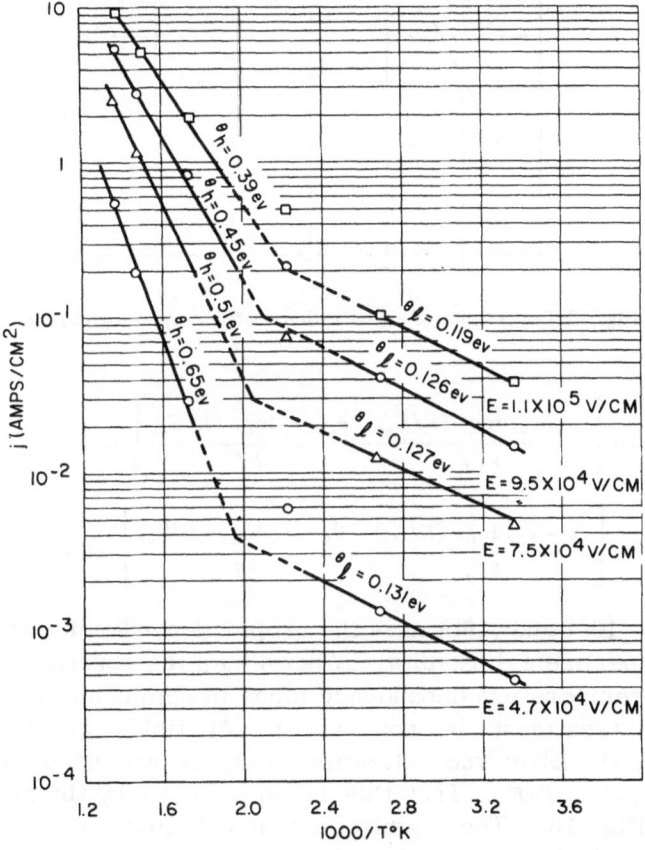

Fig. 14. Plot of current density vs. reciprocal temperature at different applied fields (B163).

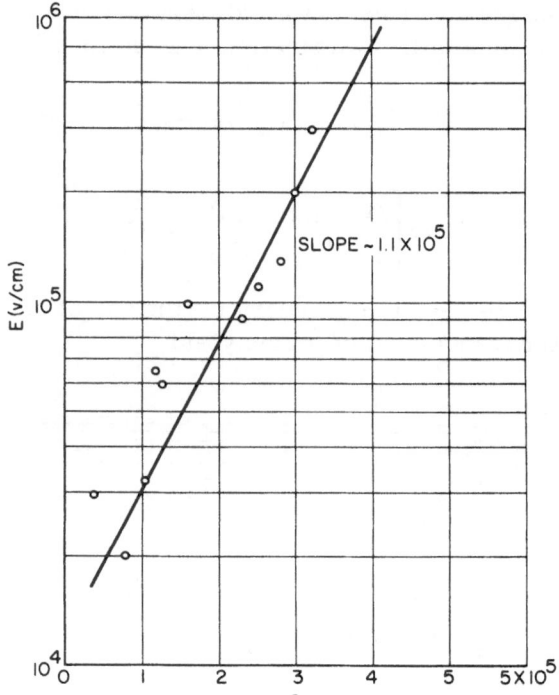

Fig. 15. Dependence of $\alpha$ (ionization coefficient) on log of field.

$$j = C_l \exp\left[\frac{(\chi - \epsilon_l)\,\log_e\,(E/E_0)}{kT_t} - \frac{\theta_l - \epsilon_l\,\log_e\,(E/E_0)}{kT}\right] \text{ for } T < T_t \qquad (10a)$$

and

$$j = C_h \exp\left[\frac{(\chi - \epsilon_h)\,\log_e\,(E/E_0)}{kT_t} - \frac{\theta_h - \epsilon_h\,\log_e\,(E/E_0)}{kT}\right] \text{ for } T > T_t \qquad (10b)$$

Equations (10a) and (10b) have the proper dependencies as required by the experimental results. When $E = E_0$, the equation reduces to the usual temperature dependency found in semiconductors. (When $E < E_0$, the treatment is not valid.) At fields greater than the threshold, the observed activation energies are seen to be equal to $(\theta_{l,h} - \epsilon_{l,h}\log_e E/E_0)$. That this is so is shown by the plot of $\theta$ vs. $\log_e E$ in Fig. 16. The significance of the constants $T_t$, $\chi$, and $\epsilon_{l,h}$ must await further research.

The activation energies, $\theta_l$ and $\theta_h$, may in principle be determined from the curves of Fig. 16 by assigning a value to $E_0$. However, to be consistent with the optical measurements and the activation energies reported in the literature, one must let $\theta_h \simeq$ 0.65 eV.    (This gives a minimum band gap between valence and conduction band of 2 $\theta_h \simeq$ 1.3 eV.)  With use of this value of $\theta_h$, one finds $E_0 = 1.7 \cdot 10^4$ V/cm and $E_0 = 5.6 \cdot 10^4$ V/cm, for samples B86 and B163, respectively. The $E_0$ value for sample B86 agrees reasonably well with the value of $1.3 \cdot 10^4$ V/cm determined by extrapolating $\alpha$ to zero in Fig. 15.

An estimate of the intrinsic resistivity of amorphous boron films may be obtained by extrapolating the appropriate curve of Fig. 14.    One obtains at low fields ($E_0$) and 25°C a resistivity of approximately $10^{12}$ $\Omega$–cm.    This is considerably higher than that

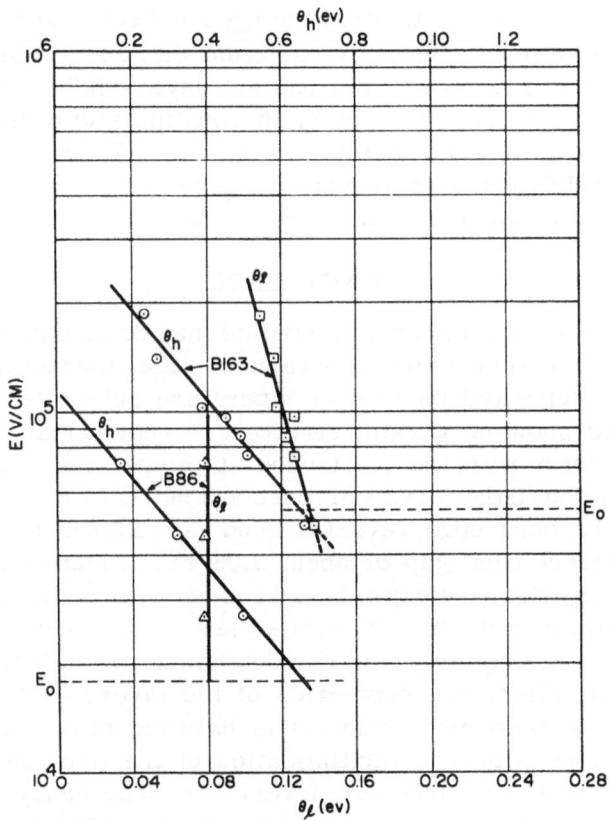

Fig. 16. Activation energies $\theta_\ell$ and $\theta_h$ for boron samples B86 and B163 as functions of electric field.

reported previously [1,16]. The value of the activation energy and the temperature at which intrinsic conductivity begins are, however, in agreement with the prior measurements [15, 16]. This implies, as one might expect, that the difference between crystalline and amorphous boron lies chiefly in the mobility values.

Using the above values of $E_0$, one finds for the activation energy $\theta_i$ values of 0.13 and 0.08 eV for samples B163 and B86, respectively. It is probable that $\theta_i$ corresponds to acceptor energy states lying just above the valence band.    It is difficult at present, however, to determine the origin of these low energy levels. Contamination by traces of gases, such as oxygen, and other impurities during deposition is always a possibility.    Similar levels have been reported for deposited films of amorphous selenium ( $\theta \approx 0.14$ eV) [25,26] and amorphous germanium ( $\theta \approx 0.18$ eV) [27]. In the case of selenium, the low activation energy has been associated with a mobility activation.  A mobility variation with energy also has been postulated for amorphous germanium layers [27].   By analogy, one could easily assign a term in equation (10) to mobility, but this would be pure speculation at present.   Additional research, including mobility measurements, is required to elucidate further the electrical behavior of amorphous boron films.

## CONCLUSIONS

This report of experimental studies may be summarized by the following physical description of vacuum-deposited boron thin films. The films, deposited on room-temperature substrates, are fine-grained and show no crystal structure.    They remain amorphous until they react with the electrodes at temperatures between 500 and 700°C or until they crystallize at about 900°C. Optical measurements in the band edge reveal a band gap of about 1.32 eV, and a possible direct band gap of about 2.08 eV. Electrical measurements support the band gap of 1.3 eV and show, in addition, acceptor states approximately 0.1 eV above the valence band. These low-energy acceptor states primarily determine the high-field, room-temperature electrical properties of the layers.  The nonlinear behavior with field and thickness is believed to be due primarily to impact ionization and multiplication of the electron-hole pairs associated with the acceptor levels.   A dependency of mobility on energy may also be present.  At low fields ( $<10^4$ V/cm) and low temperatures ($<200°C$), the films exhibit $p$-type ohmic behavior.

At higher temperatures, $n$-type conduction predominates. Questions as to the origin of the acceptor levels, the relative role of numbers of carriers and mobilities, and the form of the ionization coefficient must await further research.

## ACKNOWLEDGMENTS

It is gratifying to acknowledge the continued support of the U.S. Department of the Navy, Bureau of Naval Weapons in all phases of this study, and the aid of the National Aeronautical and Space Agency for theoretical studies supporting the effort. Sincere acknowledgment is given to Mr. Kenneth Hoggarth for his work in fabricating the samples and to Mr. A.D. McMaster for his effort in the structural studies.

## REFERENCES

1. Talley, C.P., L.E. Line, Jr., and Q.D. Overman, Jr., Boron—Synthesis, Structure, and Properties, Plenum Press (New York), 1960, p. 94.
2. Feldman, C., Nature 203: 964 (1964).
3. Moorjani, K. and C. Feldman, Rev. Mod. Phys. 36: 1042 (1964).
4. Armington, A.F., W.P. Potter, and L.E. Tanner, J. Appl. Phys. 35: 730 (1964).
5. Kohn, J.A., W.F. Nye, and G.K. Gaulé, Boron—Synthesis, Structure, and Properties, Plenum Press (New York), 1960.
6. O'Bryan, H.M., Rev. Sci. Instr. 5: 125 (1934).
7. Muggleton, A.H.F., and F.A. Howe, Nucl. Instr. Methods 13: 211 (1961).
8. Longequeue, J.P., N. Longequeue, H. Beaumevieille, E. Ligeon, F. Demartiny, and J. Fontenille, J. Phys. Radium 23: 141A (1962).
9. Erdman, K.L., D. Axen, J.R. MacDonald, and L.P. Robertson, Rev. Sci. Instr. 35: 122 (1964).
10. Hoard, J.L., and A.E. Newkirk, J. Am. Chem. Soc. 82: 70 (1960).
11. Decker, B.F., and J.S. Kasper, Acta Cryst. 12: 503 (1959).
12. Moss, T.S., Optical Properties of Semiconductors, Academic Press (New York), 1959, p. 102.
13. Morita, N., J. Sci. Res. Inst. (Tokyo) 48:8 (1954).
14. Dash, W.C., and R. Newman, Phys. Rev. 99: 1151 (1955).
15. Brungs, R.A., and V.P. Jacobsmeyer, J. Phys. Chem. Solids 25: 701 (1964).
16. Greiner, E.S., and J.A. Gutowski, J. Appl. Phys. 28: 1364 (1957).
17. Tauc, J., Proceedings of the International Conference on the Physics of Semiconductors, Exeter (1962), Institute of Physics and Physical Society, London.
18. Ehrenreich, H., H.R. Philipp, and J.C. Phillips, Phys. Rev. Letters 8: 59 (1962).
19. Weimer, P.T., and A.D. Cope, RCA Rev. 12: 314 (1951); W.E. Spear, Proc. Phys. Soc. B69: 1139 (1956).
20. Bowlt, C., Proc. Phys. Soc. 80: 810 (1962).
21. O'Connor, O.R., and J. Smilters (eds.), Silicon Carbide, Pergamon Press (New York), 1960.
22. Malter, L., Phys. Rev. 50: 48 (1936).
23. Jacobs, H., Phys. Rev. 84: 877 (1951).
24. Gunn, J.B., Progress in Semiconductors, Vol. 2, John Wiley and Sons (New York), 1957. p. 213.

25. Lanyon, P. D., and W. E. Spear, Proc. Phys. Soc. 77: 1157 (1961).
26. Hartke, J. L., Phys. Rev. 125: 1177 (1962).
27. Grigorovici, R., N. Croitorie, A. Devenyi, and E. Teleman, Proceedings of the International Conference on the Physics of Semiconductors, Paris (1964), p. 423.

# Low-Temperature Thermal Conductivity of Boron*

## J. C. Thompson and W. J. McDonald[†]

*Department of Physics, The University of Texas*
*Austin, Texas*

The thermal conductivity of a single crystal of boron has been measured between 4 and 300°K. The crystal structure is beta-rhombohedral; the Debye temperature 1200°K. An exponential temperature dependence, $\exp \Theta/aT$, was observed in the neighborhood of 150°K; the value of $a$ was 2.4. This temperature dependence is characteristic of umklapp processes. The maximum conductivity is approximately 3 W/cm-deg near 50°K. At low temperatures, the conductivity obeys a $T^{1.8}$ law and has a magnitude one-tenth that expected for boundary scattering. The latter effect may be associated with the presence of dislocations. The data are analyzed by the phenomenological model of Callaway.

## INTRODUCTION

Heat conduction by dielectric crystals is reduced by a number of scattering mechanisms [1,2]. Even in the purest of crystals the presence of several different isotopes will lead to a thermal resistance. However, in a material with a high Debye temperature, the effect of the isotopic variety is diminished as the phonons effective in energy transport are shifted to longer wavelengths and, thus, are relatively unaffected by point defects, i.e., the isotopes. Other scattering agents include boundaries, dislocations, strains due to deionized impurities, etc. Each will contribute a characteristic temperature dependence to the conductivity in a given temperature range.

Interactions also occur between phonons due to anharmonic terms in the interionic potential. The phonon–phonon scattering may be divided into two classes—normal processes wherein crystal momentum is conserved, and umklapp processes wherein the crystal

*Assisted by the Office of Naval Research. Reprinted from Phys. Rev. 132: No. 1, 82–84, October 1, 1963.
†Texas Instruments Predoctoral Fellow.

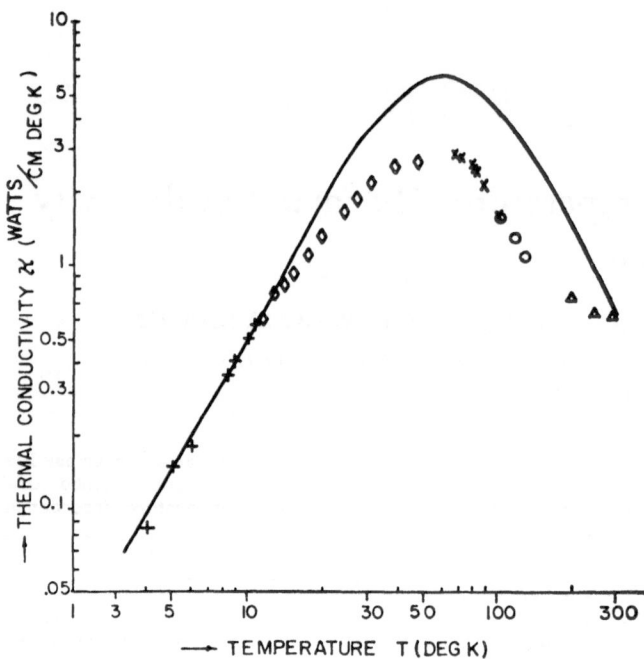

Fig. 1. Thermal conductivity as a function of temperature. The symbols refer to various runs. The solid line is based on isotope, boundary, and phonon–phonon scattering processes.

momentum is not conserved. Normal processes do not produce a thermal resistance, but rather redistribute energy among the modes.

Quite recently, single crystals of boron have become available [3], and these provide an opportunity for the investigation of the various scattering mechanisms in an elemental semiconductor with properties intermediate between silicon and diamond. Boron has a Debye temperature over 1200°K and an energy gap of 1.35 eV. Unfortunately, the presence of dislocations obscures the low-temperature properties of the sample investigated.

## EXPERIMENTAL

Boron has several allotropic forms. The sample used here was a single crystal of the beta-rhombohedral phase grown by Texaco Experiment, Inc.* Boron in this form is a semiconductor. The sample was a cylinder 3.8 cm in length and averaged 0.26 cm in

*The crystal was supplied to us through the good offices of C. P. Talley and T. S. Teasdale of Texaco, Inc.

diameter. As it was grown by the float-zone method, it lacked a uniform cross section. From studies of electrical transport [3] made on similar specimens, we may infer the band gap to be 1.35 eV. Carbon was the major impurity (0.1%); the carbon acts as a donor with a level about 0.32 eV below the conduction band. The room temperature resistivity exceeds $5 \cdot 10^6 \, \Omega$-cm. The density is $2.342 \pm 0.005$ g/cm$^3$ and the Debye temperature is 1219°K [4].

Measurements were made in the standard way [5,6] using gold–cobalt versus manganin thermocouples for temperature and temperature difference. The temperature difference across the sample was held to a few percent of the average temperature.

The results are shown in Fig. 1, which also shows a theoretical curve based on isotope, dislocation, and boundary scattering, as well as the usual anharmonic effects.

## DISCUSSION

The presence of two abundant isotopes with a 10% mass difference would be expected to preclude the observation of umklapp scattering effects. However, Klemens [1] has shown that the relaxation time $\tau_1$ for isotopic scattering can be written

$$\tau_1^{-1} = (V/N) (4\pi c^3)^{-1} \Gamma \omega^4 = A\omega^4 \tag{1}$$

where we follow the notation of Callaway [7], with $c$ the speed of sound, $V/N$ the atomic volume, and

$$\Gamma = \sum_i f_i \left(1 - \frac{M_i}{M}\right)^2 \tag{2}$$

In equation (2), $f_i$ is the fraction of isotopes of mass $M_i$ present in a crystal of average atomic mass $M$. The presence of $c^{-3}$ in the coefficient of $\omega^4$ implies that a wide variety of isotopic masses can be offset by a large sound speed. As $c$ is proportional to the Debye temperature, we expect a large $c$ in boron. Slack [8] has surveyed the variation of isotopic effects, and Table I shows a comparison among several elements. One sees that boron is no more affected by isotope scattering than the enriched germanium sample used by Geballe and Hull [9]. It is nevertheless surprising that umklapp scattering should be observed here since the value of $A$ is no smaller than in silicon. However, one cannot make a prediction without detailed knowledge of the vibration spectrum [9]. Of the materials

## TABLE I

### Isotope Scattering Effects in Several Materials

| Material | $\Theta$ (deg K) | V/N (cm³) ($\times 10^{-24}$) | c (cm/sec) ($\times 10^5$) | $\Gamma$ ($\times 10^{-4}$) | A (sec³) ($\times 10^{-44}$) |
|---|---|---|---|---|---|
| Diamond | 1960 | 5.68 | 11.8 | 0.76 | 0.00216 |
| Boron | 1219 | 7.7 | 8.1 | 12.9 | 0.150 |
| Sapphire | 1010 | 42.6 | 8.0 | 0.025 | 0.0016 |
| Silicon | 668 | 19.9 | 6.2 | 2.64 | 0.133 |
| Germanium | 390 | 22.6 | 3.8 | 5.72 | 1.87 |
| Germanium [9] (enriched) | 390 | 22.6 | 3.8 | 0.38 | 0.124 |

Data taken in part from Slack [8].

listed in Table I, only diamond and sapphire have heretofore been shown to exhibit an exponential thermal conductivity.

Figure 2 shows the mean free path $\Lambda$ of boron as a function of $\Theta/T$ at temperatures above the maximum in the conductivity. An exponential dependence is indicated. The mean free path is calculated from the thermal conductivity by $\kappa = \frac{1}{3}Cc\Lambda$, where $C$ is the specific heat expected from the Debye theory. Though simple arguments lead one to expect a value of 2 for $a$, the value of 2.4 reported here is not out of line with the value of 2.6, for example, reported for diamond [2].

It should be noted that the 0.1% carbon present may be included as an "isotope" but alters the value of $\Gamma$ by less than 1% since the carbon mass is so close to that of boron.

We have analyzed our data by using the phenomenological model of Callaway [7]. We assume that the thermal resistance is due to the following effects: (a) boundaries, (b) dislocations, (c) phonon–phonon scattering, (d) isotope scattering. Each process is described by a relaxation time according to the following scheme:

Boundaries:

$$\tau_B^{-1} = C/L \tag{3}$$

where $L$ is the sample diameter.

Dislocations:

$$\tau_D^{-1} = n\gamma^2 b^2 \omega = H\omega \tag{4}$$

where $n$ is the density of dislocations, $\gamma$ the Grueneisen constant, and $b$ the "diameter" of the dislocation.

Phonon–phonon:

$$\tau_N^{-1} = B_2 T^3 \omega^2 \tag{5}$$

for normal processes, with $B_2$ a constant; and

$$\tau_U^{-1} = B_1' T^3 [\exp(-\Theta/aT)]\omega^2 = B_1 T^3 \omega^2 \tag{6}$$

for umklapp processes, with $B_1$ taken to be a constant. The isotope scheme has already been discussed. We now combine the relaxation

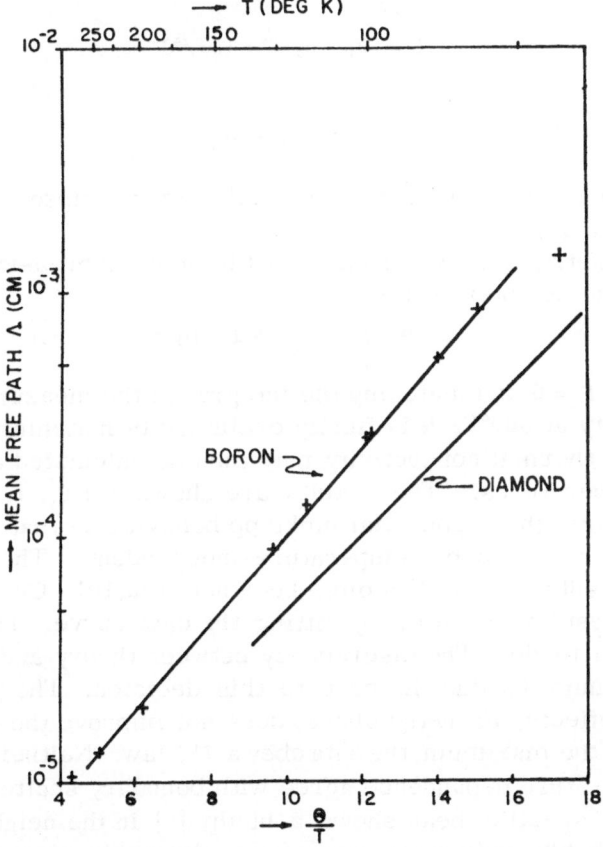

Fig. 2. The phonon mean free path as a function of temperature in the umklapp region. The mean free path is defined by $\kappa = \frac{1}{3}Cc\Lambda$, where $\kappa$ is the thermal conductivity, $C$ the heat capacity, and $c$ the speed of sound. A similar curve for diamond is shown for comparison.

times and obtain equation (7) for the thermal conductivity. The details of the calculation are given in the article by Callaway, who does not discuss dislocations, though their effect may be included by a straightforward process:

$$\kappa = \left(\frac{k}{2\pi^2 c}\right)\left(\frac{kT}{\hbar}\right)^3 \int_0^{\theta/T} (e^x - 1)^{-2}\left(Dx^4 + Ex^2 + Fx + \frac{c}{L}\right)^{-1} x^4 e^x dx \qquad (7)$$

In equation (7), $k$ is Boltzmann's constant,

$$x = \hbar\omega/kT$$

$$D = A(kT/\hbar)^4$$

$$E = (B_1 + B_2)T^3 \times (kT/\hbar)^2$$

and

$$F = H(kt/\hbar)$$

The equation was evaluated numerically using Simpson's rule and a CDC 1604 computer.

The quantity $A$ is evaluated from Klemens' expression, equation (1). A value is obtained for

$$B = B_1 + B_2 = 5.4 \times 10^{-24}$$

by setting $F = 0$ and matching the integral to the measured thermal conductivity at 300°K; $H$ is finally evaluated by matching the data at 5°K. The thermal conductivity may then be calculated at intermediate temperatures. The results are shown in Fig. 1. One cannot hope to obtain the exponential umklapp behavior, as that is removed when $B_1$ is taken to be temperature-independent. The value of $A$ is undoubtedly low, as has often been noted [2,10]. One might then seek a "best" value for $A$ by fitting the data curve. This we have not chosen to do. The discrepancy between theory and data above the maximum is due in part to this decision. The presence of umklapp effects, as noted above, does not improve the agreement.

Below the maximum, the data obey a $T^{1.8}$ law. Neither magnitude nor temperature dependence agree with boundary scattering. However, the specific heat shows a hump [4] in the neighborhood of 25°K, probably indicative of a phase change in view of the many allotropic forms of boron. One is thus led to expect a considerable number of dislocations to be present. Dislocations should produce a $T^2$ temperature dependence [2]. If we determine the dislocation density at 5°C by fitting the data and equations (4) and (7), we obtain a dislocation density of $10^9 \text{cm}^{-2}$. This result depends upon assumed values of 2 and $3 \times 10^{-8}$ cm for the Grueneisen $\gamma$ and for the dislocation core size $b$, respectively. The result is reasonable if a phase change occurs.

We do not expect to explain our result in terms of strains induced by the deionization of the carbon donor atoms as has been done by Keyes [11] for germanium. This effect leads to a temperature dependence stronger than $T^3$, as does the presence of mobile defects [12].

## CONCLUSIONS

We have reported here the third observation of umklapp scattering processes among the elements, helium and diamond being the other two. The coefficient in the exponential term $a$ has a value of 2.4 which compares well with the values of 2.3 and 2.6 for helium and carbon, respectively. The presence of two abundant isotopes of boron, with masses differing by 10%, does not lead to the large isotope scattering observed in germanium, as the Debye temperature is high in boron.

The thermal conductivity below the maximum is limited by the presence of dislocations. Heat capacity measurements lead one to believe that a phase change occurs in this temperature range. Thus, the presence of $10^9$ dislocations per square centimeter is not unreasonable.

## ACKNOWLEDGMENT

The authors would like to acknowledge a stimulating conversation with P. G. Klemens.

## REFERENCES

1. Klemens, P.G., in F. Seitz and D. Turnbull (eds.): Solid State Physics, Vol. VII, Academic Press (New York), 1958, p. 1.
2. Carruthers, P., Rev. Mod. Phys. 33:92 (1961).
3. Talley, C.P., private communication.
4. Johnston, H.L., H.N. Hersh, and E.C. Kerr, J. Am. Chem. Soc. 73:1112 (1951).
5. Thompson, J.C., and B.A. Younglove, J. Phys. Chem. Solids 20:146 (1961).
6. Holland, M.G., and L.G. Rubin, Rev. Sci. Instr. 33:923 (1962).
7. Callaway, J., Phys. Rev. 113: 1046 (1959).
8. Slack, G.A., Phys. Rev. 105:829 (1957).
9. Geballe, T.H., and G.W. Hull, Phys. Rev. 110:773 (1958).
10. Agrawal, B.K., and G.S. Verma, Phys. Rev. 126:24 (1962).
11. Keyes, R.W., Phys. Rev. 122:1171 (1961).
12. Granato, A., Phys. Rev. 111:740 (1958).

# Mechanical and Micromechanical Behavior of Bulk Polycrystalline Boron

## C. J. Speerschneider and J. A. Sartell

*Honeywell Research Center*
*Hopkins, Minnesota*

The room-temperature mechanical and micromechanical behavior of bulk polycrystalline boron was characterized in this study. Material for the test program was produced with a casting technique. It was found that in all cases fracture was initiated at cracks or steps on the sample surfaces. A study of the fracture mode shows that structural features can influence the fracture path. This provides a basis for future improvements of the mechanical properties. The micromechanical behavior of bulk polycrystalline boron shows a completely elastic response to stress at all levels up to fracture.

## INTRODUCTION

Although a large amount of work has been done on amorphous or "glassy" boron [1,2] for use as a fiber in composite materials, very little effort has been directed toward the use of bulk crystalline boron in structural applications. Many of the properties which have been measured suggest that this material would be ideal for certain structural applications. Low density, good stiffness, high hardness, and high melting point are such properties. However, the brittle nature of bulk boron and the difficulty of preparing it have thus far prevented its use as a structural material. Consequently, any investigation of the mechanical behavior of boron should include studies which will clearly define the extent of its ductility.

This study was undertaken to characterize the mechanical and micromechanical behavior of bulk crystalline boron. Particular emphasis has been on properties, such as fracture stress and microstrain behavior, related to the material's response to stress. In addition to the fracture strength, the fracture character was determined through observations of the fracture surfaces. Study in the microstrain region provides information on possible dislocation movement at room temperature, and permits observation of the stress–strain behavior with a sensitivity much greater than

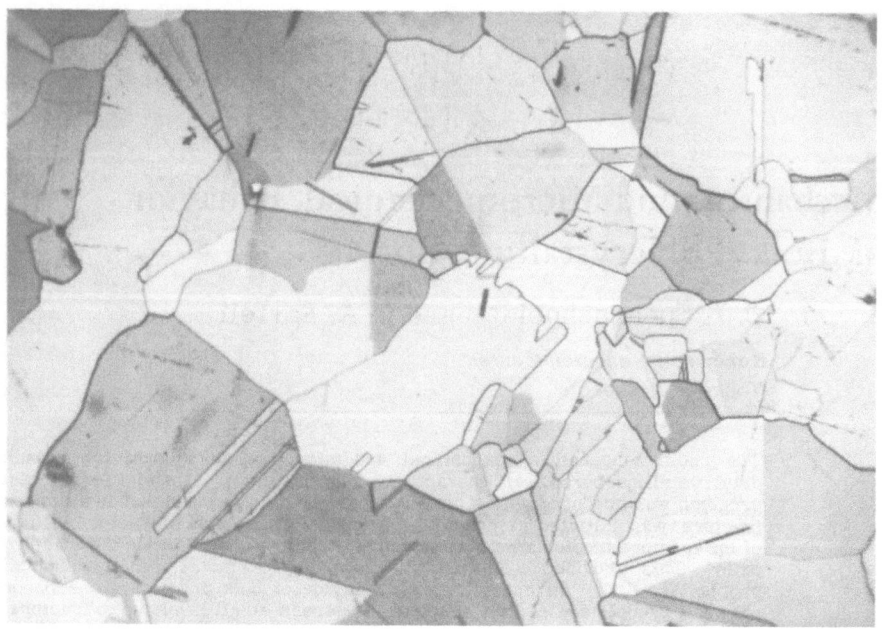

Fig. 1. Cast boron etched 15 min in a boiling $H_2SO_4$ solution which has been used previously (100x).

normally obtained.    First, however, it was necessary to develop a technique to produce bulk crystalline boron of sufficient size to provide specimens for study.

## CAST POLYCRYSTALLINE BORON

Several techniques were investigated to produce polycrystalline boron.  A casting procedure gave the best results and was selected as the source of samples.   Melting was done in a boron nitride crucible using a simple resistance furnace with a carbon element 3 in. in diameter and 6 in. long.   This heater element is not as susceptible to corrosion by the boron as was a tungsten element used in earlier experiments.   The heater is surrounded by a sufficient number of radiation shields (approx. 10–15) so that the outermost shield does not reach red heat.   Melting is done in an evacuated enclosure at a temperature of approximately 2050°C under a pressure of $1.0\mu$.  Since the system operates with a power as high as 30 kW, the low-voltage heating element and the conductors must handle currents up to several thousand amperes.  Temperature is measured optically and is controlled through powerstats.  Cooling is performed under automatic control at a rate of 150°C/hr. Slow cooling is necessary to reduce ingot cracking due to thermal shock. With use of the above technique, crack-free ingots 1 in. in diameter and 1 in. high were cast.

The quality of the cast material was assessed not only by its mechanical behavior, but also through observations of its micro-structure. Figure 1 is a photomicrograph of polished and etched cast boron showing dense polycrystalline material containing a substantial number of twins. It is felt that the occurrence of twinning may be related to the impurity content of the boron. Table I gives the chemical analysis of the cast boron material. The etchant used for the sample of Fig. 1 was a "used" solution of boiling $H_2SO_4$. As reported earlier by Ellis [3], there is considerable staining with $H_2SO_4$; however, the use of an old solution produces very good results, as shown. X-ray diffraction examination using the Laue back-reflection technique also shows the quality of the cast material. Figure 2 shows a Laue pattern of a single large crystal in a polycrystalline casting with a triad axis of symmetry for the rhombohedral crystal with no indication of internal strain, as the diffraction spots show no asterism. The absence of internal strain is not totally unexpected in view of the controlled slow cooling. Similar examination of material grown with the floating-zone technique (in which a sharp temperature gradient occurs) does show extreme asterism of the spots, due to internal strain. A limited

TABLE I

Chemical Analysis of Cast
Polycrystalline Boron

| Element | Amount, % |
|---|---|
| Tungsten | 0.2 |
| Copper | 0.05 |
| Iron | 0.05 |
| Manganese | 0.05 |
| Silicon | 0.05 |
| Cobalt | 0.01 |
| Aluminum | 0.001 |
| Beryllium | 0.001 |
| Magnesium | 0.001 |
| Boron | Balance |

Fig. 2. Laue back-reflection pattern of a large grain in cast boron. The absence of asterism indicates a low level of internal strain.

number of mechanical tests were made on zone-melted boron rods, in addition to the tests on cast boron.

Samples for testing were cut from the ingots with a diamond saw. The sawing introduces surface cracks; since boron is very notch-sensitive, it is necessary to polish the samples just prior to mechanical testing. A large number of chemical polishes were tried with only one solution — 20% KOH + 20% $K_3Fe(CN)_6$, as reported by Niemyski et al. [4] — showing the required nonpreferential polishing action. It was necessary to polish at the boiling temperature of this solution and rinse particularly carefully. Furthermore, since conventional handling techniques may introduce surface flaws, certain precautions were taken to reduce the chance of critical areas of the sample coming into contact with hard surfaces. For example, when samples were to be moved, they were always gripped lightly at their ends, never within the gauge length.

## STRENGTH AND FRACTURE CHARACTER OF BULK POLYCRYSTALLINE BORON

### Influence of Surface Imperfections on Fracture Strength

In order to understand more thoroughly the behavior of poly-crystalline boron, the fracture stress was determined and a study made of the fracture characteristics. Although the fracture stress for boron has been previously reported [2], the work was on fibers rather than on bulk polycrystalline material. In addition, in the present study a determination was made of the fracture source and the influence of surface defects on fracture strength. Bend tests were run on the polycrystalline boron specimens using an Instron testing machine at a crosshead speed of 0.002 in./min with the test bars having a length approximately ten times their thickness. The importance of surface condition is illustrated in the fracture stress of specimens tested in various surface conditions—as-sawed, as-sawed and mechanically-polished, and as-sawed and chemically-polished. Samples in the as-sawed condition with the corresponding poor surface failed at stress levels of 15–25,000 psi. Although an increase in the fracture stress was observed with the mechanically-polished samples, it was not possible to completely eliminate the surface damage introduced during the diamond sawing procedure. The most promising polishing procedure employed a 20% KOH + 20% $Fe_3K(CN)_6$ solution, and, though not 100% effective, it did result in an increase in the fracture stress to 45,000 psi in bending. All other etchants employed resulted in preferential attack with a subsequent decrease in fracture stress.

Studies of the fracture surface show that fracture initiated at the surface in all samples. Figure 3 is a typical fracture surface from a bend sample with failure initiated at a flaw on the tension surface and spread throughout the sample. These surface markings will be discussed more fully later; the brittle behavior of the material suggests that the Griffith theory [5,6] will apply. A calculation using the Griffith equation gives a critical crack length of quite small

Fig. 3. A macrophotograph of the fracture surface of polycrystalline boron, direction of crack front indicated by arrow (20 x).

Fig. 4. Photomicrograph of an enlarged view of the fracture source area from the sample shown in Fig. 3. (100×).

dimensions in agreement with our observations of the surface. It must be mentioned, however, that no quantitative study of this particular aspect was made.

Surface cracks are not the only fracture source in polycrystalline boron. Although the crack has received the most attention, Marsh [7] has shown that a surface step common to many crystal surfaces is comparable to a crack of similar dimensions. Boron rods grown by the floating-zone technique quite often contain flaws of this nature, which result in failure at very low stress levels (<10,000 psi). For example, a surface step which initiated failure at 6,000 psi was observed at a twin interface in a boron rod. Surface steps which initiate failure are also formed during chemical polishing of the samples prior to testing. With use of the 20% KOH + 20% $Fe_3K(CN)_6$ solution, a polish could be obtained with subsequent increase in fracture stress; however, in many instances, the polish was not uniform over the entire sample surface and resulted in rough areas. The surface steps in these rough areas initiate failure at points of stress concentration in the manner described above.

## Influence of Stress Field on Crack Propagation

The study of fracture surface markings has shown not only the importance of surface condition, but also the influence of the stress field and various structural features on crack propagation. Such observations have shown that in many instances boron polycrystals correspond more nearly to metallic materials than to brittle materials, such as glass and ceramics, in that the fracture behavior can be influenced by microstructure and is not dependent only upon the stress field. The fracture surface shown in Fig. 3 is typical for polycrystalline boron; it shows a conchoidal area on the tension surface near the fracture source, tear markings or striations as the compression surface is approached, and more complex regions in the finer-grained areas. As mentioned previously, all failures were initiated at a surface flaw such as shown here. The fracture characteristics can be understood more clearly if the crack front is observed as it proceeds across the sample.

Examination of these surfaces at higher magnification shows the fracture markings associated with brittle behavior, along with the

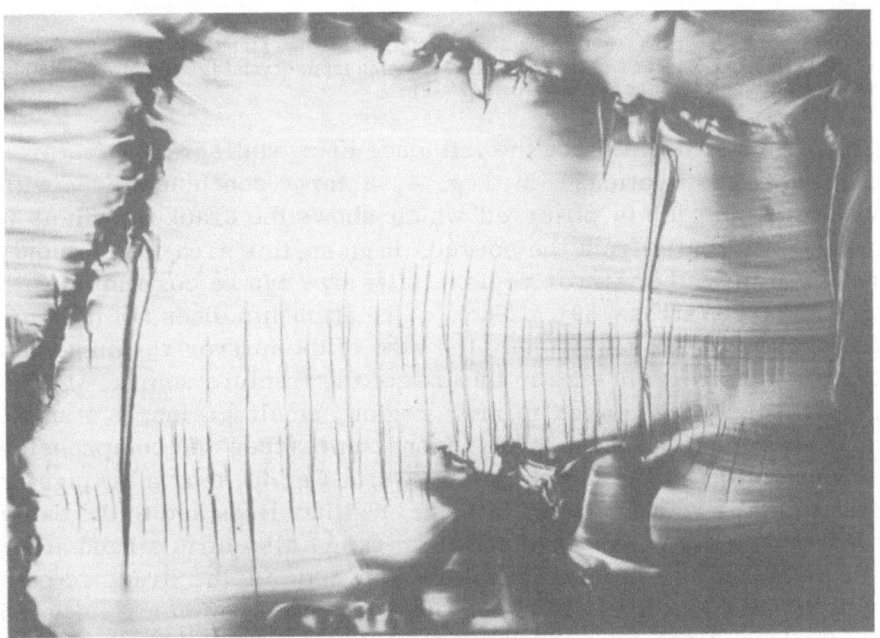

Fig. 5. Photomicrograph taken of the compressive region of the sample shown in Fig. 3, showing the appearance of tear lines or striations (200×).

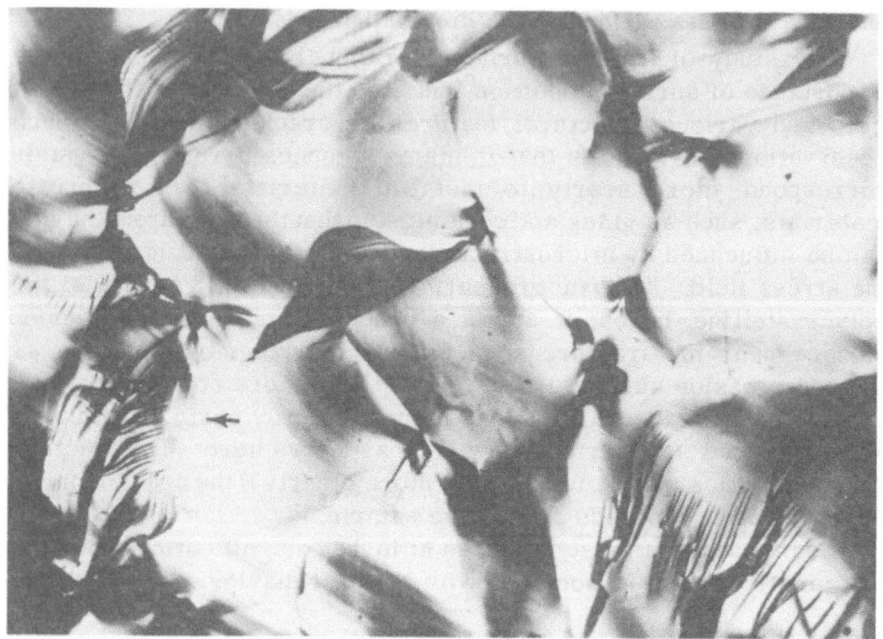

Fig. 6. Photomicrograph of the fracture surface showing striations initiating at twin interface (see arrow) and change in crack direction from crystal to crystal, as noted by the orientation shift of the Wallner lines (200×).

markings which illustrate the influence of crystallographic features on crack propagation. In Fig. 4, a large conchoidal area with Wallner lines [8] is observed which shows the crack growth as it proceeds radially from the source. In glass, this area is commonly referred to as the mirror region and its size can be correlated with the fracture stress [9]. Such a relationship does not hold for polycrystalline boron; rather, the size of the mirror region seems to depend upon grain size in the area of the fracture source. Large grains result in a large mirror region, small grains, in a small mirror region. As the crack front approaches the compression surface of the same sample (as shown in Fig. 5), tear markings or striations appear orthogonal to the Wallner lines and in the same direction as the crack front. Such markings also form a band along and adjacent to the compression surface as the fracture path approaches this free surface. These striations often are initiated at Wallner lines, as can be observed in Fig. 5; however, such markings, though more frequent nearer the compression surface, are not restricted to this area. In the regions such as shown in

Figs. 4 and 5, the propagation of the crack front is determined only by the stress field; the structure of the material has no effect.

### Influence of Structure on Crack Propagation

While in areas near the fracture source or in the large conchoidal regions no apparent interaction exists between the fracture path and the structure of the material, the movement of the crack front into the compressive region of the sample or into the finer-grained regions marks the beginning of such interactions with structural features.    This is illustrated in Fig. 6, where the crack front shifts direction from crystal to crystal as it progresses through a polycrystalline region, as indicated by the change in orientation of the Wallner lines.    It is seen also in Fig. 6 that crystallographic boundaries, such as twin interfaces, can initiate striations.    These striations are unusual when compared to the earlier illustrations in that they are not parallel to the crack front as indicated by the Wallner lines, but rather are orthogonal to the twin interface.    The complexity of fracture and the influence of structure on the progress of the crack front are manifested further

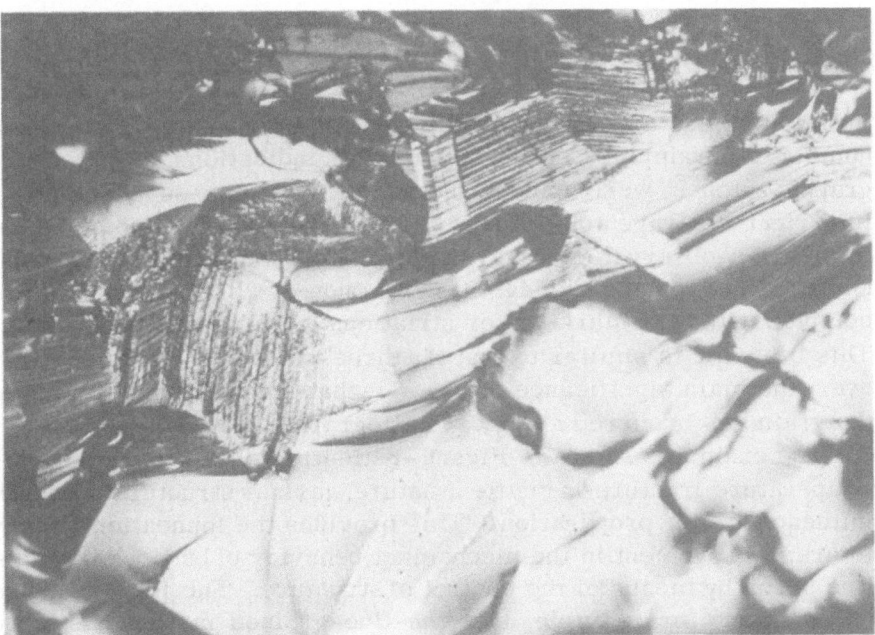

Fig. 7. Photomicrograph showing the lamellar appearance noting the influence of crystallographic planes (200×).

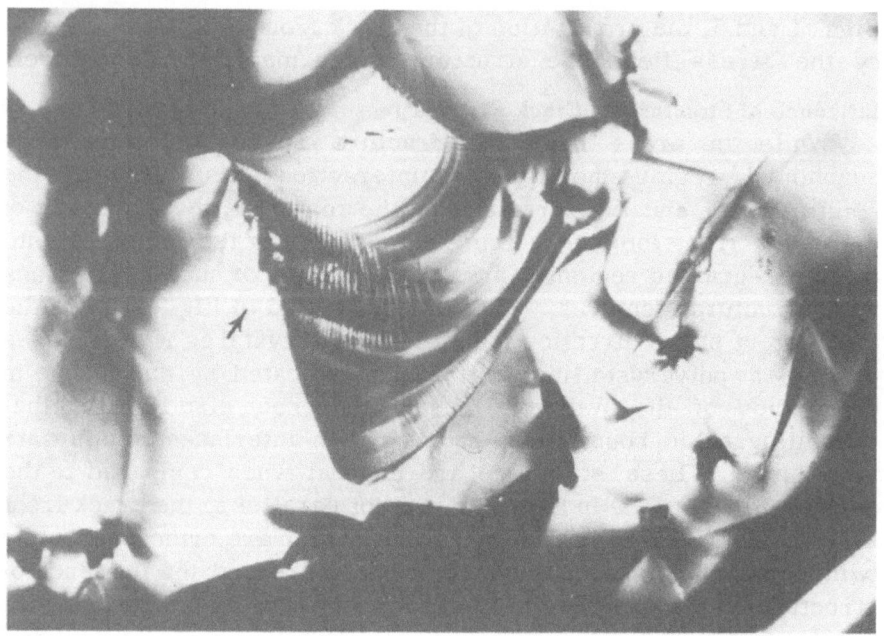

Fig. 8. Photograph showing the fracture as it occurs along crystallographic planes (see arrow). Also, the overall appearance of this grain is similar to entire three-point bend sample (200×).

in the lamellar appearance of the surface as shown in Fig. 7. In a number of examples, such regions with indications of crystallographic planes were found. These observations suggest that an imperfect cleavage took place. Figure 8 illustrates another most interesting observation on a single grain. Where the crack front enters, a simple conchoidal region is noted, which becomes more complex with tear markings or striations as the crack progresses. This behavior is similar to that of a three-point bend sample. However, the main significance of Fig. 8 is that fracture at the edges of the grain has occurred along crystallographic planes.

The examples shown in Figs. 6–8 illustrate that, although room-temperature fracture is brittle in nature, various structural features influence crack propagation. This provides the foundation for any future improvement in the mechanical behavior of bulk polycrystalline boron by means of the control of structure. The initial studies have shown, for example, that the fine-grained material is more desirable, since the influence of structural features on crack propagation is more pronounced in these regions. The observations of a transition in fracture behavior under certain conditions suggest that the influence of structural features on crack propagation will become more marked with an increase in temperature.

## MICROMECHANICAL BEHAVIOR OF POLYCRYSTALLINE BORON

### Analysis of Response to Stress

Microstrain measurements as employed by Brown et al. [10–12] are one of the newest techniques used to determine the initial response of materials to stress. Recent work by Brown and co-workers [10–12] and by Bonfield and Li [13] used measurements of the mechanical hysteresis loops in the anelastic region to study the initial response of various metals to stress. Although the studies by the above authors led to certain conclusions on the nature of dislocation movement and interaction, the chief aim of the studies on polycrystalline boron was to determine whether dislocation movement occurs at room temperature, and, if so, at which stress level.

For most metals, there is a mechanical hysteresis, as shown by the closed loop in the load–unload curve. The hysteresis is associated with dislocation movement. The stress required to move the first dislocations can be determined from the analysis of closed hysteresis loops. It has been shown [10,14] that the irreversible work $\Delta W$ done during each load–unload cycle is simply

$$\Delta W = 2 \sigma_F \Delta \gamma$$

where $\Delta W$ is the enclosed area of the mechanical hysteresis loop, $\Delta \gamma$ is the maximum strain amplitude (width) of the hysteresis loop, and $\sigma_F$ is the stress required to move the first dislocations. With techniques similar to those employed for metals, it was possible to examine the microdeformation characteristics of polycrystalline boron.

### Material Response to Stress in the Microstrain Region

Properties in the microstrain region were determined in four-point bend tests using an Instron testing machine at a crosshead speed of 0.002 in./min. Strain gauges having a sensitivity of $2 \cdot 10^{-6}$ in./in. were attached using Duco cement as specified by the gauge manufacturer. The experimental sequence in these tests was to observe the continuous variation in microstrain with stress during the entire load–unload cycle in order to establish the nature

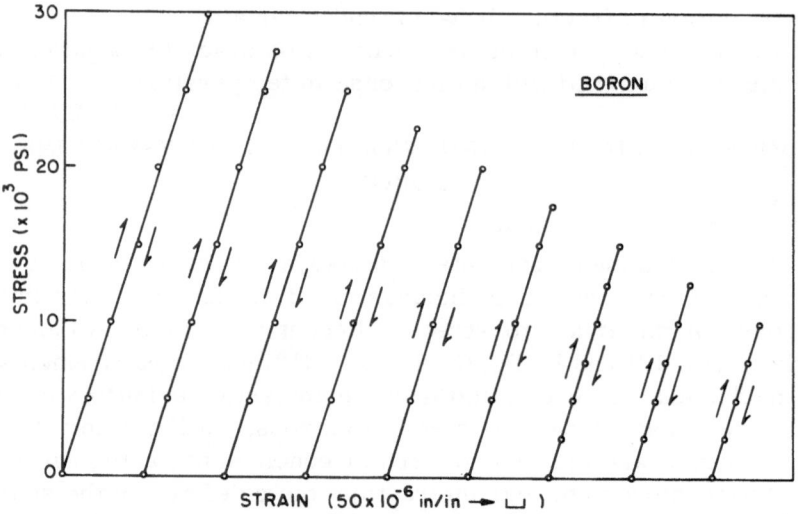

Fig. 9. Continuous load–unload, stress–strain curves for polycrystalline cast boron, tested in four-point bending at a strain sensitivity of 2 · $10^{-6}$ in./in. It should be noted that the points of loading and unloading (see arrows) are coincident.

of the room-temperature response to stress. Normally, such tests are run at successively increasing stress levels until a number of open hysteresis loops are formed or up to fracture.

The response to such load–unload cycles for polycrystalline boron was a straight line with no measurable hysteresis up to fracture. Figure 9 shows the typical microstrain behavior in bending. It should be noted that the points of loading and unloading in Fig. 9 coincide. The usual analysis of this response to stress cannot be applied, as no measurable hysteresis exists at stress levels up to failure. Such a response indicates the lack of any measurable dislocation movement; any movement would result in a hysteresis effect. It can be seen from Fig. 9 that the dimensional response to stress will be entirely predictable, and after removal of stress, the material will return to its original shape. Furthermore, the lack of dislocation movement at stresses greater than 30,000 psi suggests that the stress required to move the first dislocations will be very high. Additional work in this area will require material having flaw-free surfaces to allow tests at higher stress levels. In addition, the influence of temperature in the microstrain region should be investigated in order to understand more thoroughly the mechanical behavior of the material.

## CONCLUDING REMARKS

The casting technique developed in this investigation not only produced the unstressed, crack-free bulk polycrystalline boron ingots required for testing, but more significantly demonstrated that boron of high quality can be cast. Such a technique makes available for structural applications the bulk crystalline material, in addition to the fibers which are presently available.

The study of the fracture behavior has shown that, as expected, boron is very sensitive to surface imperfections which reduce the fracture stress of bulk material. This effect has not been eliminated as yet, although the fracture stress can be increased substantially through careful handling and polishing. As an example, bend tests have shown a three-fold increase in strength through careful sample preparation. Two types of surface imperfections that initiated failure, cracks and steps, were found in bulk material. Such failures emphasize the need for flaw-free surfaces; even though progress has been made, additional work is necessary in this area.

Surface markings associated with brittle failures are present in bulk crystalline boron; yet, observations of the fracture characteristics have shown that various structural features, such as twin interfaces, grain boundaries, and crystallographic planes, can influence the fracture path. This provides a foundation for future improvement in the mechanical behavior of boron by means of control of structure, and this should be investigated more thoroughly. The initial results have shown that fine-grained material is more desirable, since the influence of structural features on crack propagation is more pronounced in fine-grained regions. Furthermore, since the observations of the fracture surface show a transition in fracture mode under certain conditions, the effect of a temperature increase may be significant and should be investigated.

One of the potential uses of boron is for inertial guidance devices and other precision instruments which demand that materials used in their structural elements not only be dimensionally stable, but also that dimensional response to stress be entirely predictable. Boron satisfies such requirements, as the nature of the response to stress in the microstrain region shows no mechanical hysteresis at stress levels up to failure. That is, the dimensional response to stress is entirely elastic; the material returns to its original dimensions after being strained, and no measurable dislocation movement occurs.

## ACKNOWLEDGMENTS

The authors wish to acknowledge the capable assistance of R. G. Zurn. They also wish to thank Dr. J. N. Dempsey, Director of Research, Honeywell Research Center, for his interest and for permission to publish this work. The work was partially supported by the Navy, Bureau of Weapons.

## REFERENCES

1. Talley, C. P., L. E. Line, Jr., and O. D. Overman, Jr., Boron—Synthesis, Structure, and Properties, Plenum Press (New York), 1960, p. 94.
2. Talley, C. P., J. Appl. Phys. 30: 114 (1959).
3. Ellis, R. C., Boron—Synthesis, Structure, and Properties, Plenum Press (New York), 1960, p. 135.
4. Niemyski, T., I. Pracka, R. Szczerbinski, and Z. Frukacz, Proceedings of the Seventh International Conference on the Physics of Semiconductors, Paris, 1964, Dunod (Paris), p. 1295.
5. Griffith, A., Phil. Trans. Roy. Soc. London A 221: 180 (1921).
6. Griffith, A., Intern. Cong. Appl. Mech., Delft, 1924, p. 61.
7. Marsh, D.M., Fracture of Solids, Interscience Publishers (New York), 1963, p. 119.
8. Wallner, H., Z. Physik 114: 368 (1939).
9. Anderson, O. L., Fracture, John Wiley & Sons (New York), 1959, p. 344.
10. Roberts, J. M., and N. Brown, Trans. Am. Inst. Min. 218:454 (1960).
11. Brown, N., and K. F. Lukens, Acta Met. 9: 106 (1961).
12. Brown, N., and R. A. Ekvall, Acta Met. 10: 1101 (1962).
13. Bonfield, W., and C. H. Li, Acta Met. 13: 317 (1965).
14. Cottrell, Dislocations and Plastic Flow in Crystals, Clarendon Press (Oxford), 1953, p. 111.

# The Effect of Chemical Polishing on the Strength and Fracture Characteristics of Amorphous Boron Filaments

F. E. Wawner, Jr.

*Texaco Experiment Incorporated*
*Richmond, Virginia*

The effects of surface and internal flaws on the tensile and flexural strengths of amorphous boron filaments are discussed. It is shown that surface flaws can be eliminated by chemical polishing, greatly enhancing strength properties. Strength values exceeding 500,000 psi in tension and 2,000,000 psi in flexure have been obtained from the tests. Analysis of the stress distribution and fracture characteristics shows that the crystalline boride core is responsible for the large difference recorded in the two types of tests. Theoretical considerations indicate the maximum attainable flexural strength of amorphous boron is 3,900,000 psi.

## INTRODUCTION

Filaments of amorphous or glassy* boron possess extremely interesting mechanical properties. Talley [1] has given the modulus of rupture (flexural strength) as 350,000 psi and Young's modulus as 64,000,000 psi for glassy boron. These values coupled with its low density, 2.34 $g/cm^3$, make it quite attractive as a structural material for space-age applications. Previous difficulties of fabricating useful shapes from boron due to its lack of ductility have been circumvented by the advent and advancement of composite technology. Consequently, boron produced in filament form can be used advantageously as a reinforcement for structural composites.

The Griffith criterion and the statistical flaw theories assume that pre-existing flaws of some nature (surface cracks, notches, inclusions, etc.) are present in glass and other brittle solids, and that the ultimate strength and fracture are related to the number and severity of these flaws. Many experimental studies have supported these predictions and are described in the literature [2–5].

*This form of boron probably consists of microcrystalline aggregates of one of the crystalline polymorphs, and will be described more adequately in a paper to be published.

The fact that amorphous boron is a somewhat vitreous and brittle material suggested that surface defects probably influence its strength. A study was initiated, therefore, to determine the nature of the flaws present in the filaments and how they influence the strength properties and fracture characteristics. The approach taken in this study was a comparison of the strength of the filaments in the as-produced condition and after various degrees of polishing, which was correlated with microstructural features and fracture characteristics.

## EXPERIMENTS AND RESULTS

### Boron Filaments

Boron filaments used in this study were prepared by the chemical vapor plating of amorphous boron from a volatile halide in the manner described by Talley, Line, and Overman [6]. Tungsten wire, 0.5 mil in diameter, was used as a substrate for all samples. The filaments produced for this study were of two diameter ranges, 3–5 and 10–15 mils. The diameter of each test specimen was measured with a microscope using a filar micrometer eyepiece. The accuracy of the measurement was ±0.05 mil. Filaments were tested in the as-produced and chemically polished conditions. As a precise method of comparing the original and polished filaments, the samples were extracted alternately from a given piece of material. In this manner, one section of the filament was used for the polishing experiments, while its adjacent section was used to represent the as-produced state. The samples that were polished were mounted in a silicone rubber holder and immersed in a solution of water and red fuming nitric acid for the desired amount of time. The filaments were then rinsed in boiling water for several minutes to remove any oxides which formed on the surface. In order to maintain relatively constant conditions, a fresh polishing solution was prepared for each group of filaments.

### Selection of Polishing Solution

Preliminary experiments were performed to select a polishing solution of optimum concentration and to determine the rate at which the boron was polished away. Solutions of water and red fuming nitric acid in the proportions 0:1, 1:1, 2:1, 3:1, and 10:1 (water:acid) were evaluated. The 1:1 mixture produced the best results, removing the boron at a rate of approximately 0.1 mil/min at a bath temperature of 100°C. In addition to a slower etching rate, filaments polished in mixtures other than 1:1 consistently

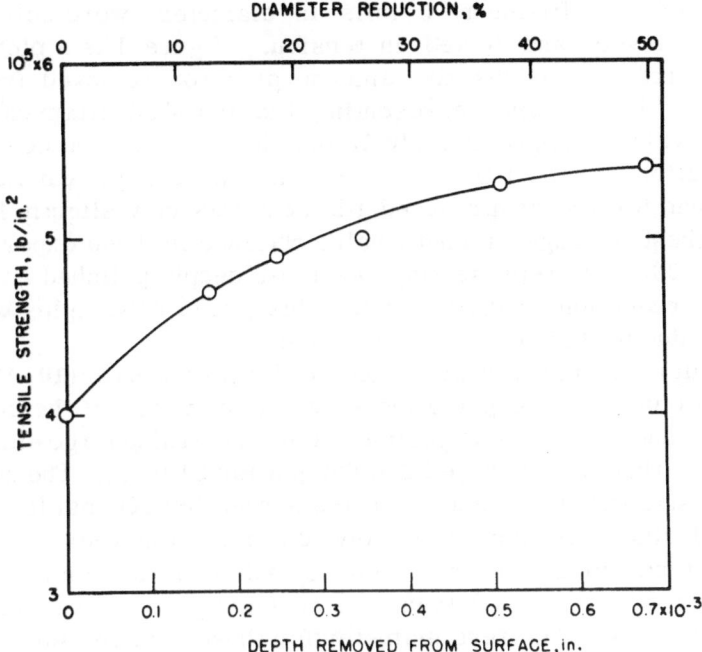

Fig. 1. Effect of surface polishing on tensile strength.

gave lower strength values. The selected etchant produced a very uniform surface on the filaments with no signs of preferential etching except in the cases with extremely long polishing times. Sulfuric and nitric acid mixtures and potassium periodate solutions are reported to be etchants for boron [7,8]; however, they did not yield results as good as those obtained with the nitric acid solution.

Mechanical Tests

Two types of strength tests, tensile and flexure, were used to evaluate the filaments. The tests were performed on an Instron tester at a loading rate of 0.05 in./min. Filaments tested in tension were mounted in tabs with an epoxy adhesive at $1/4$-in. gage length. Flexure tests were made with the 3-point loading method. The test jig consisted of two lower knife-edge supports, adjustable with a micrometer head, and a single knife edge applying the load from above. In keeping with ASTM standards [9], a gage length was selected which gave a span-to-depth ratio of 16–20/1.

Tensile Tests

The samples used in this study gave an average tensile strength of 400,000 psi for 76 individual tests in the original state. Six groups of the filament, 3 mils in diameter, were polished to different levels and tested in tension. Figure 1 is a plot of the tensile strength versus the amount of boron removed from the surface. Each point representing the polished filament is the average value of approximately 15 individual tests. The coefficient of variation for the polished groups averaged approximately 5%, while that for the as-produced filament was only slightly higher. The highest average strength value obtained in these experiments was 535,000 psi, representing the most deeply polished filament. Further reduction in diameter for this group of samples was not feasible due to their extreme smallness.

Similar experiments were made on large-diameter (10–12 mils) filaments in order to get a more complete picture of the curve at higher reductions. The experiment was made using large-diameter filaments that had averaged 280,000 psi for 24 tests. The average tensile strength recorded in 15 tests was 480,000 psi for lightly polished filaments (approximately 0.3 mil removed). Further tensile tests made on the samples polished to a much greater degree (4 mils removed from the surface) gave an average value of 560,000 psi. The test diameter for these samples was 4 mils. The increase in strength after polishing shows that surface flaws are present in the filaments and that they do influence the strength.

Flexural Tests

Flexural-strength determinations were made primarily on large-diameter filaments. The samples selected to represent the as-produced state were 10–15 mils in diameter and gave an average flexural strength of 730,000 psi for all groups tested (53 individual

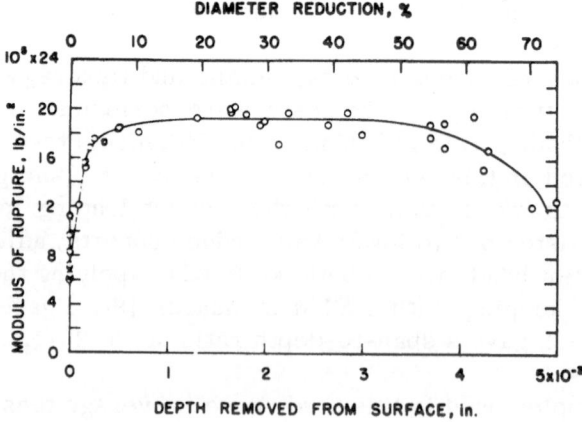

Fig. 2. Effect of surface polishing on modulus of rupture.

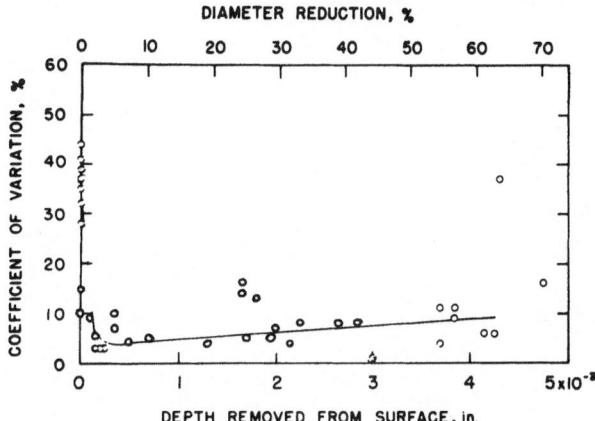

Fig. 3. Effect of surface polishing on coefficient of variation.

tests). Alternate samples were taken from each group and polished, removing 0.1–5.0 mils of boron from the surface. This range represented an approximate diameter reduction of from 1% to 73%.

In Fig. 2, the modulus of rupture or flexural strength for the above groups is plotted as a function of the surface removed. Each of the points on the graph represents at least five tests. As was the case with the tensile experiment, the flexural strength increased to a level of approximately 1,900,000 psi as the surface layers were etched away.

Figure 3 is a companion curve showing the coefficient of variation as a function of the surface removed for the same tests. The relatively high coefficient of variation for tests on the as-produced samples dropped from 40 to 4% after 0.3 mil had been removed and remained at this level until the higher reductions were reached, where it began to rise again. The high coefficient of variation in the as-produced case indicates that the surface flaws are not all equally severe. However, their ability to concentrate stresses is reduced to the same magnitude after slight polishing. If the surface flaws are microcracks, this reduction is caused by the acid solution penetrating the crack and attacking its more energetic tip, rounding it off.

It was interesting to note that out of 164 individual tests in this series of experiments, 36 gave modulus of rupture values exceeding 2,000,000 psi. The highest value obtained was 2,700,000 psi. These room-temperature strength values, recorded with no special precautions other than acid-polishing the filaments, are among the highest ever witnessed for a bulk material.

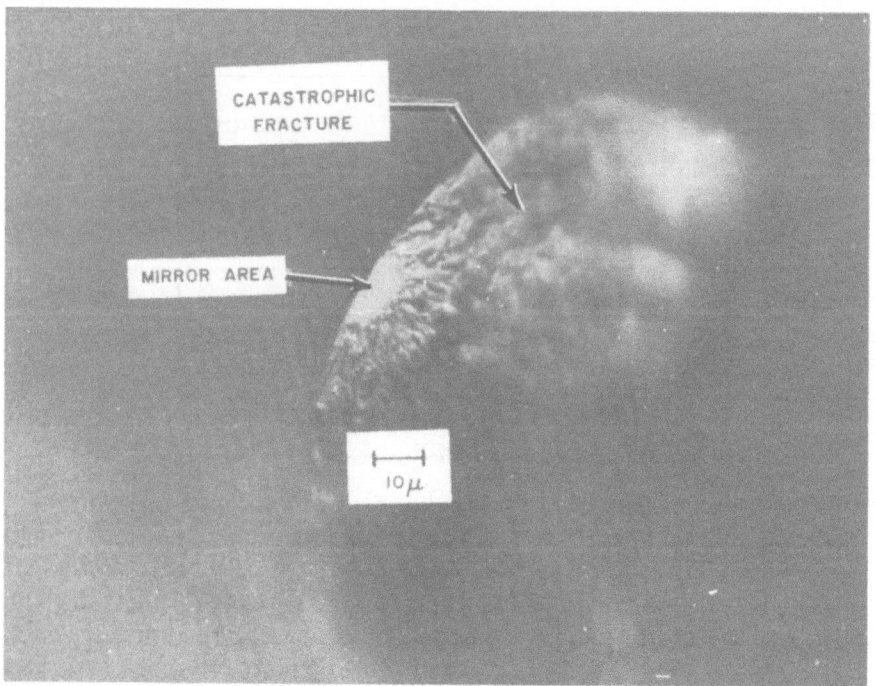

Fig. 4. Typical fracture surface in boron filaments. Nucleation site of typical surface fracture in unpolished filament.

## Examination of Fracture Surfaces

Whenever possible, broken ends of the filaments were captured during the above tests for examination of the fracture surfaces. This could be done with regularity in tensile tests; fracture in bend tests, however, was so explosive that the ends were not obtainable. Examination of the surface of as-produced filaments tested in tension showed that fracture always nucleated on the surface (as denoted by the smooth mirror area or region of slow crack growth). After polishing to the optimum depth, the origin of fracture for filaments tested in the same manner was adjacent to the interface between coating and tungsten boride core. Figures 4 and 5 depict typical fracture surfaces for as-produced and etched filaments showing the origin fracture.

## Surface Observations

Visual observations of the surface of many filaments were made before and after polishing in an attempt to detect irregularities, such as cracks, inclusions, notches, or anything that might be deemed a typical surface flaw and to follow changes in the microstructural features. Figures 6-8 are a series of optical micrographs showing an as-produced surface and polished surfaces. As-produced boron filaments exhibit a characteristic surface of nodules or areas of preferred growth separated by boundaries. The microstructure is generally similar for filaments of all diameters, differing only in the size of the nodules. Upon chemical polishing, the change is slow until approximately 0.45 mil has been removed. At this point, the characteristic features are destroyed, yielding a somewhat smooth surface that is difficult to focus optically. After long times in the solution, the polishing effect is quite pronounced, indicating some preferential attack.

Similar experiments were made with the electron microscope, utilizing replication techniques to follow the surface changes at higher magnifications. Figure 9 is an electron micrograph of an as-produced filament showing the characteristic appearance. Figure 10 shows the surface after approximately 0.15 mil has been removed. The surface is considerably smoother and the boundary outlines

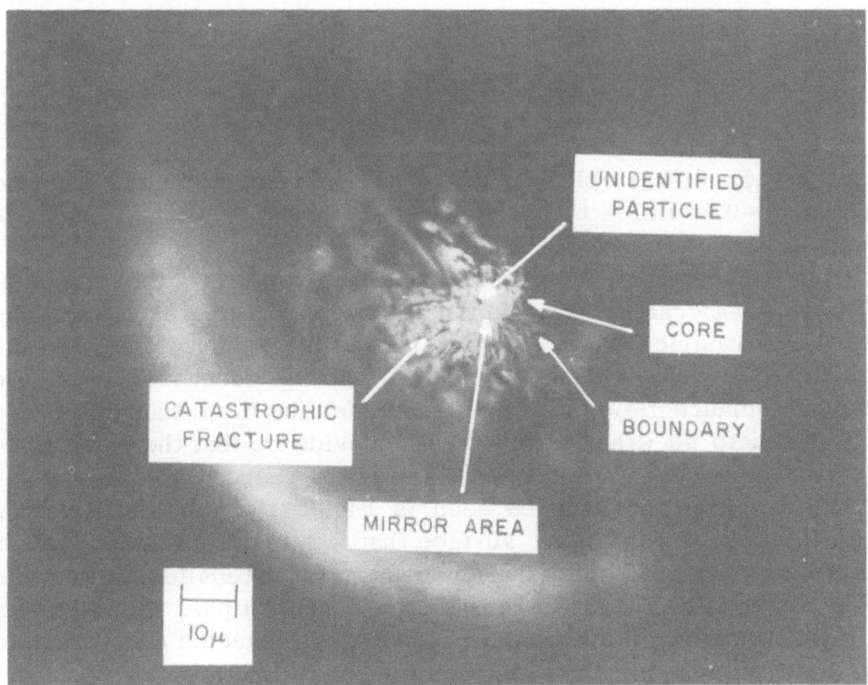

Fig. 5. Typical fracture surface in boron filaments. Nucleation site of typical interfacial fracture on polished filament.

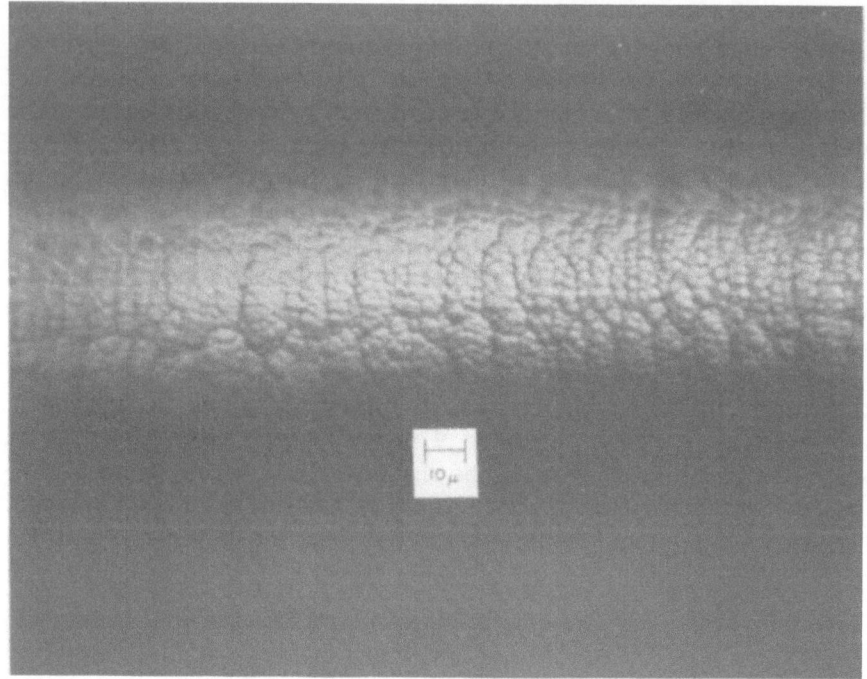

Fig. 6. Optical micrograph of filament surface—unpolished.

are barely discernible. After the optimum amount of surface removal—0.45 mil — the surface is completely smooth, as shown in Fig. 11.

In this series of observations, no obvious flaw was detected. The only noticeable change in the surface structure was elimination of the boundaries which characteristically appeared on every as-produced surface. This suggested that the boundaries were the typical surface flaws or were masks for the flaws that influenced the strength so drastically. Further evidence that the boundaries were the surface flaws responsible for the reduced strength of as-produced filaments can be seen in Fig. 12, which is an electron micrograph of a filament surface that was unusually smooth after deposition and had quite shallow boundaries. Tensile tests on this filament gave an average strength value of 480,000 psi. Similar tests on the filaments after they were polished revealed an average of 490,000 psi or essentially no increase. Examination of the fracture surfaces showed that fracture nucleated adjacent to the interface in every case and not on the surface. Hence, it appears that the strength is dependent upon the depth of the boundaries and that the boundaries are the inherent surface flaws.

## DISCUSSION

Analysis of the curves in Figs. 1 and 2 shows that they are generally similar in shape. The increase in strength demonstrates that surface defects do exist on amorphous boron filaments and that they are removed or rendered ineffective after approximately 0.5 mil of the surface has been polished away. The strength curves at this point enter a plateau region, whereupon deeper polishing does not enhance the strength. After reaching the plateau region, the strength is apparently controlled by some other flaw site. The primary difference in the curves is the strength level they attain at the plateau. In tensile tests, 500,000–600,000 psi appears to be the upper limit. However, flexural tests of polished filament give strength values exceeding 1,900,000 psi.

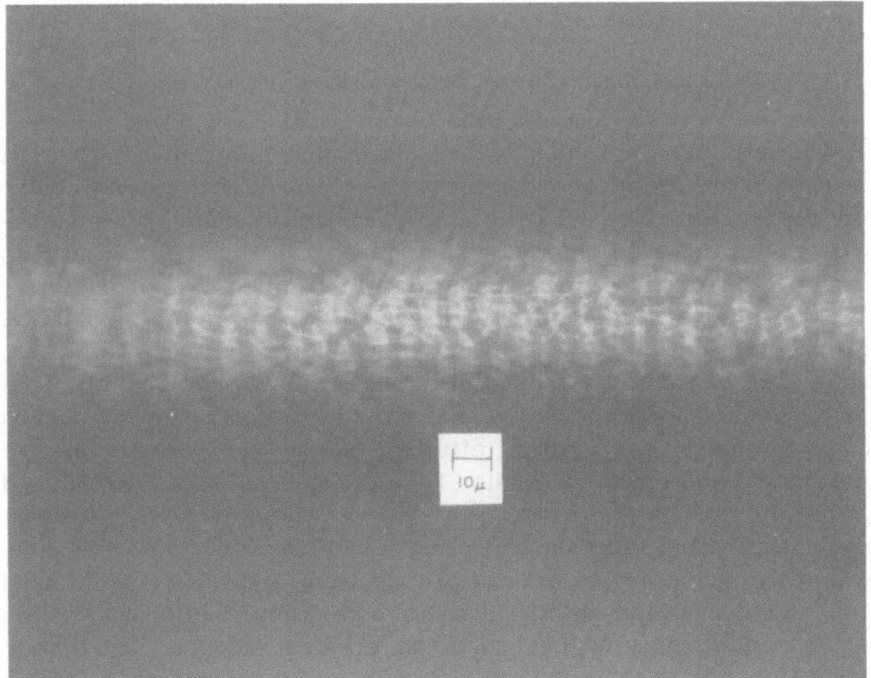

Fig. 7. Optical micrograph of filament surface—0.45 mil removed.

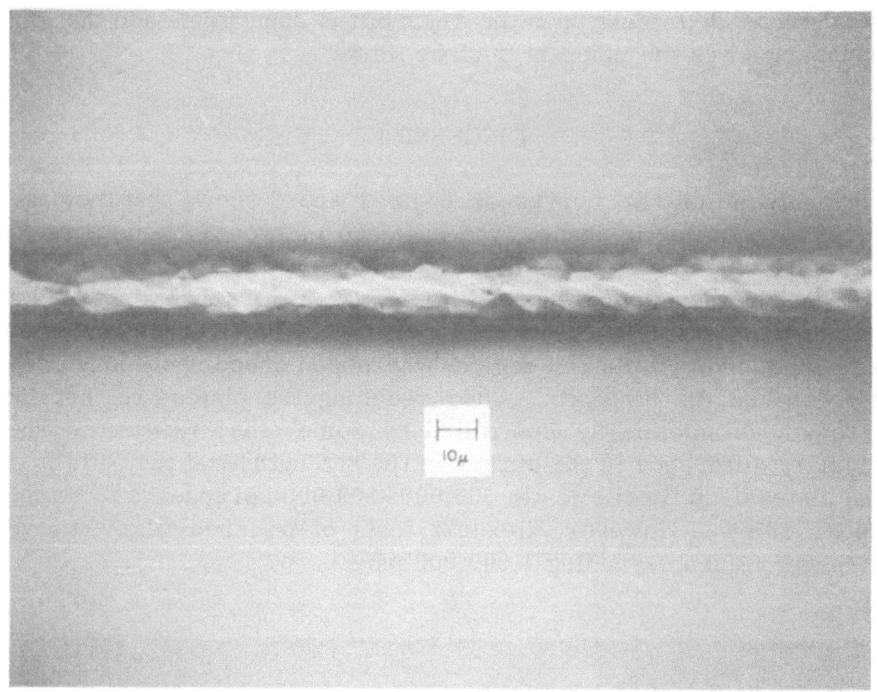

Fig. 8. Optical micrograph of filament surface—1.68 mil removed.

Weibull [10] derived a theoretical expression showing that the ratio of bend strength to tensile strength for a rectangular homogeneous material is 1.41. He, as well as others [11], tested the relationship experimentally and found excellent agreement. This ratio for as-produced amorphous boron filaments, approximately 1.6, is quite similar in spite of the difference in test geometry. After polishing away the surface flaws, however, the ratio is 3.8, much greater than that predicted by the Weibull relationship. This very large ratio suggests that severe stress concentration occurs in the tensile case after elimination of the surface defects. Uniform bodies tested in tension have the applied stress distributed uniformly across the cross section. If a severe flaw is present within this cross section, the stress will concentrate at this flaw, promoting crack nucleation and propagation. In the three-point bend test, the test sample is placed under both compression and tension with the outermost edges under the greatest stress. The tensile and compressive stresses become less as the neutral axis is approached and are zero along this axis. Consequently, a potential stress concentration site near the neutral axis in a bend test contributes little toward the fracture of the sample.

Boron filaments are nonhomogeneous structures composed of a tungsten boride core, approximately 0.5 mil in diameter, surrounded by a coating of bulk amorphous boron. It is reasonable to assume that the strength of the filament is affected not only by the inherent properties of the bulk boron and the condition of its surface, but also by the presence of the core material and the interface which exists between the core and the boron. In flexural tests, the core, being along the neutral axis, contributes little toward the fracture of the filament. Hence, a strength value representative of the bulk amorphous boron is obtained from this type of test. A filament tested in tension, however, has the core and interfacial region as an integral part of its cross section tending to concentrate applied stresses and nucleate fracture. This analysis is consistent with the strength data obtained from the two types of tests and with observations of the fracture surfaces showing crack initiation adjacent to the core in polished filaments. The analysis also

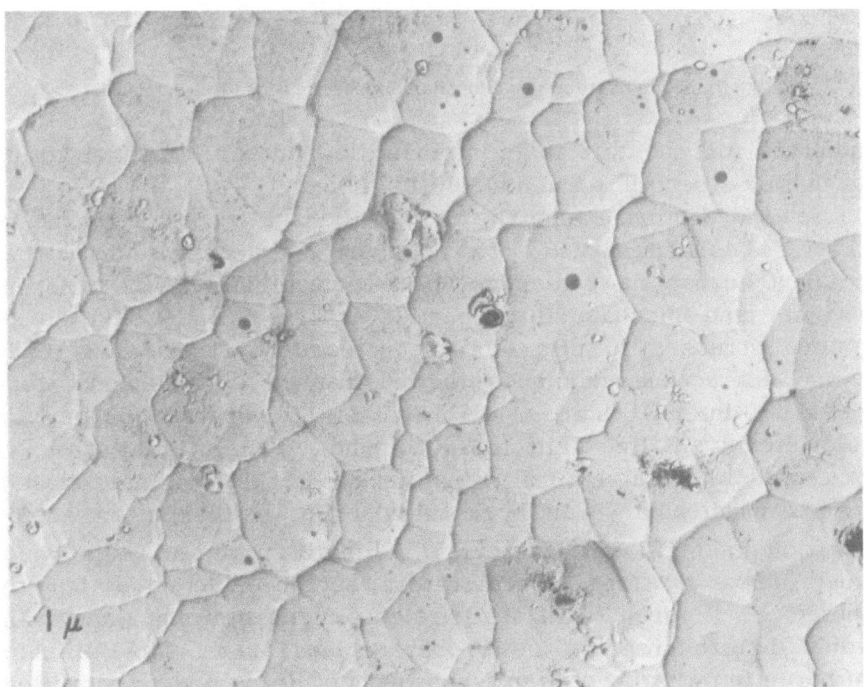

Fig. 9. Electron micrograph of filament surface—unpolished.

Fig. 10. Electron micrograph of filament surface—0.15 mil removed.

indicates that the core is an intrinsic flaw in the filament, limiting its tensile strength to the 500,000-psi level.

Several factors cause the region adjacent to the core to act as a stress concentrator. Firstly, there is a definite structural gradient across the interface with two materials having different physical and mechanical properties. Hence, one would expect applied stresses to pile up here. Secondly, stress analysis by X-ray diffraction techniques showed that the core is under considerable internal stress. A value of 150,000–200,000 psi was calculated from line-shift measurements. This internal stress is generated by the structural transformation of the original tungsten wire to tungsten boride and by the mismatch of the thermal expansion coefficients of the two materials. Finally, pores or voids, which could serve as stress concentrators, have been detected along the interface. Figures 13 and 14 are electron micrographs showing the voids along this region. They probably arise from the rapid diffusion of boron into the tungsten and subsequent vacancy condensation along the interface between the two materials, a phenomenon called the

Kirkendall effect [12].    With all these effects present near the
center of the filaments, it is not difficult to visualize this region
as deleterious to the tensile properties. That fact that the core has
a definite size implies that it should especially influence the
flexural strength and fracture properties of extremely small
filaments.    This, along with experimental inaccuracies due to
excessive deflection of the small-diameter filaments, explains the
observed decrease in the flexural strength when the diameter has
been greatly reduced (Fig. 2).

Theoretical Considerations

The extremely high values obtained in bend tests on polished
boron filaments raised the question of what the maximum strength
of the material might be. Theoretical strength values for materials
are generally approximated from a rule of thumb that gives the range
of values $E/10 - E/20$, where $E$ is Young's modulus of the material.
Therefore, Gilman [13], using the 5% strain method, has calculated
the maximum strength for boron as 2,600,000 psi. Indications are
that the theoretical strength of boron is closer to $E/10$ , since an

Fig. 11. Electron micrograph of filament surface—0.45 mil removed.

Fig. 12. Electron micrograph of smooth unpolished filament surface.

experimental value from the present study exceeded Gilman's value. Hence, a different approach was taken to get an indication of the theoretical strength of boron.

Proctor [4], in his study on the effects of etching on the strength of glass, proposed a crack model for the typical surface flaw. In this study, he modified an equation originally proposed by Inglis [14] to fit his model. The equation in its modified form is

$$\sigma_T = \sigma\left(1 + 2\sqrt{l_0/r_0 + \delta}\right) \tag{1}$$

where $\sigma_T$ is stress at the crack tip or maximum stress of the material, $\sigma$ is the overall applied stress or breaking stress, $l_0$ is the crack depth, $r_0$ is the radius of the crack tip, and $\delta$ is the depth removed from the surface at any instant. Using this model, he predicted the shape of the strength versus depth removal curve and gave an estimate of the values for the parameters $l_0$ and $r_0$ for his glass samples. Wilkinson and Proctor [15] extended this work by making some conclusive observations which verified the proposed crack model.

Amorphous boron has many properties similar to glass; hence, it was felt that the theoretical model proposed by Proctor could be used to give an indication of the cohesive strength of the material and to plot the theoretical strength versus surface removal curve. The experiments made in this study indicated that the boundaries were the surface flaws causing premature failure; hence, measurement of the boundary dimensions from electron micrographs gave the approximate values of 0.0024 mil (0.06 $\mu$) for the width and 0.012 mil (0.3 $\mu$) for the depth* for the most severe ones. If the boundaries are the surface flaws, then the measured values can be interpreted as $2r_0$ and $l_0$, respectively, of the equation. The average

*The measurement for the depth of the boundary was made from the shadow cast at a known shadowing angle and is actually a measurement of the height of the nodule above the boundary. In reality, the boundaries probably extend deeper into the filament.

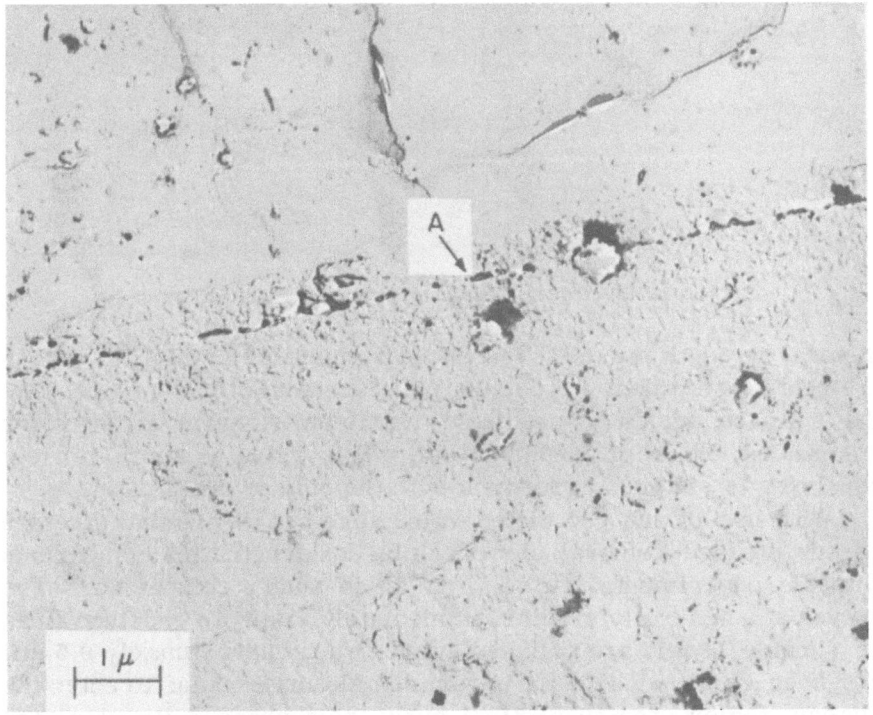

Fig. 13. Electron micrograph of pores along interface.

Fig. 14. Electron micrograph of pores along interface.

flexural strength for the as-produced filament from which the measurements were taken was 530,000 psi. Substitution of the values into
the equation yields a predicted maximum flexural strength for
amorphous boron of 3,900,000 psi.    This value is approximately
equal to $E/15$, in good agreement with the rule of thumb.

With use of the calculated value for $\sigma_T$, a theoretical strength
versus surface removal curve can be constructed for comparison
with the experimental curve; Fig. 15 is such a comparison. The
curves are generally similar, immediately rising to greater values
as surface layers are polished away. When approximately 0.5 mil
has been removed and the experimental curve begins to enter the
plateau region, the theoretical curve also changes its slope and
approaches the ultimate value of 3,900,000 psi very slowly in a

semiplateau fashion. The difference in strength at any point on the curves is to be expected, since consideration has been given only to surface flaws in the theoretical model used. In reality, internal flaws and the core of the filaments influence the strength considerably and must be included in the theoretical case for a closer match.

## SUMMARY AND CONCLUSIONS

It has been shown that surface flaws are present in boron filaments produced by chemical vapor plating and are deleterious to strength properties. The flaws appear to be boundaries separating areas of preferred growth on the surface of the filaments. Chemical polishing with a nitric acid solution will eliminate the flaws, giving an increase in strength equivalent to over 100,000 psi in tension and 1,200,000 psi in flexure. Although the flaws appear to be relatively small, their influence is not completely removed until 0.5 mil of the surface has been polished away.

Analysis of the filament fracture characteristics shows that the relatively small increase in strength in the tensile case can be explained by the interaction of the crystalline core. This active stress concentrator limits the tensile strength to the 500,000-psi level. Strength values from bend tests are more indicative of the true strength of amorphous boron. The bend values obtained in

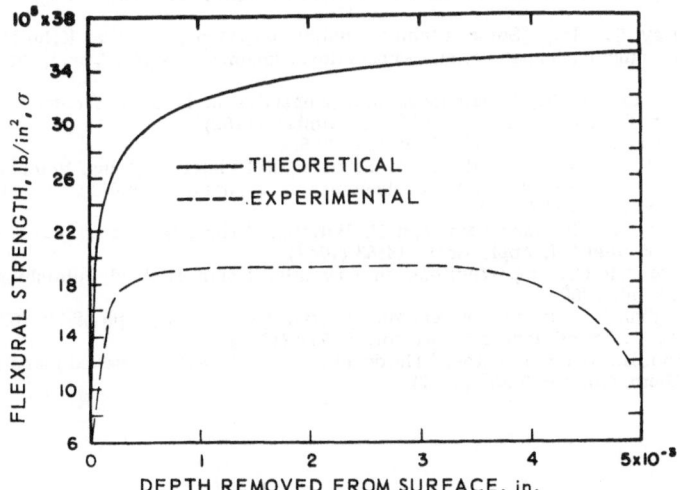

Fig. 15. Comparison of theoretical and experimental strengths. The theoretical curve was calculated from equation (1) where $\sigma_T$, $l_{0}$, and $r_0$ were 3,900,000 psi, 0.012 mil, and 0.0012 mil, respectively.

this study demonstrate that amorphous boron is one of the strongest bulk materials known. Theoretical considerations indicate that 3,900,000 psi is the maximum cohesive strength of amorphous boron.

## ACKNOWLEDGMENTS

The author wishes to acknowledge the Air Force Materials Laboratory, Advanced Filaments and Composites Division, which sponsored this work under Contract AF 33(616)-8067. Appreciation is also expressed to Dr. Homer A. Hartung for his helpful discussions and to the Ernest F. Fullam Company, which prepared the electron micrographs.

## REFERENCES

1. Talley, Claude P., "Mechanical properties of glassy boron," J. Appl. Phys. 30: 1114-1115 (1959).
2. Anderson, O. L., "The Griffith Criterion for Glass Fracture," in: B. L. Auerbach et al. (eds.), Fracture, John Wiley and Sons (New York), 1959, pp. 331-353.
3. Charles, R. J., and J. C. Fisher, "Strength of Amorphous Solids," in: V. D. Frechette (ed.), Non-Crystalline Solids, John Wiley and Sons (New York), 1960, pp. 491-507.
4. Proctor, B., "The effects of hydrofluoric acid etching on the strength of glass," Phys. Chem. Glasses 3: 7-27 (February 1962).
5. Johnson, T. L., C. H. Li, and R. J. Stokes, "Strength of ionic solids," Honeywell Research Center, HR-60-558 (November 1960).
6. Talley, Claude P., Lloyd E. Line, Jr., and Quinton D. Overman, Jr., "Preparation and Properties of Massive Amorphous Elemental Boron," in: J. A. Kohn et al. (eds.), Boron—Synthesis, Structure, and Properties, Plenum Press (New York), 1960, pp. 94-104.
7. Ellis, Ray C., Jr., "Some Etching Studies on Boron," in: J. A. Kohn et al. (eds.), Boron—Synthesis, Structure, and Properties, Plenum Press (New York), 1960, pp. 135-139.
8. Fasolino, Ludwig G., "Thermodynamic properties of bimetallic compounds," Natl. Res. Corp. Quarterly Tech. Rept. 31, December (1962).
9. ASTM Standards Book of 1961, Part 9, p. 358.
10. Weibull, W., "A statistical theory of the strength of materials," Ing. Vetenskaps Akad. Handl. NR 151 (1939); and "The phenomenon of rupture in solids," Ing. Vetenskaps Akad. Handl. NR 153 (1939).
11. Davidenkov, N., E. Shevandin, and F. Wittman, "The influence of size on the brittle strength of steel," J. Appl. Mech. 14:63 (1947).
12. Van Buren, H. G., Imperfections in Crystals, North Holland Publishing Company (Amsterdam), 1960.
13. Gilman, John J., "Strength of ceramic crystals," NBS Monograph 59:79-102 (1963).
14. Inglis, C. E., Trans. Inst. Nav. Archit. 55:219 (1913).
15. Wilkinson, B., and B. Proctor, "The development of defects on etched glass surfaces," Phys. Chem. Glasses 3:203 (1962).

# Boron Semiconductor Devices

## Wolfgang Dietz and Hermann Helmberger

*Consortium für elektrochemische Industrie*
*Munich, Germany*

The current vs. voltage characteristic of boron makes it possible to produce simple boron semiconductor devices which in principle act as contactless switches. The switching operation can be triggered by varying the current through the boron, the temperature of the boron, or the ambient temperature. Measurements of the current vs. voltage characteristic are reported. These characteristics and the possible operating points are also calculated under the assumption of a pure temperature effect. Various examples of circuits employing boron semiconductor devices are given.

## INTRODUCTION

When a small and slowly increasing voltage is applied to a boron sample (which is fitted with two contacts or clamped between electrodes) in series with a resistor, we observe at first only a slight increase in current. Above a certain voltage, the current suddenly rises sharply, while the voltage drop at the boron sample diminishes. The current is then limited only by the resistor in the circuit. The current–voltage characteristic so obtained is similar to that of a gas discharge. This breakdown effect in boron has been known for quite a long time. For instance, Weintraub [1], who noted the current–voltage characteristic of boron as early as 1913, explained the breakdown effect by the high resistance-temperature coefficient of boron and the heat generated by the passage of the current. His measurements were made with direct current. In 1918, Lyle [2] described the breakdown effect somewhat more precisely and compared it with the breakdowns occurring in insulators. He, too, attributed the breakdown to the heat generated by the passage of current and the temperature-dependent resistance. Lyle also established correlations between the breakdown voltage, the ambient temperature, and the geometric dimensions of the boron sample. He made his measurements with direct current and believed that similar conditions would apply to alternating current. In 1939, Bruce and Hickling [3] again investigated the breakdown effect,

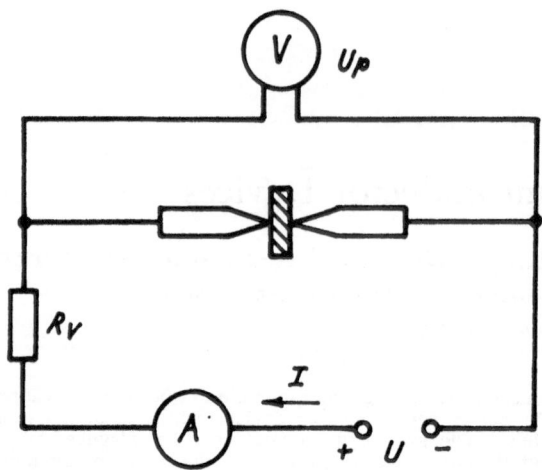

Fig. 1. Circuit for measuring current–voltage characteristic with DC.

using direct current and 50-cycle alternating current. These authors came to the conclusion, because of the short times involved, that the effect could by no means be a pure temperature effect, but they failed to produce an alternative explanation.

In recent years, a similar effect was observed primarily in ZnO and $BaTiO_3$, and was termed the assistor effect. Schnupp [4] explained this effect as a process preliminary to the thermal breakdown and obtained quite good agreement between experimental results and his theory for the current–voltage characteristic, the breakdown voltage, etc. Schnupp also calculated the case of alternating currents and arrived at limit frequencies of several megacycles per second for his samples, which were cubes of 10 $\mu$ on each side.

## MEASUREMENT OF THE CURRENT–VOLTAGE CHARACTERISTIC

For the measurement of the current–voltage characteristic of the boron manufactured by us, we used disks that were cut from zone-melted rods and then ground down to about 50–100 $\mu$. The disks, initially without contacts, were clamped between brass electrodes. The characteristic was then determined with direct current, using the circuit shown in Fig. 1. It proved expedient to use a pentode with a constant screen-grid voltage, in preference to an ohmic series resistor. With this, the curve could be recorded almost completely.

The curve (obtained by plotting the individual operating points) for increasing current is not identical with the curve for decreasing current. (See Fig. 2.) This is probably a pure temperature effect, since the measuring times were relatively short. If the time for the measurement of the individual points is lengthened, the return curve approaches the initial curve more closely.

The difference between the two curves is especially marked when AC is used. Our AC measurements were made at 50 cps, using the circuit shown in Fig. 3. The display obtained in this case is shown in Fig. 4.

If the measurements are carried out as described on samples without attached contacts, the individual measured values fluctuate sometimes considerably, presumably because of the varying contact resistance, and also possibly due to variations in heat dissipation. To achieve the reproducibility required for the

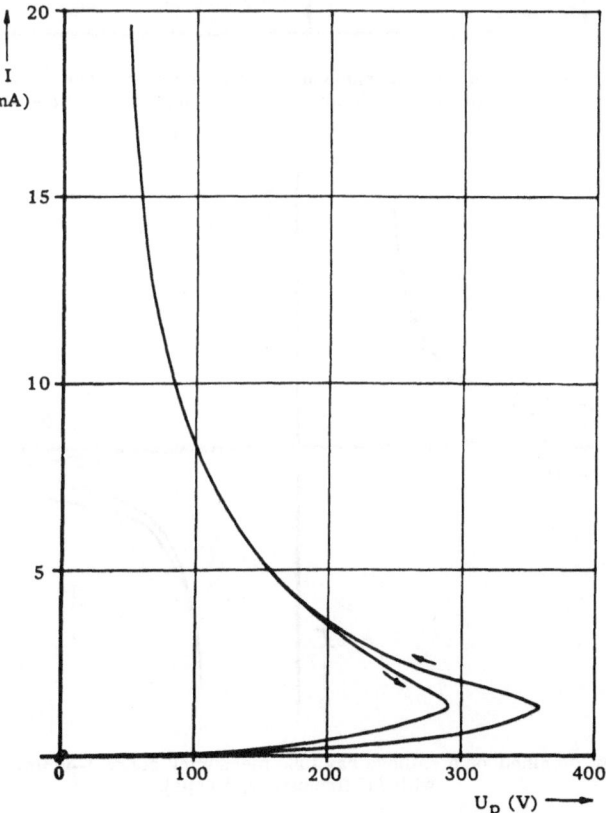

Fig. 2. Current–voltage characteristic of boron disk.

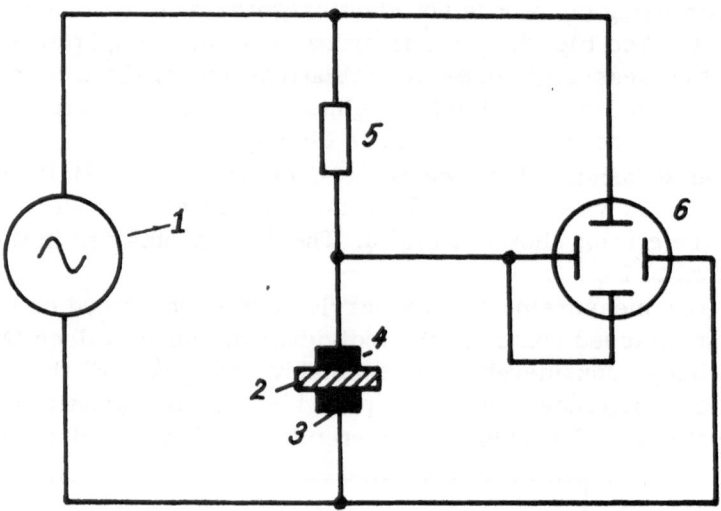

Fig. 3. Circuit for measuring current—voltage characteristic with AC. (1) AC power supply; (2) boron disk; (3 and 4) electrodes; (5) series resistor; and (6) oscilloscope.

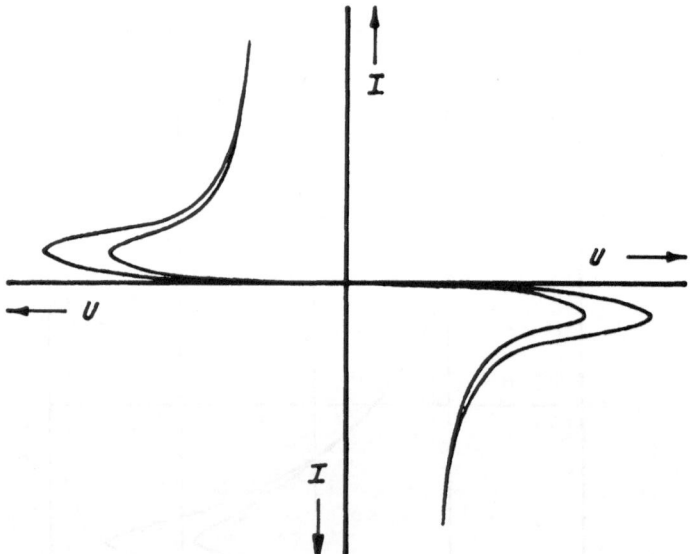

Fig. 4. Display obtained on a cathode ray tube measuring current—voltage characteristic with AC (frequency, 50 cps).

manufacture of components, either flawless pressure contacts on metallic coatings on the boron or soldered or welded terminals are necessary.

In our opinion, the breakdown effect is predominantly a heat effect. In a pure heat effect, the breakdown is produced by the Joule heat acting on the strongly temperature-dependent resistance. A basic circuit for the case of a pure heat effect is given in Fig. 5.

The thermal and electrical quantities are related as follows. The power input $N_1$ is

$$N_1 = \frac{U^2}{R(1 + R_v/R)^2} \tag{1}$$

where $U$ is the total voltage, $R$ is the temperature-dependent resistance of the boron sample, and $R_v$ is a constant series resistor. The power output $N_2$ for the temperature range of about 20–300°C (with only heat conduction assumed) is

$$N_2 = (\lambda/l)F(\theta - \theta_U) \tag{2}$$

where $\lambda$ is the thermal conductivity of the leads, $l$ is the length of the leads, $F$ is the cross section of the leads, $\theta$ is the temperature of the boron sample, and $\theta_U$ is the ambient temperature (assumed constant).

Figure 6 shows curves calculated from equations (1) and (2). Curves 1, 2, and 3 show the input power $N_1$ as function of tempera-

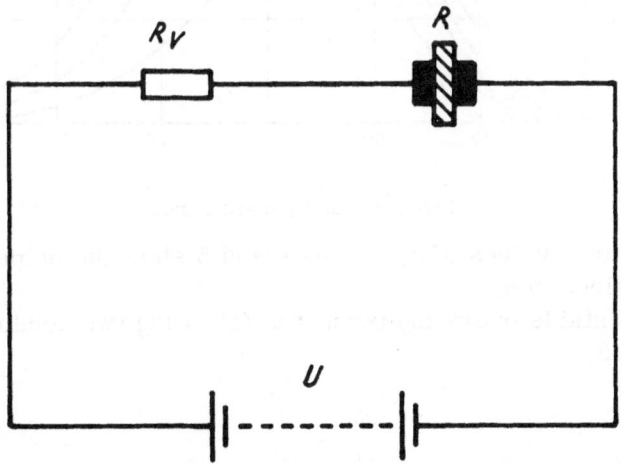

Fig. 5. Basic circuit for calculating power balance. $R$ is resistance of boron semiconductor device; $R_v$ is series resistor; and $U$ is voltage of power supply.

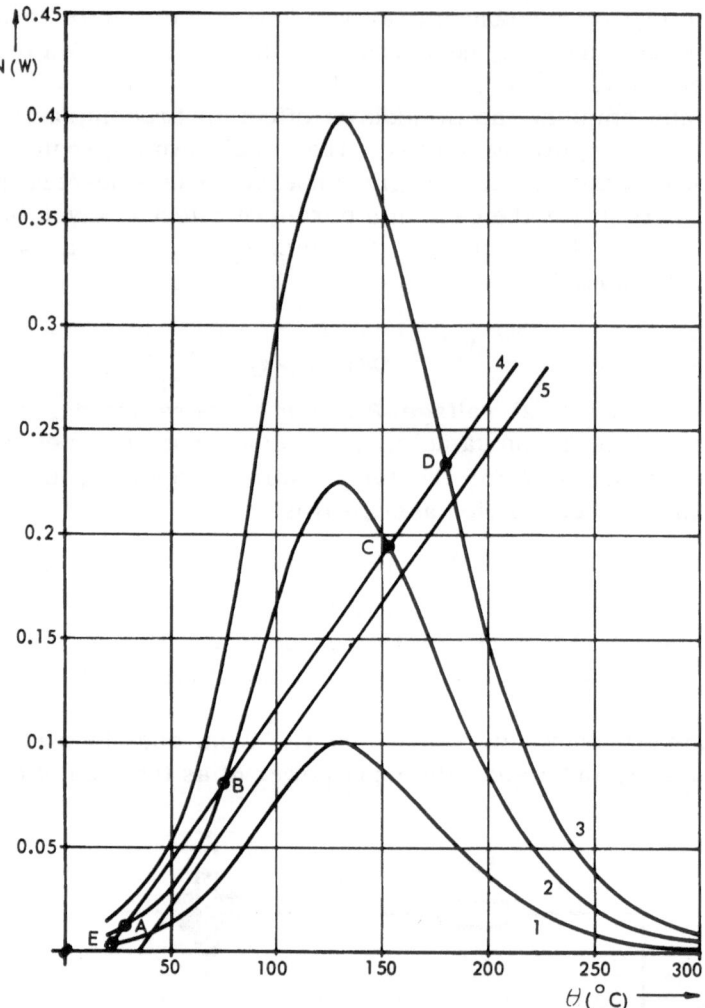

Fig. 6. Power input and output.

ture for three values of $U$; curves 4 and 5 show the output power $N_2$ for two values of $\theta_U$.

For a stable operating point, the following two conditions must be satisfied:

$$N_1 = N_2 \tag{3a}$$

$$\frac{dN_1}{d\theta} < \frac{dN_2}{d\theta} \tag{3b}$$

If the output $N_2$ is given by curve 4, we obtain only one stable operating point (E) for voltage $U_1$ (curve 1), two stable operating points (A and C) for voltage $U_2$ (curve 2), and one stable operating point (D) for voltage $U_3$ (curve 3). If we apply the voltage $U_2$, operating point A is obtained (low temperature, high resistance). If the voltage is raised to $U_3$, we obtain operating point D (higher temperature, smaller resistance). Then, if we subsequently revert to $U_2$ without breaking the circuit, we obtain operating point C. Hence, temporary application of a higher voltage to boron transforms a state of poor conductivity into one of higher conductivity, which is then retained even at a lower voltage. This arrangement operates as a switch. If now, with the sample in a conducting state, the voltage is dropped from $U_2$ to $U_1$, we first obtain operating point E, and, after stepping up again to $U_2$, we finally obtain operating point A. The switch is off again. Instead of only reducing the voltage to $U_1$, the circuit may, of course, be opened for a short time.

The process described above constitutes the mode of operation of a basic circuit which leads to a variety of possibilities for the application of boron semiconductor devices. The switching process is triggered by a brief stepping-up of the current through the boron. This can be achieved not only by increasing the voltage as described, but also by reducing $R_v$.

A second method of triggering the switching process also can be derived from the diagram of the power curves (Fig. 6). When

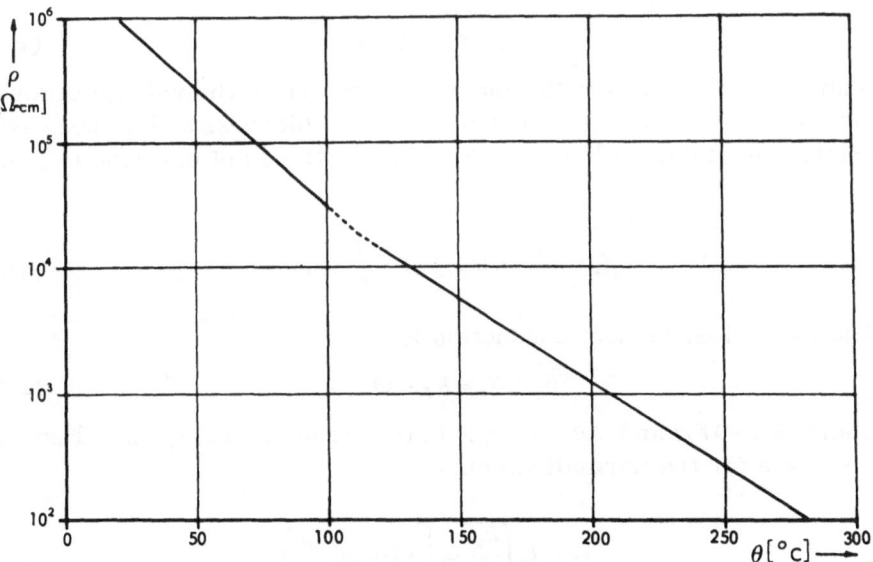

Fig. 7. Resistivity of boron as a function of temperature.

the ambient temperature is increased such that $N_2$ is now given by curve 5 (instead of curve 4), only one stable operating point will exist, if the voltage is properly chosen. This operating point will be at a high temperature; thus, the device has been switched on. Conversely, if the ambient temperature is reduced with the device switched on, only one stable operating point is possible within the range of lower temperatures; the device switches off. This mode of triggering the switching action leads to a second group of applications of boron semiconductor devices.

## CALCULATION OF THE CURRENT–VOLTAGE CHARACTERISTIC

We assume again that we are dealing with a pure temperature effect and calculate the current vs. voltage characteristic according to Schnupp's [4] treatment of the assistor.

The resistivity of boron is plotted versus temperature in Fig. 7. A good approximation of the temperature dependence for the range $20 < \theta < 110°C$ is given by

$$\rho = \rho_0 \cdot \exp\left(-\frac{\theta - \theta_0}{\Theta}\right) \tag{4}$$

where $\Theta = 20.75°C$, $\rho_0 = 1.05 \cdot 10^6 \, \Omega\text{-cm}$, and $\theta_0 = 20°C$. For the range $110 < \theta < 300°C$, $\Theta = 32.8°C$, where $\rho_0 = 1.9 \cdot 10^4 \, \Omega\text{-cm}$ and $\theta_0 = 110°C$. The power fed to the boron disk is

$$N_1 = I^2 R = U_p^2/R \tag{1a}$$

where $I$ is the current through the boron and $U_p$ the voltage across the boron. With $R_0 = \rho_0 d/q$ (where $d$ is the thickness and $q$ the cross section of disk), with $\theta_N = \theta - \theta_0/\Theta$, and with use of equation (4), we obtain

$$N_1 = I^2 R_0 \cdot \exp(-\theta_N) = \frac{U_p^2}{R_0} \exp(\theta_N) \tag{1b}$$

The power lost by heat conduction is

$$N_2 = K_E \cdot \Delta\theta \tag{2a}$$

where $K_E = \lambda F/l$ and $\Delta\theta = \theta - \theta_U$. In the steady state, $N_1 = N_2$. Hence, we obtain for the normalized current

$$i_N = I\sqrt{\frac{R_0}{K_E}} = \sqrt{\Delta\theta \cdot \exp\left(\frac{\theta_N}{2}\right)} \tag{5}$$

and analogously we obtain for the normalized voltage $u_N$

$$u_N = U \cdot \sqrt{\frac{1}{K_E \cdot R_0}} = \sqrt{\Delta\theta \cdot \exp\left(-\frac{\theta_N}{2}\right)}$$ (6)

If $\theta_U = \theta_0$, $i'_N = i_N/\sqrt{\Theta}$, and $u'_N = u_N/\sqrt{\Theta}$, we obtain

$$i'_N = \sqrt{\theta_N} \cdot \exp\left(\frac{\theta_N}{2}\right)$$ (5a)

$$u'_N = \sqrt{\theta_N} \cdot \exp\left(-\frac{\theta_N}{2}\right)$$ (6a)

The incremental (normalized) resistance then becomes

$$r_N = \frac{du'_N}{di'_N} = \frac{1 - \theta_N}{1 + \theta_N} \exp(-\theta_N)$$ (7)

For the relaxation point, where $r_N = 0$, we obtain $\theta_N = 1$. Since $\theta_N = \theta - \theta_0/\Theta$ and $\Theta = 20.75°C$, we conclude that $\theta - \theta_0 = 20.75°C$.

Thus, the relaxation point is attained by a current through the boron disk which causes a rise in temperature of approximately 21°C. In several series of measurements, we found values of the same order of magnitude, but always somewhat below the calculated values. This is so probably because the measurements record only the surface temperature, while the temperature inside the disk is certainly higher.

## EXAMPLES OF APPLICATIONS FOR BORON SEMICONDUCTOR DEVICES

Figure 8 shows an example of a circuit using a boron semiconductor device as a DC switch. A boron disk (1) with two electrodes (2 and 3) is connected in series, with an incandescent lamp (4) and a pushbutton switch (5) with normally-closed contact, to a DC supply (6). In parallel with the boron disk is a DC supply (7) connected in series with an operating pushbutton switch (8) and a capacitor (9). The voltage (6) is such that in the beginning only a small leakage current flows via the switch (5) through lamp (4) and the boron disk. At this stage, the boron is in a state of low conductivity. When switch (8) is actuated, a higher voltage [from source (7)] is applied to the boron disk for a short time. As a result, the boron disk is transformed into a state of greater conductivity; the circuit with source (6) and the lamp is closed. By pressing the button (5), the lamp can be switched off again.

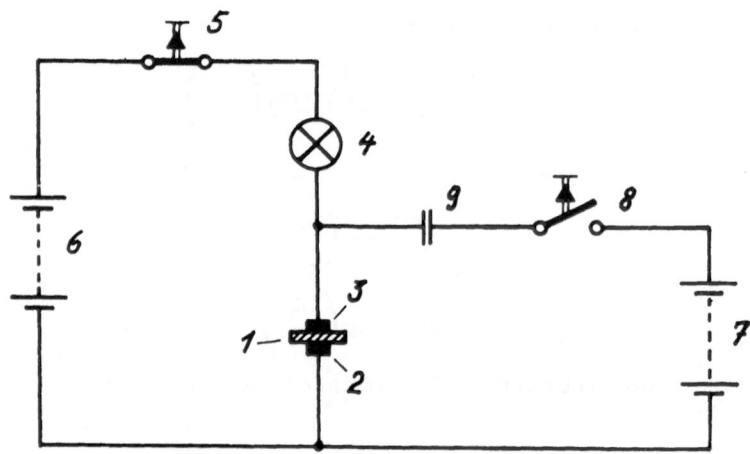

Fig. 8. Boron semiconductor device operating as DC switch. (1) Boron device; (2 and 3) electrodes; (4) incandescent lamp; (5) pushbutton with normally-closed contact; (6) DC supply; (7) DC supply; (8) operating pushbutton; and (9) capacitor.

A 7-W pilot lamp could be switched with boron disks about 60 $\mu$ thick and with cross sections of several square millimeters. The ignition voltage [source (7)] was roughly 500 V, the operating voltage [source (6)] about 220 V.

The semiconductor element can also be switched on without a second power source, as illustrated in Fig. 9. In this example, a

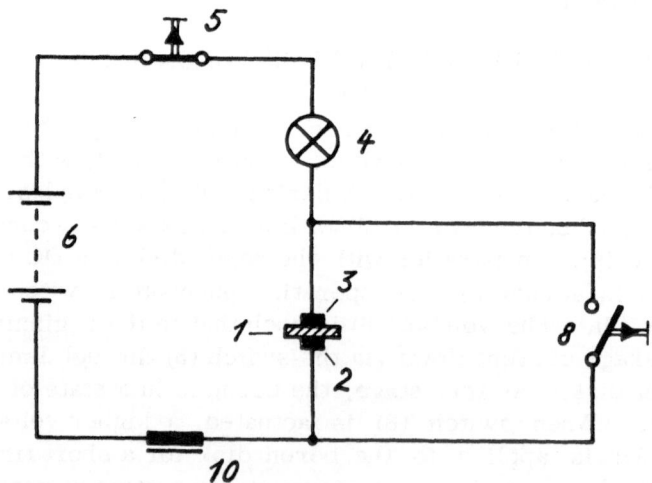

Fig. 9. Boron semiconductor device operating as DC switch. (1) Boron disk; (2 and 3) electrodes; (4) incandescent lamp; (5) pushbutton with normally-closed contact; (6) DC source; (8) operating pushbutton; (10) choke.

choke (10) is in series with the boron disk, while a pushbutton switch (8) is parallel to the disk. As in the previous example, only a small leakage current flows at first. When the on button (8) is pressed, the boron switch is short-circuited and the full current flows. When the circuit is now broken again by releasing the button (8), a state of greater conductivity is induced in the boron disk by the induction pulse of the choke (10), and the circuit remains closed until it is opened again with the off button (5).

Figure 10 is a schematic of a circuit where a boron semiconductor device controls an alternating current. The semiconductor device (2) is in series with a load resistance (3) and an AC supply (1). This AC source also feeds a pulse shaper (5) in series with a phase shifter (4). In the simplest case, the phase shifter may consist of a capacitor and an adjustable resistor. A boron semiconductor device may be used for the pulse shaper.

For every positive half-wave of the alternating current, the pulse shaper produces a short positive pulse; and for every negative half-wave, a short negative pulse. The phase relation of these pulses can be shifted in relation to the voltage of source (1) by the phase shifter. The pulses are fed to the semiconductor element via a coupling capacitor (6) or a coupling transformer. The semicon-

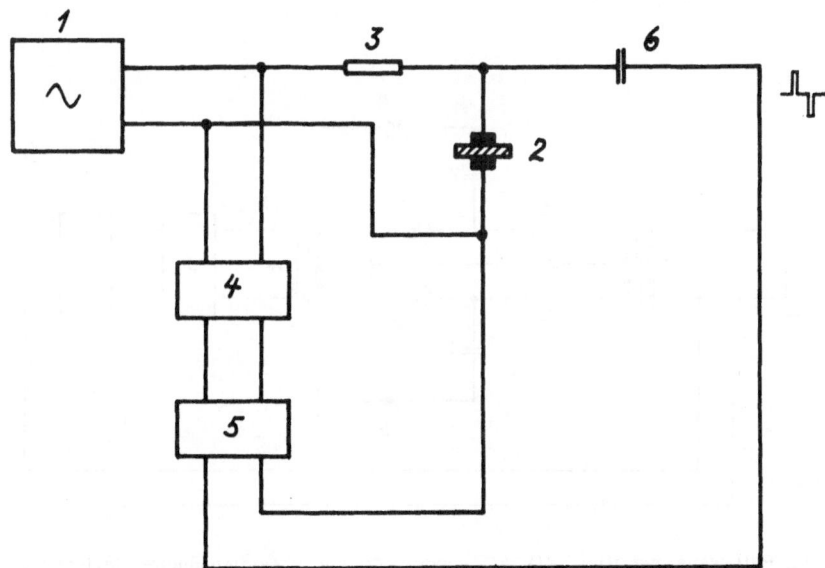

Fig. 10. AC control circuit. (1) AC power supply; (2) boron semiconductor device; (3) load resistance; (4) phase shifter; (5) pulse shaper; and (6) capacitor.

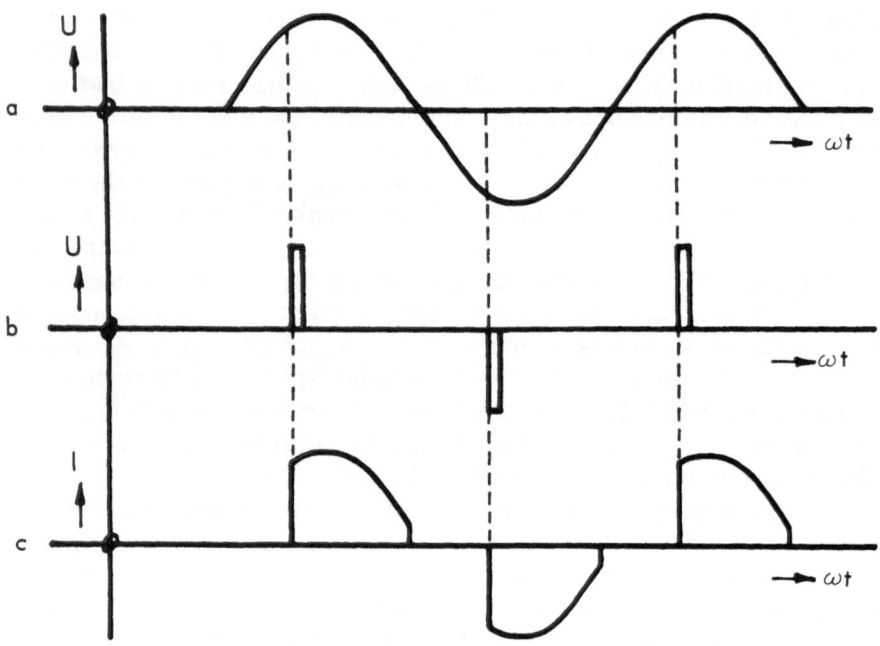

Fig. 11. Voltages in the circuit of Fig. 10: (a) voltage of power supply; (b) pulses supplied by pulse shaper; and (c) current through loading resistance.

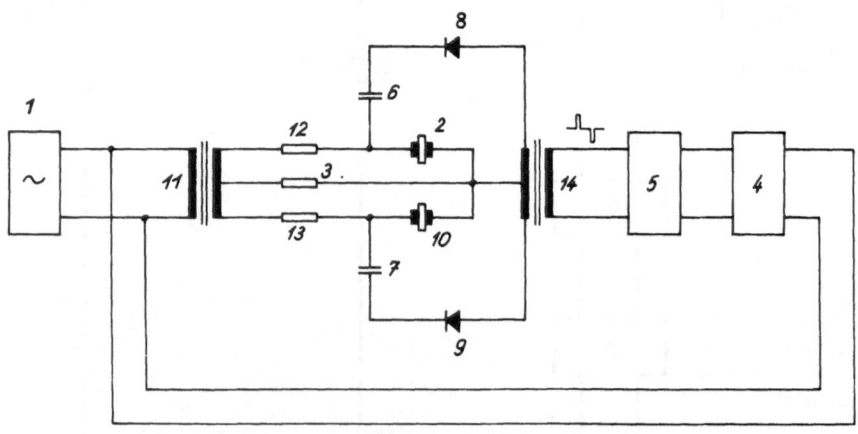

Fig. 12. Full-wave rectifier. (1) AC supply; (2) boron semiconductor device; (3) load resistance; (4) phase shifter; (5) pulse shaper; (6 and 7) capacitors; (8 and 9) diodes; (10) boron semiconductor device; (11) transformer; (12 and 13) resistors; and (14) transformer.

ductor element is so designed that it is in a state of low conductivity when the AC voltage alone is applied.

When a pulse from the pulse shaper reaches the semiconductor
element, it becomes more conductive and current passes through
the load resistance until the momentary value of the alternating
current drops below that required to maintain the conductive state.
The ignition of the next half-wave is triggered by the next pulse.
Figure 11 shows the currents and voltages as a function of time:
Fig. 11a shows the voltage from source (1); Fig. 11b shows the
pulses supplied by the pulse shaper, which in this case have a 60°
phase shift relative to voltage (1); and Fig. 11c shows the current
through the load resistance. By taking two semiconductor elements
of the type described and adding two diodes, a controlled full-wave
rectifier may be built (Fig. 12). The mode of operation is the same
as in the system of Fig. 10, but because of the diodes (8 and 9) the
semiconductor devices (2 and 10) are triggered alternately, so that
the positive half-wave passes one element and the negative, the
other.

In addition to the examples given above, it is also possible to
build, for instance, bistable, monostable, and free-running multi-
vibrators, ratchet circuits, and memories, with boron semiconductor devices.

Figure 13 shows electrically heated equipment which is protected
against overheating by a boron device. The boron device (1) is in
good thermal contact with the component (2). If the temperature

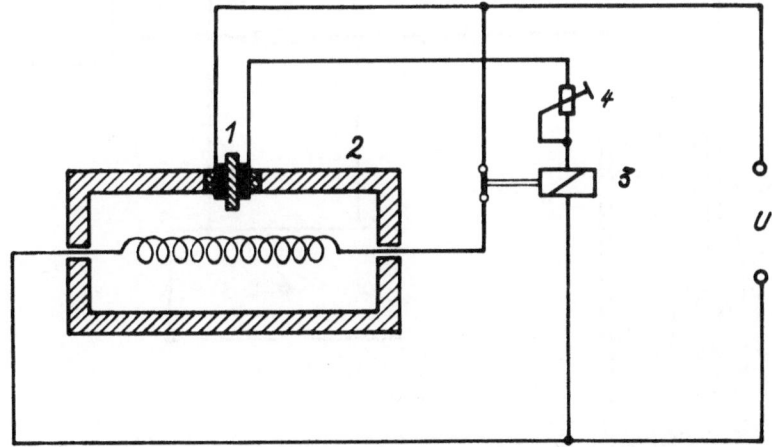

Fig. 13. Protective circuit. (1) Boron semiconductor device; (2) heater; (3) relay; (4)
variable resistor; and (U) power supply (AC or DC).

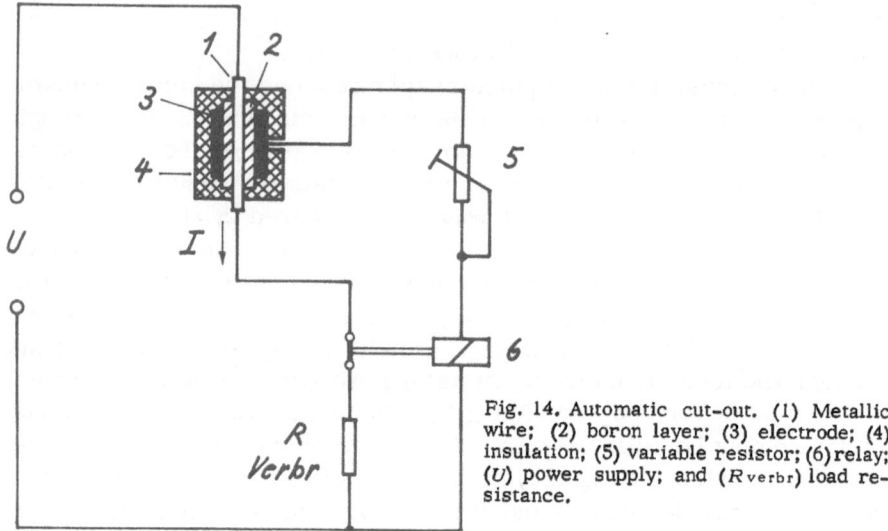

Fig. 14. Automatic cut-out. (1) Metallic wire; (2) boron layer; (3) electrode; (4) insulation; (5) variable resistor; (6) relay; ($U$) power supply; and ($R_{verbr}$) load resistance.

rises above a certain value, the boron becomes conductive. The relay (3) is then actuated and breaks the heating circuit. The cut-off temperature is set with the rheostat (4).

Figure 14 shows a boron semiconductor device controlling a circuit breaker. The boron coating (2) is applied directly to a metal wire. The operating current flows through the wire, which is

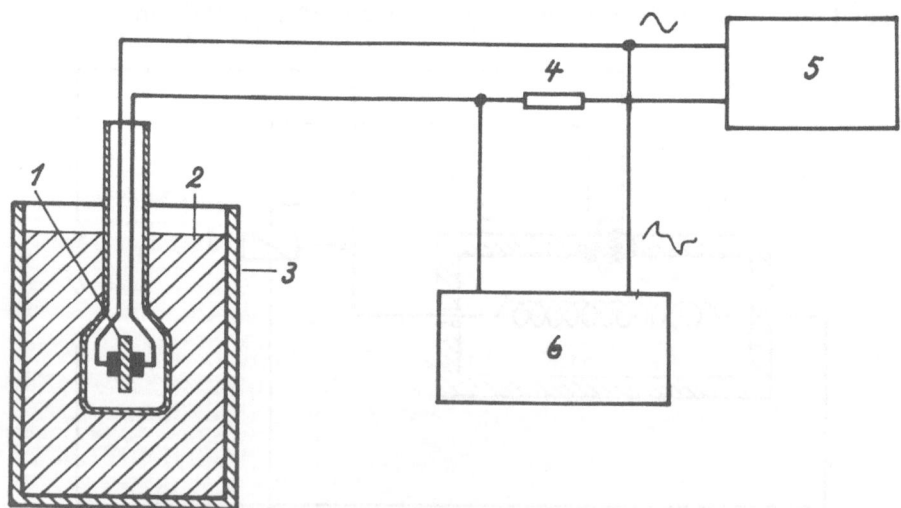

Fig. 15. Circuit for temperature measurement. (1) Boron semiconductor device; (2) liquid; (3) container; (4) series resistor; (5) power supply; and (6) phase meter.

simultaneously one of the electrodes of the semiconductor device. A second electrode (3) is connected to the boron coating. This electrode is thermally and electrically insulated by the coating (4). The semiconductor device is so designed that at the rated current $I$ only a small leakage current flows through the boron layer. If the current $I$ through the wire is excessive, the temperature is raised both at the electrodes and in the boron layer. Eventually, the boron element triggers the relay (6), breaking the main circuit. Sensitivity and time delay of the relay circuit are set with rheostat (5).

In the arrangement of Fig. 15, a boron semiconductor device serves to measure temperature. The enclosed boron device (1) is in good thermal contact with the liquid (2). It is in series with a resistor (4) and an AC power source (5). The AC amplitude is such that, at the lowest rated temperature, the boron device switches on slightly below the peak voltage. The potential across the boron then varies with time, as shown in Fig. 16a. When the temperature of the liquid rises, the semiconductor element switches on, even at lower momentary values of the AC voltage (Fig. 16b). The phase difference between the source voltage and the breakdown voltage is a measure of temperature.

Figure 17 is a basic circuit diagram for the temperature control of an electrically heated appliance, using two boron semi-

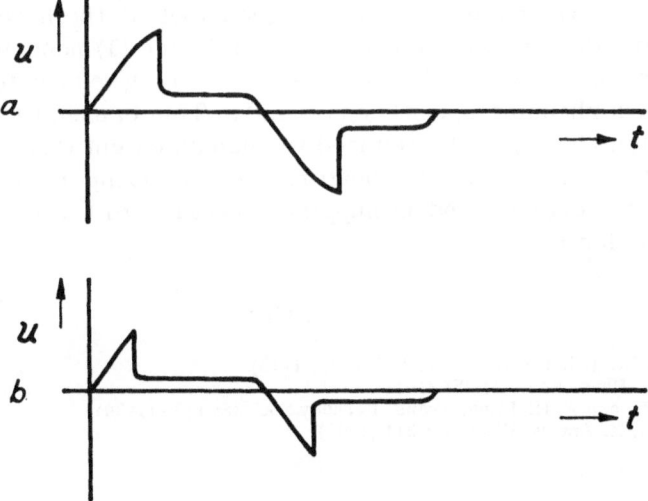

Fig. 16.  Voltage at input of the phase meter: (a) at lowest temperature to be measured, and (b) at higher temperature.

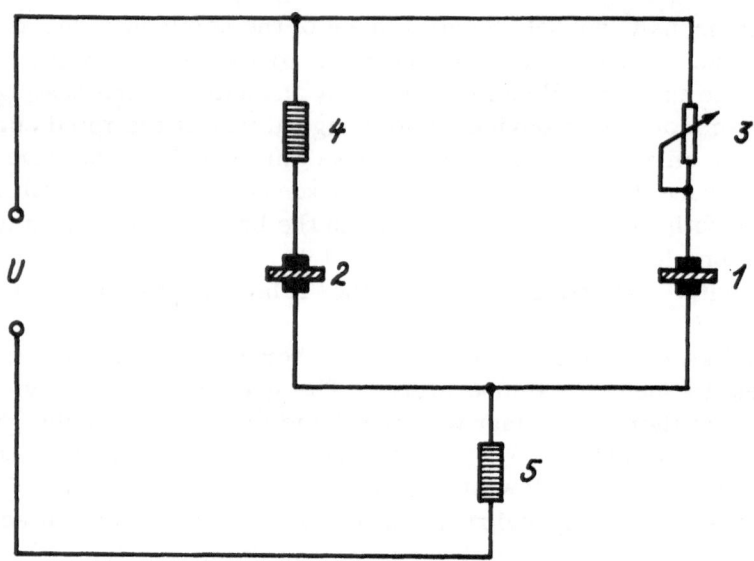

Fig. 17. Temperature-control circuit. (1 and 2) Boron semiconductor devices; (3) variable resistor; and (4 and 5) heaters.

conductor devices. The semiconductor device (1) is here in good thermal contact with the appliance, while semiconductor element (2) is either at the ambient temperature (room temperature) or at another fixed temperature. While the main heater (5) is constantly on, the supplementary heater (4) is controlled by boron device (2). The entire arrangement is so designed that at the lowest rated temperature device (2) is conductive and device (1) nonconductive. If the temperature rises, device (1) switches on when the temperature set with rheostat (3) has been reached. This causes the current through element (2) to be reduced to such an extent that the latter cuts off the supplementary heater. If the temperature drops, device (1) switches off and the supplementary heater (4) is turned on again by device (2).

## REFERENCES

1. Weintraub, E., J. Ind. Eng. Chem. 109, (Feb. 1913).
2. Lyle, F. W., Phys. Rev. 11: 253 (1918).
3. Bruce, J. H., and A. Hickling, Trans. Faraday Soc. 35: 1436 (1939).
4. Schnupp, P., Z. Angew. Phys. 16: 311 (1963).

# $B^{10}/B^{11}$ Thermistor Pairs and Their Application

## G. K. Gaulé and R. L. Ross

*Institute for Exploratory Research*
*U. S. Army Electronics Command, Fort Monmouth, New Jersey*

## and J. L. Bloom*

*Nuclear Division, Martin Marietta Corp.*
*Baltimore, Maryland*

Previous work [1] described the semiconductor properties of high-purity, crystalline, beta-rhombohedral boron. It was shown that the temperature coefficient of the resistivity is high and that ohmic contacts can reliably be attached to the material. This led to the development of boron thermistors, devices which sense changes in temperature as changes in resistance. The present work uses pairs of thermistors, where the two members of a pair are matched with respect to their semiconductor properties, but differ in their nuclear properties. One thermistor is made of $B^{11}$, the other of $B^{10}$. Both isotopes are stable, but, when exposed to neutrons with thermal energies, only $B^{10}$ atoms undergo nuclear reactions according to

$$B^{10} + n \rightarrow Li^7 + He^4 + 2.8 \text{ MeV}$$

The energy released causes the $B^{10}$ thermistor to become warmer than the $B^{11}$ thermistor. This temperature difference is converted into an electrical signal as usual. Experiments in a water-moderated research reactor showed that a neutron flux of $10^{11} \text{n/cm}^2$-sec or more gives a reproducible electrical signal in this fashion. Signals caused by other nuclear effects, such as transmutations, nuclear damage, and gamma-radiation heating, were either negligible or could be eliminated by comparing the results from a $B^{10}$ thermistor, a $B^{11}$ thermistor, and a blank. These results were obtained with simple prototypes [2] carrying crystalline boron layers with platinum and silver contacts on alumina plates. Preliminary experiments with $B^{10}/B^{11}$ thermistor pairs made of very high-purity boron rods and the detection of the thermal neutron part of a nuclear radiation pulse are also discussed.

## INTRODUCTION

An investigation of the semiconductor properties of crystalline beta-rhombohedral boron of the highest purity then available was completed by one of the present authors and his collaborators in 1960 [1]. The material was found very suitable for thermistors, semiconductor resistors which sense changes in temperature as changes in resistance. In 1961, one of the present authors [3]

*Present Address: U. S. Atomic Energy Commission, Washington, D. C.

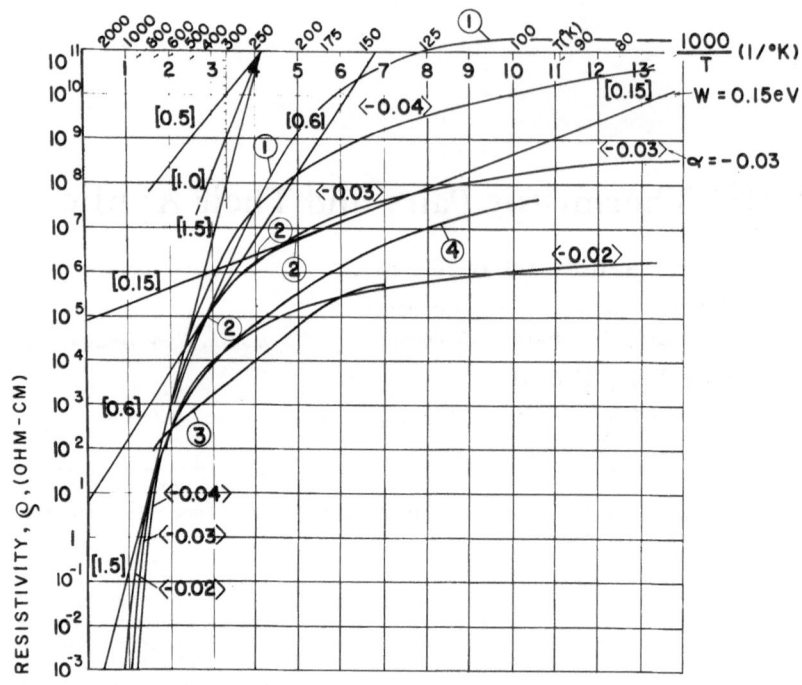

Fig. 1. Resistivity $\rho$ as function of temperature $T$. $< -0.03 >$ represents the ideal curve of a thermistor with $\alpha = -0.03$ for all $T$. It is closely approximated under the assumption of high-resistivity $p$.-type material with a few ($10^6$ cm$^{-3}$) shallow (activation energy, 0.15 eV) acceptor levels. [1.5] indicates hypothetical characteristic with activation energy $W = 1.5$ eV for all $T$. Curves 1, 3, and 4 obtained experimentally from $p$-type material [1].

proposed to expose such thermistors to a flux of neutrons with thermal (or slightly higher) velocities. Nuclear reactions caused by the thermal neutrons in the boron crystal would cause the temperature of the thermistor to rise, thus making the thermistor a sensor for thermal neutrons. To distinguish the effect of the neutrons from influences of, for example, ordinary temperature changes of the environment, the ensuing experimental work was performed with pairs of boron thermistors. One thermistor was made of the isotope $B^{11}$, which does not capture thermal neutrons; the other thermistor was made of $B^{10}$, an isotope with a large capture cross section for thermal neutrons. Enriched $B^{10}$ and $B^{11}$ raw materials were purified, and experimental $B^{10}$ and $B^{11}$ film thermistors were made by the Chemical and Metals Division of the Eagle-Picher Company, Miami, Oklahoma [2], and tested in the steady neutron flux of a nuclear reactor by the authors. As expected, a B$^{10}$/B$^{11}$ thermistor pair gave a differential signal proportional to the neutron flux. In a more recent development, high-purity, isotope-enriched rods were used to build more sensitive B$^{10}$/B$^{11}$ thermistor pairs. The application of such pairs for the detection of the thermal neutron part of a short nuclear pulse was also begun.

Topics discussed in the sections that follow are: essential semiconductor properties of boron, dominant nuclear effects, theory of the new sensor principle, preparation of the experimental thermistors, and measurements and test results.

## SEMICONDUCTOR PROPERTIES OF INTEREST

A number of properties make crystalline, beta-rhombohedral boron an attractive choice as a thermistor material. It is a hard material which resists most corroding influences and has a high melting point (approx. 2100°C). Furthermore, it is a semiconductor with a wide bandgap, giving a large intrinsic activation energy, $W_i = 1.5$ eV. This intrinsic activation energy dominates the resistivity vs. temperature characteristic of a semiconductor at high temperatures, as seen in Fig. 1, where curve 1 represents high-purity $p$-type material such as used in thermistors made with boron rods [1], while curve 4 represents boron films deposited on alumina substrate, as will be discussed later. In the intrinsic temperature range, which, for these two curves, extends from $\approx 400°$K ($\approx 127°$C) upward, the resistivity is given by

$$\rho \propto \exp (W_i/2kT) \tag{1}$$

where $k$ is Boltzmann's constant. The effectiveness of a thermistor material is determined by the relative change $a$ of the resistivity with temperature. Equation (1) leads to [1]:

$$a = (1/\rho) (d\rho/dT) = - W_i/2kT^2 \tag{2}$$

The formula indicates that the $a$ values become successively smaller as the temperature increases, but for pure boron, represented by curve 1 of Fig. 2, $a$ is still near -0.01 even at 1000°K. A 1°C temperature change would thus be sensed as a 1% resistance change by a thermistor operating near 700°C. At lower temperatures, the magnitude of $a$ is generally greater, which tends to make the thermistor still more sensitive. For temperatures below the

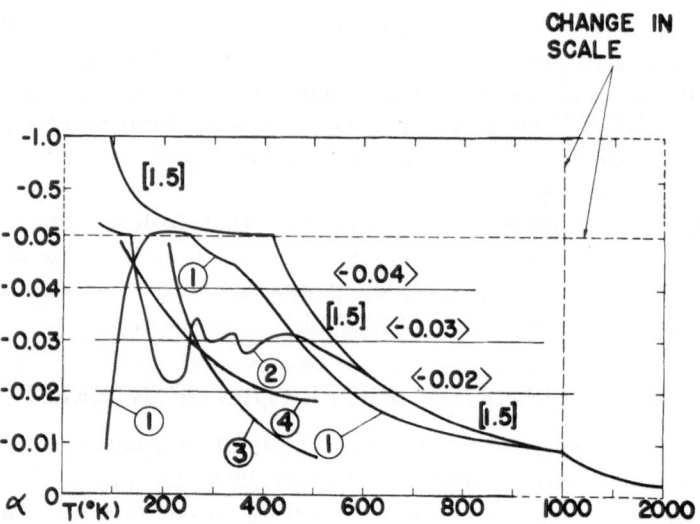

Fig. 2. The relative change $\alpha$ of resistivity $\rho$ with temperature $T$ as function of $T$, as derived from the characteristics of Fig. 1.    Curves 1, 3, and 4 are empirical [1].

intrinsic range, the resistivity is usually given by

$$\rho \propto \exp\left(-W_e'/kt\right) \tag{1a}$$

where $W_e'$ represents the extrinsic activation energy, the depth of the dominant acceptor level in the boron of interest.* For a consistent treatment of the intrinsic and the extrinsic range, the parameter $W_e = 2\,W_e'$ is used below.   For highly purified boron, the extrinsic range extends from 400°K down and is characterized by $W_e$ values that are high in comparison with those of other semiconductors. Near room temperature, for instance, one finds $W_e \approx 0.6$ eV ($W_e' \approx 0.3$ eV) for curve 1, and $W_e \approx 0.4$ eV for curve 4 of Fig. 1. This and the $1/T^2$ factor in equation (2) secure high sensitivities also in the low-temperature range down to 150°K, as demonstrated in Fig. 2.   From the point of view of thermistor performance, the contact properties of high-purity beta-rhombohedral boron are superior to those of most other crystalline semiconductors. The resistance of a properly-made contact is usually much smaller than the resistance of the bulk material.   Also, the contacts are usually ohmic, even at low temperatures [1].   Figure 3 gives the resistance vs. temperature characteristics of typical rods and typical films of $B^{10}$ and of $B^{11}$ enriched boron, and of natural boron.

*See "Semiconductor Properties of Boron," this volume, p. 143.

The rods, as a group, are more temperature-sensitive than the films, as a group, because of higher purity. Differences within each group are probably not related to the isotopic composition, but are caused by differences in the impurity level of the raw material.

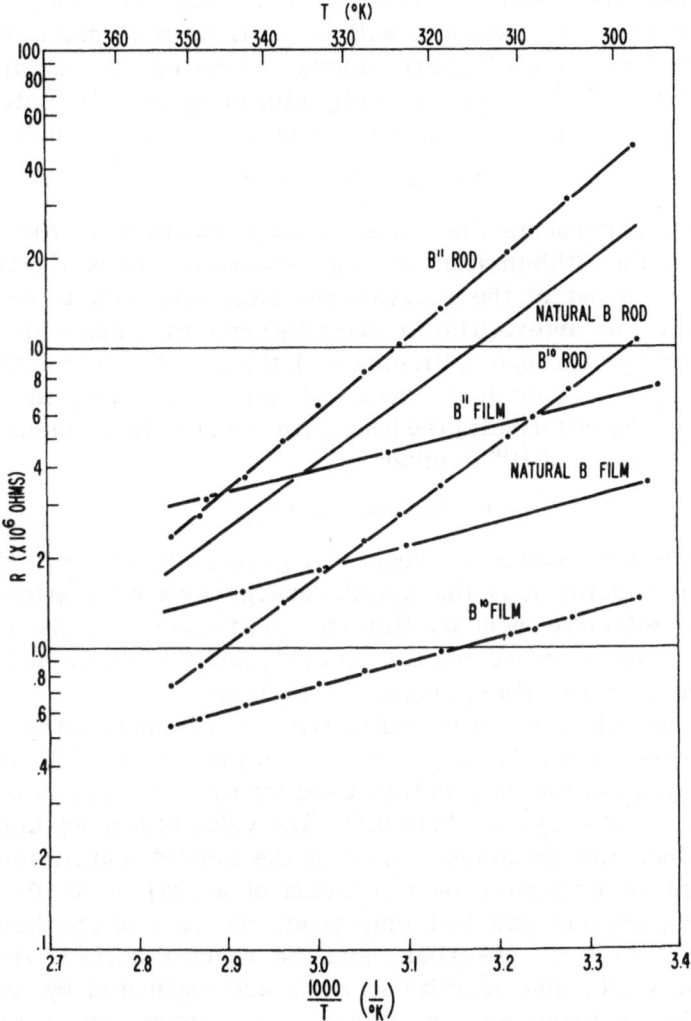

Fig. 3. Resistance vs. temperature characteristics of typical film and rod thermistors. Because of higher purity, the rods, as a group, are more sensitive than the films, as a group. Differences within each group are probably not related to the isotopic composition, but are caused by differences in the impurities in the isotope-enriched raw material.

## NUCLEAR PROCESSES OF INTEREST

Natural boron is a mixture of two stable isotopes, namely, $B^{10}$ ($\approx$ 19%) and $B^{11}$ ($\approx$ 81%) [4]. Enriched material with 93% $B^{10}$ is available and was used for the experimental $B^{10}$ thermistors. The material employed in the $B^{11}$ thermistors was 98.6% enriched. The only noticeable difference between a $B^{10}$ and a $B^{11}$ thermistor, if made by identical methods, will be in their response to neutrons of thermal (or slightly higher) velocity $v$. Whereas the capture cross section of the $B^{11}$ atoms is negligible, many of the $B^{10}$ atoms will capture a neutron and undergo a nuclear reaction according to

$$B^{10} + n = Li^7 + He^4 + E_{kin} \tag{3}$$

where $E_{kin}$ represents the kinetic energy acquired by the reaction products, the lithium atom and the $\alpha$-particle, during the reaction. In a solid, most of the reaction products will come to rest in the solid in either interstitial or substitutional sites after losing their kinetic energy through collisions with the lattice. Thus, the kinetic energy is converted into lattice vibrations, or heat, in less than $10^{-9}$ sec. Quantitatively, the heat $\dot{Q}$ (in W/g) produced in one second within one gram of $B^{10}$ is given by

$$\dot{Q} = fK\phi\sigma N E_{kin} = F E_{kin} \tag{4}$$

where $F$ is the number of reactions/g-sec, $f$ is the self-shielding correction factor, $K$ is the kinetic-energy loss correction factor, $\phi$ is the effective neutron flux (in n/cm$^2$-sec), $\sigma$ is the effective neutron reaction cross section (in cm$^2$), and $N$ is the atomic density of the neutron absorbing target (in atoms/g).

The quantity $f$ is estimated by the method employed by Reynolds and Mullins [5] for thin foils, or as shown by Zweifel [6]. Since the boron layers in the thermistors used for this work are of the order of 30 $\mu$, $f$ is found to be about 0.95. The value of $K$ is assumed to be 0.67, since the maximum range of the helium and lithium atoms produced is estimated by Cervellati et al. [7] to be 10 $\mu$ and the reaction products are lost only from one face of the thermistor. Since $B^{10}$ is a 1/$v$ absorber, and the reactor neutron flux is not monoenergetic, the quantities $f$ and $\sigma$ are estimated by averaging over the neutron energy range. For simplicity, a Maxwell–Boltzmann distribution is assumed. The effective absorption cross section is then given by

$$\sigma = (\pi T_0/4T)^{1/2} \sigma_0$$

with $\sigma_0 = 3850 \cdot 10^{-24}$ cm$^2$ for the monoenergetic absorption cross section corresponding to the standard temperature $T_0 = 293.59°$K [8]. The reactor temperature in the experiments discussed here was $T = 310.95°$K (37.8°C), giving for the thermal cross section $\sigma = 3315 \cdot 10^{-24}$ cm$^2$. Boron enriched to 93% B$^{10}$ was employed in fabricating the B$^{10}$ thermistors; hence, $N$ is computed to be $5.4 \cdot 10^{22}$ atoms B$^{10}$ per gram of boron. $E_{kin}$ is 2.79 MeV or $4.5 \cdot 10^{-13}$ W-sec. Therefore, $\dot{Q} = 6.2 \cdot 10^{-11} \phi$ (in W/g).

Use of boron enriched to 100% B$^{10}$, instead of 93% as in the actual case, would result in only a 7.5% higher heat input $\dot{Q}$. The residual B$^{10}$ atoms in the (nominal) B$^{11}$ thermistor present a more serious problem. Fortunately, the actual B$^{11}$ material contained only 1.4% B$^{10}$, so that the heat input per gram of actual B$^{11}$ material is only 1.5% of that for actual B$^{10}$ material. Material which is B$^{11}$-enriched to an even higher degree would, however, be desirable for future work.

The transmutations described by equation (3) could presumably influence the semiconductor properties of a B$^{10}$ crystal by the creation of boron vacancies or through the introduction of lithium and helium atoms. With equation (4) and the assumptions made above, the fraction of lithium atoms in the crystal $\lambda$ after exposure to a neutron flux $\phi$ for a time $t$ becomes

$$\lambda = Ft/N = fk\phi t \approx 2.5 \cdot 10^{-21} \phi t \qquad (5)$$

As shown below, a typical neutron dose for one experiment would be $\phi \cdot t = 10^{16}$ n/cm$^2$, resulting, for example, from a day's (approx. $10^5$ sec) exposure to a flux of $\phi = 10^{11}$ n/cm$^2$-sec, giving $\lambda \approx 2.5 \cdot 10^{-5}$. The concentration of the reaction products then corresponds to the concentration of impurities already present in the boron, which is in the 10-ppm range [1,2]. It should be remarked, however, that most impurities are very ineffective as dopants in boron [1] and that, furthermore, impurities have generally little influence at the high ambient temperatures at which neutron sensors usually operate, as shown previously. The experiments performed so far indicate a dose tolerance of $10^{16}$ n/cm$^2$ or higher.

Frequently gamma radiation of high intensity is present during experiments with neutron sensors. This radiation will be absorbed by the thermistor material and will cause a temperature increase. Furthermore, it will ionize gases in the environment, thus causing an electrical leak across a thermistor. It is possible, however, to make an independent determination of the influence of the gamma radiation and to correct the thermistor current accordingly, as shown below.

It is known from the work on the semiconductor properties of boron [1] that the mean free path for the majority carriers, as well as the minority carrier lifetime, is extremely small. The influence from nuclear damage through the introduction of additional scattering or recombination centers thus can be expected to be very small also, in contrast to the influence of nuclear damage on, for example, high-purity, nearly perfect germanium crystals.

## THEORY OF NEUTRON RATEMETERS AND DOSIMETERS USING THERMISTORS

According to Shive [9], the equivalent circuit for the thermal system representing a thermistor can be given by a diagram such as the one shown in Fig. 4, where $C$ (in W-sec/°C) represents the effective heat capacity of the system and $R$ (in °C/W) represents the heat leak through which energy is lost to the environment. The power input $P$, which in this case is caused by and is proportional to the neutron flux, divides accordingly into $P_1$, leaving the system through $R$, and $P_2$, which accumulates heat in $C$.

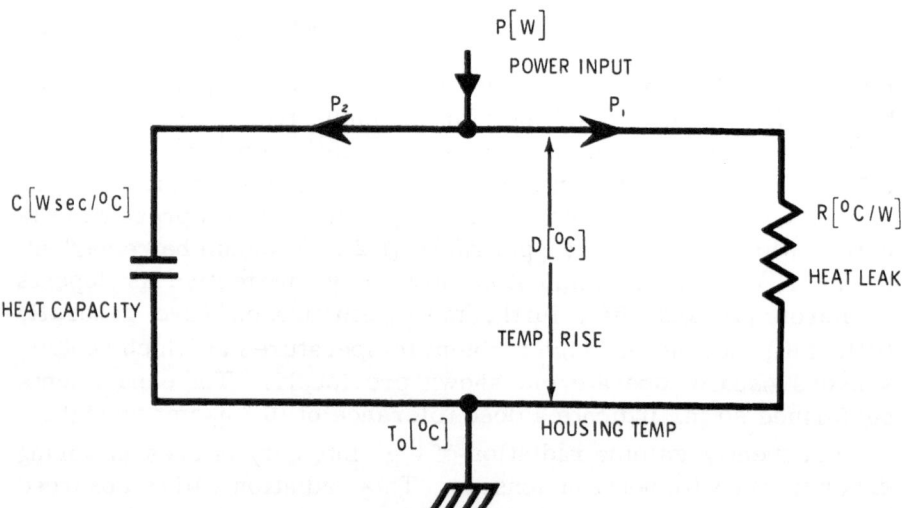

Fig. 4. Equivalent circuit representing the flow of heat in a boron thermistor. The heat leak $R$ represents losses, mainly through lead wires. The heat capacity $C$ is given by the mass of the boron and (if applicable) the substrate.

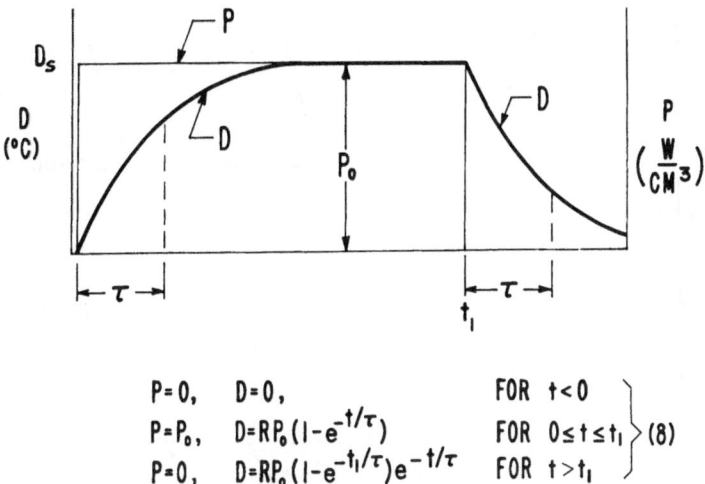

$$P=0, \quad D=0, \qquad\qquad\qquad \text{FOR } t<0$$
$$P=P_0, \quad D=RP_0(1-e^{-t/\tau}) \qquad\quad \text{FOR } 0\leq t \leq t_1 \quad \bigg\} \ (8)$$
$$P=0, \quad D=RP_0(1-e^{-t_1/\tau})e^{-t/\tau} \quad \text{FOR } t>t_1$$

Fig. 5. Thermal response of a B$^{10}$ thermistor to a long neutron pulse. After the initial buildup, the thermistor yields a signal $D_s$, which is proportional to the neutron-flux rate.

The pertinent equations are

$$D = T - T_0 = P_1 R = {}_0\!\int^t P_2 \, dt'/C \tag{6}$$

$$P = P_1 + P_2 \qquad\qquad P_1/P \leq 1$$

$$\tau dD/dt + D = RP \tag{7}$$

$$\tau = RC$$

Exact solutions of equation (7) are easily obtained for the case of a square heat (or neutron) pulse, namely,

$$P = 0, \qquad D = 0 \qquad\qquad\qquad\qquad \text{for } t < 0$$
$$P = P_0, \qquad D = RP_0(1 - e^{-t/\tau}) \qquad\qquad \text{for } 0 \leq t \leq t_1 \tag{8}$$
$$P = 0, \qquad D = RP_0(1 - e^{-t_1/\tau})e^{-t/\tau} \quad \text{for } t > t_1$$

It is evident that the system is characterized by a time constant, $\tau = RC$, which is approximately 20 sec in the experiments with films and approximately 120 sec in the experiments with boron rods. Much smaller or much higher values for $\tau$ may be obtained by changing the thermistor design [9]. The temperature increase for the B$^{10}$ thermistor builds up essentially during a time $\tau$ after the onset of a long pulse and then approaches a steady value

$$D_s = RP_0 \qquad\qquad \text{for } \tau < t \leq t_1 \tag{8a}$$

This and the exponential decay after the pulse is shown in Fig. 5. It is noted that $D_s$ is independent of $\tau$ and, thus, of $C$. The sensitivity

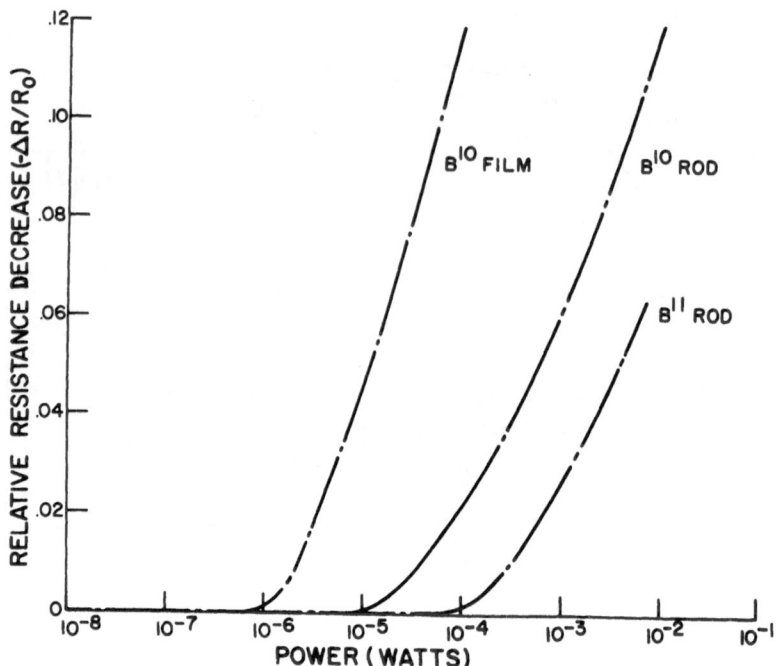

Fig. 6. Self-heating curves from encapsulated (see Fig. 9) boron thermistors. In conjunction with Fig. 3, these curves indicate (a) the highest tolerable bias current and (b) the input power necessary to raise the temperature by 1°C.

of a thermistor operating in the steady state is determined by $R$. Typical $R$ values range from $10^2$ to $10^5$ °C/W with the lower values representing simple designs, such as used in this work, and the higher values representing thermistors suspended in a vacuum by thin wires. Taking the lowest value for an unencapsulated film, as used in the first experiment, one obtains a temperature increase $D_s$ of 1°C for a power input $P_0$ of $10^{-2}$ W. Figure 6 shows that usually less power is necessary to produce the resistance corresponding to a one-degree temperature increase (compare with Figs. 2 and 3) when the thermistors are embedded in a silicone elastomer. The curves of Fig. 6 were obtained by warming the encapsulated thermistors with Joule heating, but they apply to heating from nuclear reactions as well.

The power input $P_0$ (in W) into a mass $M$ from a neutron flux $\phi$ is, according to the aforementioned nuclear processes

$$P_0 = M\dot{Q} = 6.2 \cdot 10^{-11} \, M \, \phi \qquad (9)$$

With $M \approx 10^{-2}$ g for the typical mass of the boron in a $B^{10}$ film thermistor, the flux producing a temperature increase of $D_s = 1°C$ becomes $\phi \approx 10^{11}$, in agreement with the experimental results below. Since according to the semiconductor properties discussed, a 1°C temperature difference between the $B^{10}$ and the $B^{11}$ thermistors produces a satisfactory electrical signal, this value may serve to define the sensitivity of a neutron ratemeter. The experimentally obtained sensitivity of the system used is thus $\phi = 10^{11}$ n/cm$^2$-sec. Since the thermal insulation of the boron, i.e., the value of $R$ in equation (8a), can be improved by several orders of magnitude [9], much higher sensitivities seem possible.

One easily derives from equations (6) and (7) that

$$P_1/P = \int_0^t (1 - P_1/P) \frac{dt'}{\tau} < t/\tau \qquad (10)$$

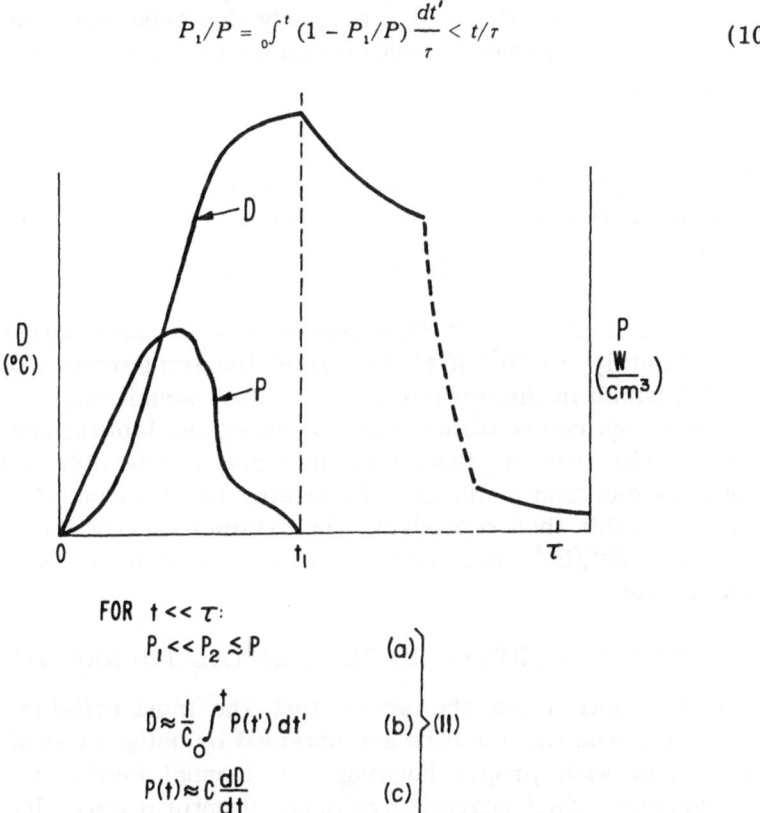

FOR $t \ll \tau$:

$$P_1 \ll P_2 \lesssim P \qquad \text{(a)}$$

$$D \approx \frac{1}{C} \int_0^t P(t')\, dt' \qquad \text{(b)} \Bigg\} \text{(11)}$$

$$P(t) \approx C \frac{dD}{dt} \qquad \text{(c)}$$

Fig. 7. Thermal response of a $B^{10}$ thermistor to a short neutron pulse. $D$ is now given by the neutron dose, but the neutron-flux rate, which is proportional to the power input $P$, can also be sensed by electronically obtaining the time derivative of $D$.

which indicates that, for times $t$ very short in comparison to the time constant of the system $\tau$, the flow of heat $P_1$ through the heat leak $R$ is negligible. The system's reaction to a short pulse (Fig. 7) for $t \ll \tau$ is

$$P_1 \ll P_2 \leq P \tag{11a}$$

$$D = {}_0\!\int^t P(t') \; dt'/C \tag{11b}$$

$$P(t) = C \; dD/dt \tag{11c}$$

The $B^{10}/B^{11}$ thermistor pair now acts as a dosimeter, with $D$ proportional to the dose. The sensitivity may again be defined by requiring that $D = 1°C$. In a properly designed boron dosimeter, most of the heat capacity $C$ will result from the mass $M$ of the boron, which yields

$$C = M \cdot c_B \tag{12}$$

where the specific heat of boron $c_B$ is approximately 1.14 W-sec/°C-g. Combination of this result with equations (9) and (11b) yields

$$D = 5.5 \cdot 10^{11} \; {}_0\!\int^t \phi \; dt' \tag{13}$$

which means the neutron dose represented by the integral must be approximately $2 \cdot 10^{10}$ n/cm$^2$ to yield the temperature difference $D = 1°C$, which in this case defines the dose sensitivity.

The temperature difference $D$ becomes available as an electrical signal. The time derivative of the signal can be obtained by conventional electronic means. According to equations (9) and (11c) and Fig. 7, this then reproduces the original neutron pulse. In this case, the $B^{10}/B^{11}$ thermistor pair is applied as a ratemeter for short pulses.

## PREPARATION OF EXPERIMENTAL THERMISTORS

It is clear from the above that the most reliable and most sensitive boron thermistors are obtained by using a rod of crystalline boron with proper housing and thermal insulation from the environment. Such thermistors made of natural boron have indeed been found satisfactory earlier [1]. The difficulties in obtaining and purifying isotope-enriched boron prevented this approach for the first experimental series of $B^{10}/B^{11}$ thermistor pairs. Crystal-

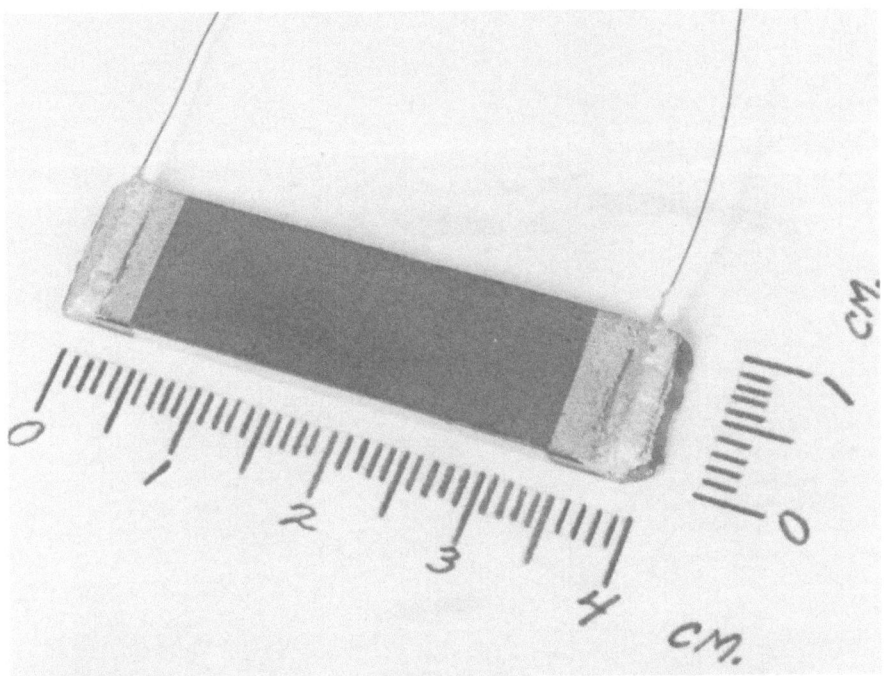

Fig. 8. Experimental thermistor made by thermal decomposition of BBr$_3$ in hydrogen on hot alumina substrate. Resulting films of crystalline B$^{10}$ (or B$^{11}$) are approximately 30 $\mu$ thick.

line boron films, formed by vapor-phase deposition and subsequent heating, were used instead. The most successful method, described in detail elsewhere [2], was to expose alumina plates to the gaseous-phase reduction of BBr$_3$ in hydrogen at 1200°C. Subsequent further heating enhanced the crystallization of the boron film, which improves and stabilizes the electrical characteristics. The films thus obtained were approximately 30 $\mu$ thick. Adherence to the alumina substrate was excellent. Platinum layers were vapor-deposited on the ends of each film, giving satisfactory ohmic contacts to the film (Fig. 8). Lead wires were then attached to the end contacts using silver-resin paste and baking. The lead wires provided electrical connections as well as mechanical support in the first series of experiments.

Figure 9 shows one of the recently developed thermistors made of a rod of highly purified, isotope-enriched boron (98% B$^{11}$ or 92% B$^{10}$) next to a film thermistor. A capsule with coaxial cable,

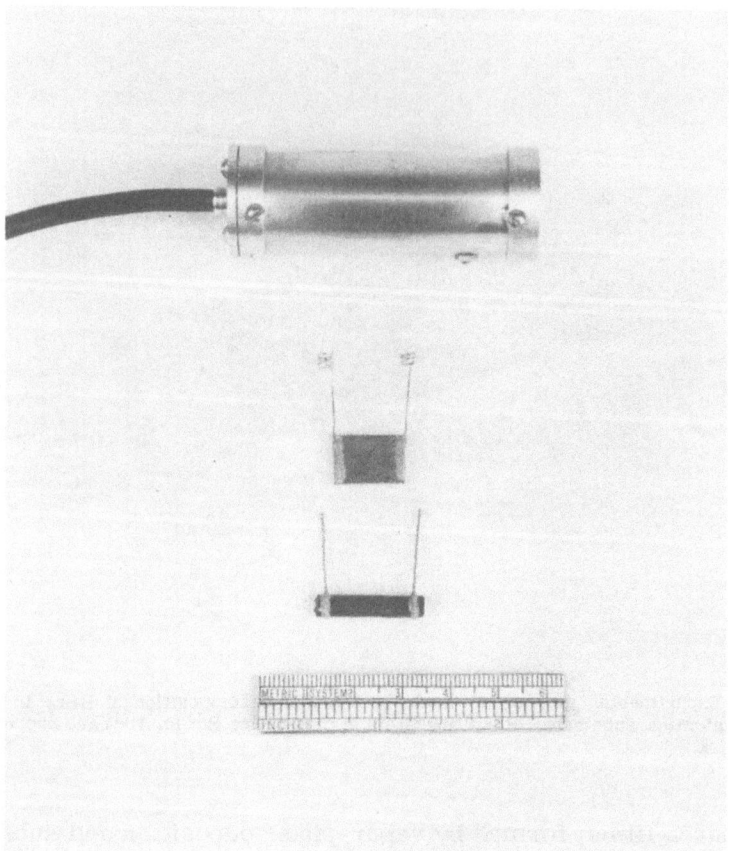

Fig. 9. Rod and film thermistor with aluminum capsule used in the nuclear pulse experiments. The thermistor is placed lengthwise in the capsule, connected to the capsule and the coaxial cable, and embedded in a silicone-based elastomer.

which contained one thermistor during the neutron pulse experiment, is also shown. The capsule was filled with a silicone elastomer.

## RESULTS WITH EXPERIMENTAL THERMISTORS

Matched $B^{10}/B^{11}$ thermistor pairs were selected according to their resistance vs. temperature characteristics between 25 and 210°C. The thermistors were mounted in a temperature-controlled oven and any significant temperature hysteresis was eliminated through long measuring times. The resistance vs. temperature characteristics of the thermistors were obtained by applying a small, constant DC potential (usually 1V) and measuring the current. The applied potential was kept small to prevent self-heating effects. (Compare with Fig. 6.)    Reproducibility of the resistance vs. temperature characteristics was within 1%. Typical characteristics are given by curves 1 and 4 of Figs. 1 and 2 and in Fig. 3.

In the first series of tests, $B^{10}/B^{11}$ film thermistor pairs were mounted in an aluminum irradiation capsule of the type shown in Fig. 10.  The capsule was filled with dry air and contained, next to the thermistors, a "blank," a device simulating the contact configuration of a film thermistor.  The blank was used to determine the ionization current.  The method of mesurement was similar to that for the purely electrical measurements and is indicated in Fig. 10.

In the first run of experiments, a pair of well-matched $B^{10}/B^{11}$ thermistors was employed, but the results were greatly falsified through moisture-induced leakage currents in part of the system. Because of the high resistance of the experimental thermistors ($10^6$ Ω and higher), moisture must be eliminated.  This was done in the second run, but the $B^{10}/B^{11}$ thermistor pair available for that run

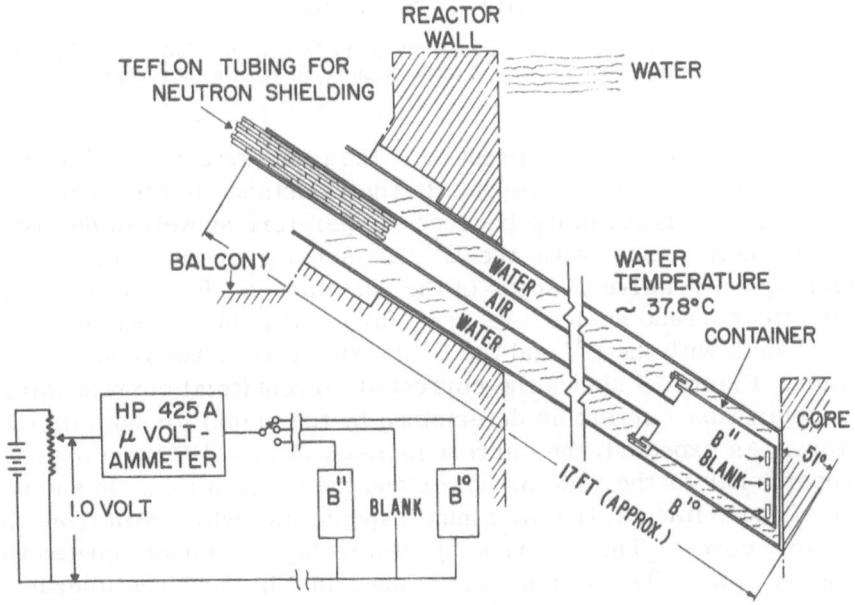

Fig. 10. Aluminum capsule containing B¹⁰ and B¹¹ thermistors and blank in position closest to reactor core.  The blank is used to determine ionization current caused by gamma radiation.

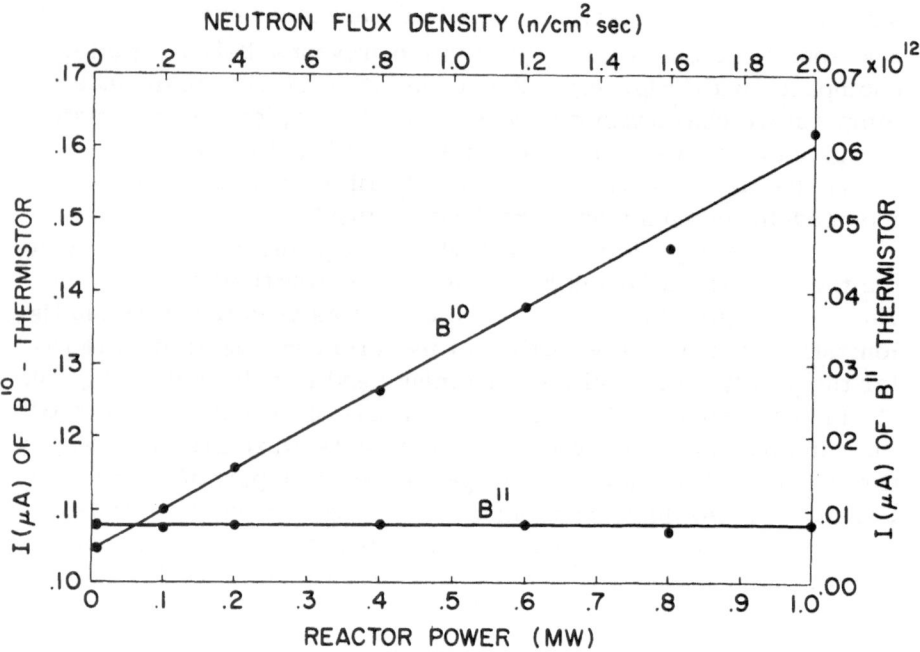

Fig. 11. Corrected current (total current minus ionization current) for $B^{10}$ and $B^{11}$ thermistors in position shown in Fig. 3 (zero distance) as function of reactor power and of neutron flux.

was matched only with repsect to $a$, the temperature coefficient of resistance, and not with respect to the resistance itself. This was not a serious shortcoming because temperature as well as neutron-flux measurements with the $B^{10}/B^{11}$ sensing system depend essentially on relative changes of the resistances. For convenience, separate current scales are used for the graphical description of the results with the $B^{10}$ and the $B^{11}$ thermistors in the final experiments. Figure 11 shows the corrected current (total current minus the ionization current as determined by the blank) for both thermistors. As expected, the current increase for the $B^{10}$ thermistor is proportional to the reactor power and the neutron flux. In another set of experiments, the aluminum capsule was withdrawn from the reactor core. The interposing water layer then attenuated the neutron flux. The results are shown in Fig. 12. The ionization current and the bias current (the current flowing when the thermistors are at a great distance from the reactor core) have been subtracted from the measured current, which gives the differential current $\Delta I$ for each thermistor. The decrease of the $B^{11}$ current, as the distance from the core is increased, is probably due to three effects that raise the temperature of the $B^{11}$ (as well as that of the $B^{10}$) thermistor when both are very near the reactor core. These three effects, listed in the estimated order of their importance, are: (1) fast neutrons, not yet slowed down by the water, (2) a somewhat higher water temperature near the core, and (3) heating by gamma radiation. Even without exact knowledge of these influences, the relative change of the $B^{10}$ current, which is much larger than that for the $B^{11}$ current, can be corrected accordingly to obtain the signal representing the thermal neutrons only. This signal decreases exponentially with distance, as expected. In all the experiments, it was necessary to measure the ionization current through the blank simultaneously with the thermistor currents,

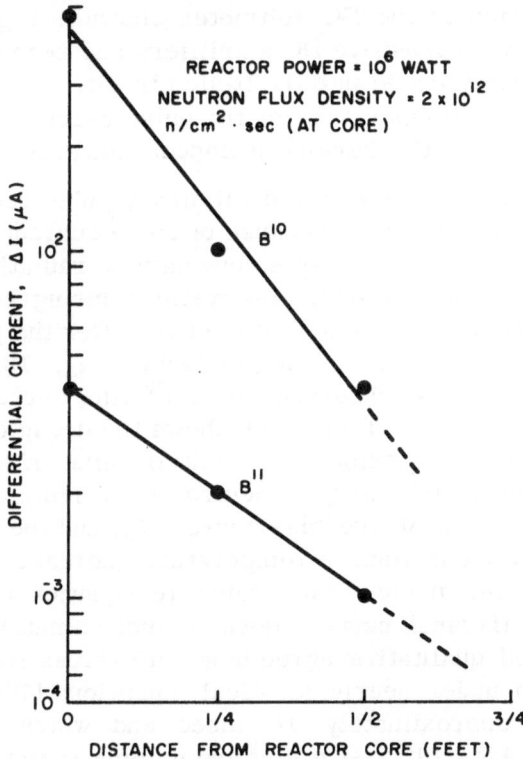

Fig. 12. Differential current $\Delta I$ (total current minus bias current minus ionization current) as function of distance from reactor core. Change in current through $B^{11}$ thermistor was probably caused chiefly by fast neutrons close to core.

because neutron flux and gamma radiation did not change concurrently. For example, the gamma radiation persisted for a while after the reactor was shut off. The ionization current was usually a fraction of the $B^{10}$ current.

For the preliminary experiments with nuclear pulses, an arrangement basically similar to that for the steady-state experiments (Fig. 10) was made. The main differences were:

1. The thermistors were not exposed to the air inside the radiation container, but were protected from it through individual aluminum capsules (Fig. 9). Each capsule was mechanically and electrically connected to the shield of a coaxial cable. Each capsule was filled with a silicone elastomer. This improved the thermal resistance ($R$ in Fig. 4) and reduced ionization effects.
2. In addition to the DC voltmeter shown in Fig. 10, oscilloscopes with sensitive DC amplifiers and recording cameras were incorporated into the basic circuit.
3. The water temperature of the pulse reactor used was only 20°C, making the thermistor impedances even higher (Fig. 3).

The circuitry applied during the preliminary pulse reactor experiments proved unsatisfactory because of cable currents generated in the coaxial cables by the very strong gamma radiation during the pulse. This prevented reliable observations during the pulse. Such observations were thus limited to the time after the pulse, corresponding to $t \geq t_1$ in Fig. 7, and to $t' \geq 0$ in Fig. 13. Although a distinct signal was also observed with a $B^{10}$ film, curves were taken only with $B^{10}$ rod A. Figure 13 shows the decay of the current increment $\Delta I$ after the pulse. The initial value of $\Delta I$, which was extrapolated, gives the thermal neutron dose from the just-completed pulse. From $\Delta I$, the bias current $I_B$, and the $R$ vs. $T$ charteristic (Fig. 3), one finds a temperature increase $D \approx 5°C$ at the completion of the pulse. According to equation (13), this corresponds to a thermal neutron dose of approximately $10^{11}$ n/cm$^2$. This is in good qualitative agreement with values from activation analyses taken under nearly identical conditions [10]. The pulse, which lasted approximately 18 msec and which also carried approximately $4.2 \cdot 10^{11}$ fast n/cm$^2$ and gamma radiation equivalent to approximately 4200 rad (absorbed), had no observable effect on the $B^{11}$ thermistors. No thermistor, $B^{10}$ or $B^{11}$, showed any lasting

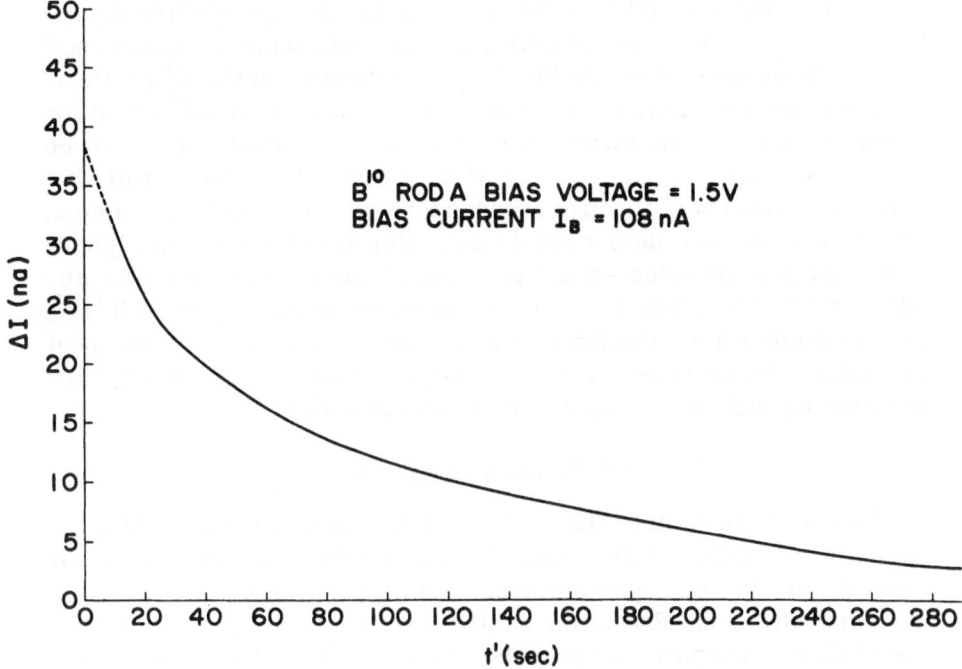

Fig. 13. Decay of the excess current $\Delta I$ of B $^{10}$ rod thermistor A after 18 msec of a pulse with $\approx 10^{11}$ thermal n/cm$^2$. Compare with Fig. 7 for $t > t_1$. The initial $\Delta I$ value corresponds to $\approx 5°C$ temperature increase. (See Fig. 3.)

change of the DC characteristic after it had undergone a number of pulses (5−11) during the experiments.

## CONCLUSION

The experimental results show that a B$^{10}$/B$^{11}$ thermistor pair can be applied as a sensor for thermal neutrons. The sensitivity of an experimental pair used as a ratemeter under steady-state conditions was $10^{11}$ n/cm$^2$-sec, but a hundred times better sensitivity should be obtainable with improved thermal insulation of the boron. Used as a dosimeter for short (up to 1 sec) pulses of thermal neutrons, a B$^{10}$/B$^{11}$ thermistor pair should have a sensitivity of $10^{10}$ n/cm$^2$, or better. In a preliminary experiment with an 18-msec nuclear radiation pulse, $10^{11}$ n/cm$^2$ produced a large signal. With better circuitry, the new sensor system also has promise as a ratemeter for nuclear pulses. Thermistors with lower impedances

(100 k $\Omega$ or less) would greatly facilitate the design of better detector circuits. The experiments gave no indication of detrimental effects from the neutron-induced transmutations in the $B^{10}$; a theoretical estimate predicts no such effect for a dose of $10^{16}$ n/cm$^2$ or possibly higher. The influence of gamma radiation through induced ionization currents across the thermistors was substantial, but could be eliminated through independent measurements of a blank. Adaption of the new device for the detection of neutrons with higher-than-thermal energies seems possible by surrounding the detector with a moderator, which would slow down the neutrons, or with $U^{238}$, which would deliver fission products that would then enter and heat the boron. An increase of the sensitivity for slow neutrons appears possible by making a similar arrangement with $U^{235}$.

## ACKNOWLEDGMENTS

The authors wish to thank Mr. J. J. O'Connor, Nuclear Reactor Division, AMRA, Watertown, Massachusetts and his staff for support in the steady-state measurements, and Mr. S. Marcus, Diamond Ordnance Radiation Facility, Silver Spring, Maryland and his staff for support in the pulse experiments. The authors also wish to express their appreciation for the contribution made by the late W. Medcalf and by Mr. R. Starks at Eagle-Picher Research Corporation in preparing the experimental thermistors. Special gratitude is due to Dr. S. Kronenberg and Mr. K. Nilson, USAEL for their invaluable help in preparing the pulse experiments. Helpful discussions with Mr. J. Mellichamp and electrical measurements by Mr. John Winter, both of USAEL, are gratefully acknowledged.

## REFERENCES

1. Gaulé, G. K., et al., "Optical and Electrical Properties of Boron and Potential Application," in: J. A. Kohn, W. F. Nye, and G. K. Gaulé, (eds.) Boron—Synthesis, Structure, and Properties, Plenum Press (New York), 1960, pp. 159–174.
2. "Research Investigation of Physical Chemistry and Metallurgy of Semiconducting Material," Chemical and Metals Division of the Eagle-Picher Company, Miami, Oklahoma, Contract Report No. DA36-039-SC-85131, Supplemental Report, December, 1963.
3. Bloom, J. L., unpublished.
4. Kaplan, I., Nuclear Physics, Addison-Wesley Publ. Co. (Reading, Mass.), 1958, pp. 369 ff and 448 ff.
5. Reynolds, S. A., and W. T. Mullins, Int. J. Appl. Rad. and Isotopes 14 (8): 421 (1963).
6. Zweifel, P. F., Nucleonics 18 (11): 174 (1960).
7. Cervellati, R., R. Gislon, and B. Rispoli, "Neutron Thermopiles for Flux Measurement in Nuclear Reactors," Neutron Dosimetry, Vol. I, IAEA (Vienna), 1963, p. 319.
8. Gray, D. E., (ed.), American Institute of Physics Handbook, second edition, McGraw-Hill Book Co. (New York), 1963, pp. 8–148.

9. Shive, J.N., Semiconductor Devices, D. van Nostrand Co. (Princeton), 1959, pp. 41 ff and 56 ff.
10. Kronenberg, S., private communication.

## DISCUSSION

BILLIG: What is the lifetime of your sample?

GAULÉ: That depends on the neutron flux density. We estimate that the B$^{10}$ thermistor tolerates a long-time dose of $10^{16}$ n/cm$^2$ and probably much more without significant changes in the characteristics. The maximum flux of $10^{12}$ n/cm-sec corresponds in our experiment to an exposure time of $10^4$ sec ($\approx$ 3 hr). The detectors, which are inexpensive, would have to be exchanged after receiving a certain dose. The low mobility and high trap density of the boron make it apparently very insensitive against nuclear damage.

SEILER: Can the device be used to measure a very high flux?

GAULÉ: When the high flux is applied for too long a time (many seconds), the B$^{10}$ temperature will become too high. It is also possible that the rapid development of helium within the B$^{10}$ (from the nuclear reaction) will do mechanical damage. One way out of this problem would be to use natural boron or boron material with an even smaller concentration of B$^{10}$. Another way would be to put the detector behind an attenuator for thermal neutrons, for example, several centimeters of the moderator substance in a reactor.

HERRMANN: What was the crystal structure of your films?

GAULÉ: The material was crystalline and predominantly of the beta-rhombohedral structure. Details are contained in the reports by Eagle-Picher Corp., which are quoted in the reference section.

NIEMYSKI: Did you observe lithium in the B$^{10}$ after the irradiation?

GAULÉ: We have not yet made such an analysis.

SEILER: Can you use your device as a detector for other particles?

GAULÉ: Since the absorption cross section of B$^{10}$ changes inversely with the neutron velocity, fast neutrons (with kinetic energies over 1 eV) can be detected only after they have been slowed down by a moderator. Similarly, a shield of uranium may be used to convert nondetectable particles to detectable particles.

NIEMYSKI: Did the detector become radioactive in your experiments?

GAULÉ: Boron itself does not become very radioactive upon radiation with neutrons, but some of the material used for contacts and for the capsule were quite radioactive immediately after the experiment. In the future, this can be largely avoided by using only a few selected materials, for example, pure aluminum and $Al_2O_3$, for the construction.

# List of Conference Participants

| | |
|---|---|
| Eberhard Amberger | Institut für Anorganische Chemie<br>Universität München<br>1 Meiserstrasse<br>Munich, Germany |
| Alton F. Armington | Air Force Cambridge Research Laboratories<br>Office of Aerospace Research<br>L. G. Hanscom Field<br>Bedford, Massachusetts |
| Hermann Josef Becher | Laboratorium für anorganische Chemie der<br>\Technischen Hochschule<br>26 Schellingstrasse<br>Stuttgart, Germany |
| E. Billig | Electrical Research Association<br>Cleeve Road<br>Leatherhead, Surrey, England |
| Robert A. Brungs | Research Institute of the Natural Sciences<br>(RINS)<br>Woodstock College<br>Woodstock, Maryland |
| J. T. Buford | The Eagle-Picher Company<br>Miami Research Department<br>200 Ninth Avenue N. E.<br>Miami, Oklahoma |
| Wolfgang H. Dietz | Consortium für elektrochemische Industrie<br>20 Zielstattstrasse<br>Munich, Germany |
| W. Dietze | Institut für Anorganische Chemie<br>Universität München<br>1 Meiserstrasse<br>Munich, Germany |
| Gerhart K. Gaulé | Institute for Exploratory Research<br>U.S. Army Electronics Command<br>Fort Monmouth, New Jersey |

D. Geist*

Physikalisches Institut
Universität Köln
1 Claudiusstrasse
Cologne, Germany

Arno K. Hagenlocher

General Telephone & Electronics
    Laboratories
Bayside, New York

H. A. Herrmann

Consortium für elektrochemische Industrie
20 Zielstattstrasse
Munich, Germany

Ingeborg Hinz

Consortium für elektrochemische Industrie
20 Zielstattstrasse
Munich, Germany

Robert E. Hughes**

Department of Chemistry
University of Pennsylvania
Philadelphia, Pennsylvania

Günter Köhl

AEG-Forschungsinstitut
235 Goldsteinerstrasse
Frankfurt am Main, Germany

Werner Neft‡

Institut für Theoretische und Angewandte
    Physik
Technischen Hochschule Stuttgart
Stuttgart, Germany

Tadeusz Niemyski

Institute of Physics (P.A.N.)
Polish Academy of Sciences
37 Zielna
Warsaw, Poland

Ulrich Rössler

Institut für Theoretische Physik (II) der
    Universität Marburg/Lahn
33 Mainzer Gasse
Marburg an der Lahn, Germany

K. Seiler‡‡

Intermetall Gesellschaft fur Metallurgie und
    Elektronik
19 Hans-Bunte-Strasse
Freiburg im Breisgau, Germany

Jürgen Smidt

Consortium für elektrochemische Industrie
20 Zielstattstrasse
Munich, Germany

---

*Present address: Institut für Angewandte Physik, Technische Hochschule Clausthal, Clausthal-Zellerfeld, Germany.
**Present address: Department of Chemistry, Cornell University, Ithaca, New York.
‡Present address: Ohm-Polytechnikum, 40 Kesslerstrasse, Nürnberg, Germany.
‡‡Present address: W. C. Heraeus, Heraeusstrasse, Hanau, Germany.

L. Sosnowski

Polska Akademia Nauk
Instytut Fizyki
37 Zielna
Warsaw, Poland

George Szasz

General Electric Company
Research Laboratory
37 Pelikanstrasse
Zürich, Switzerland

Joe Walker

Kansas State College of Pittsburg
Pittsburg, Kansas

Heinz Wirth

Consortium für elektrochemische Industrie
20 Zielstattstrasse
Munich, Germany

W. Zawadzki

Institut of Physics (P. A. N.)
Polish Academy of Sciences
37 Zielna
Warsaw, Poland

# Index *

*Unless otherwise noted, properties usually refer to beta-rhombohedral boron.